AF352907

Metal-Based Radiopharmaceuticals in Inorganic Chemistry

Metal-Based Radiopharmaceuticals in Inorganic Chemistry

Editors

Alessandra Boschi
Petra Martini

MDPI • Basel • Beijing • Wuhan • Barcelona • Belgrade • Manchester • Tokyo • Cluj • Tianjin

Editors
Alessandra Boschi
Department of Chemical,
Pharmaceutical and
Agricultural Sciences
University of Ferrara
Ferrara
Italy

Petra Martini
Department of Environmental
and Prevention Sciences
University of Ferrara
Ferrara
Italy

Editorial Office
MDPI
St. Alban-Anlage 66
4052 Basel, Switzerland

This is a reprint of articles from the Special Issue published online in the open access journal *Molecules* (ISSN 1420-3049) (available at: www.mdpi.com/journal/molecules/special_issues/molecules_MetalbasedRadiopharmaceuticals2020).

For citation purposes, cite each article independently as indicated on the article page online and as indicated below:

LastName, A.A.; LastName, B.B.; LastName, C.C. Article Title. *Journal Name* **Year**, *Volume Number*, Page Range.

ISBN 978-3-0365-6989-5 (Hbk)
ISBN 978-3-0365-6988-8 (PDF)

Contents

molecules

Editorial

Metal-Based Radiopharmaceuticals in Inorganic Chemistry

Alessandra Boschi [1,*] and Petra Martini [2,*]

1   Department of Chemical, Pharmaceutical and Agricultural Sciences, University of Ferrara, 44121 Ferrara, Italy
2   Department of Environmental and Prevention Sciences, University of Ferrara, 44121 Ferrara, Italy
*   Correspondence: alessandra.boschi@unife.it (A.B.); petra.martini@unife.it (P.M.);
    Tel.: +39-0532-455354 (A.B. & P.M.)

Citation: Boschi, A.; Martini, P.
Metal-Based Radiopharmaceuticals
in Inorganic Chemistry. *Molecules*
2023, *28*, 2290. https://doi.org/
10.3390/molecules28052290

Received: 27 February 2023
Accepted: 27 February 2023
Published: 1 March 2023

The field of radiopharmaceuticals is constantly evolving thanks to the great contribution of specialists coming from different disciplines such as inorganic chemistry, radiochemistry, organic and biochemistry, pharmacology, nuclear medicine, physics, etc. In particular, the use of radiometals has experienced a great increase as a result of the development of radionuclides production technologies. In this particular area, inorganic chemistry skills are mainly involved in developing target-specific radiopharmaceuticals based on radiometals for non-invasive disease detection and cancer radiotherapy.

The Special Issue "Metal-Based Radiopharmaceuticals in Inorganic Chemistry", which follows a similar topical Special Issue "New Trends in Production and Applications of Metal Radionuclides for Nuclear Medicine" [1–16], includes eleven research articles and one review. The production and applications of conventional and newly emerging research radiometals for diagnosis, therapy, and theranostics are the main focus of this Special Issue. Their employment in all Nuclear Medicine branches (SPECT/PET diagnostic, therapy, and theranostics) is regulated by their physical characteristics, such as half-life, radiation emission energy and type (γ, β^+, β^-, auger, α), availability and chemical ability to coordinate with ligands. The actual trend in the nuclear medicine research field is the use of radiometals for PET and SPECT, such as ^{68}Ga, ^{64}Cu, ^{89}Zr, ^{44}Sc, ^{86}Y, ^{52}Mn, ^{99}mTc, etc., for therapy, such as ^{177}Lu, ^{90}Y, ^{89}Sr, ^{223}Ra, ^{225}Ac, etc., and for theranostics, such as ^{67}Cu, ^{47}Sc, theranostics pairs, etc.

Hernández-Jiménez et al. [17] developed a ^{225}Ac-delivering nanosystem by encapsulating the radionuclide into rHDL nanoparticles, as a potential targeted radiotherapeutic agent. It is well known that reconstituted high-density lipoproteins (rHDL) specifically recognize the scavenger receptor B type I (SR-BI) overexpressed in several types of cancer cells. Furthermore, after rHDL-SR-BI recognition, the rHDL content is injected into the cell cytoplasm. They performed the synthesis of rHDL in two steps using the microfluidic synthesis method for the subsequent encapsulation of ^{225}Ac, previously complexed to the lipophilic molecule ^{225}Ac-DOTA-benzene-p-SCN. The nanosystem (13 nm particle size) showed a radiochemical purity higher than 99% and stability in human serum. In vitro studies in HEP-G2 and PC-3 cancer cells (SR-BI positive) demonstrated that ^{225}Ac was successfully internalized into the cytoplasm of cells, delivering high radiation doses to cell nuclei (107 Gy to PC-3 and 161 Gy to HEP-G2 nuclei at 24 h), resulting in a significant decrease in cell viability down to 3.22 ± 0.72% for the PC-3 and to 1.79 ± 0.23% for HEP-G2 at 192 h after 225Ac-rHDL treatment. After intratumoral ^{225}Ac-rHDL administration in mice bearing HEP-G2 tumors, the biokinetic profile showed significant retention of radioactivity in the tumor masses (90.16 ± 2.52% of the injected activity), which generated ablative radiation doses (649 Gy/MBq). The results demonstrated adequate properties of rHDL as a stable carrier for selective deposition of ^{225}Ac within cancer cells overexpressing SR-BI.

The work by Da Silva et al. [18] has been dedicated to developing new biomedical cyclotron irradiation and radiochemical isolation methods to produce ^{165}Er suitable for targeted radionuclide therapeutic studies and characterize a new agent targeting prostate-specific membrane antigen. They irradiated 80–180 mg ^{nat}Ho targets with 40 µA of

11–12.5 MeV protons to produce ^{165}Er at 20–30 MBq·µA^{-1}·h^{-1}. Radiochemical isolation yielded ^{165}Er in 0.01 M HCl (400 µL) with a decay-corrected (DC) yield of 64 ± 2%. Proof-of-concept radiolabeling studies were successfully performed synthesizing [^{165}Er]PSMA-617, which will be utilized in vitro and in vivo to understand the role of AEs in PSMA-targeted radionuclide therapy of prostate cancer.

Helal et al. [19] proposed the use of ^{213}Bi, an alpha-emitter with a short physical half-life of 46 min, to treat invasive fungal infections (IFI) with radioimmunotherapy (RIT). RIT uses antigen–antibody interaction to deliver sufficient activities of ionizing radiation to cells to induce DNA double-strand breaks and alter cell membrane and intracellular components for cell apoptosis while preserving healthy tissues. Microorganism-specific monoclonal antibodies have shown promising results in the experimental treatment of fungal, bacterial, and viral infections, including their recent and encouraging results from treating mice infected with *Blastomyces dermatitidis* with ^{213}Bi-labeled antibody 400-2 to (1→3)-β-glucan. They performed a safety study of ^{213}Bi-400-2 antibody in healthy dogs as a prelude for a clinical trial in companion dogs with acquired invasive fungal infections and later on in human patients with IFI. No significant acute or long-term side effects were observed after radioimmunoterapy (RIT) injections; only a few parameters were mildly and transiently outside reference change value limits, and a transient atypical morphology was observed in the circulating lymphocyte population of two dogs. Their results demonstrate the safety of systemic ^{213}Bi-400-2 administration in dogs and encourage to pursuit of evaluation of RIT of IFI in companion dogs.

Summer et al. [20] developed one imaging probe for hybrid imaging combining the beneficial properties of radioactivity and optical imaging. They modified the macrocyclic gallium-68 chelator fusarinine C (FSC) by conjugating a fluorescent moiety and tetrazine (Tz) moieties. The resulting hybrid imaging agents were used for pretargeting applications utilizing click reactions with a *trans*-cyclooctene (TCO) tagged targeting vector for a proof of principle both in vitro and in vivo. The evaluation included fluorescence microscopy, binding studies, logD, protein binding, in vivo biodistribution, µPET (micro-positron emission tomography), and optical imaging (OI) studies. ^{68}Ga-labeled conjugates showed suitable hydrophilicity, high stability, and specific targeting properties towards Rituximab-TCO pre-treated CD20 expressing Raji cells. Biodistribution studies showed fast clearance and low accumulation in non-targeted organs for both SulfoCy5- and IRDye800CW-conjugates. In an alendronate-TCO based bone targeting model the dimeric IRDye800CW-conjugate resulted in specific targeting using PET and OI, superior to the monomer. This proof-of-concept study showed that the preparation of FSC-Tz hybrid imaging agents for pretargeting applications is feasible, making such compounds suitable for hybrid imaging applications.

Two research articles covered the production and preparation of PET radionuclides and radiopharmaceuticals. Kazakov et al. [21] have dedicated a study to cobalt ^{55}Co, ^{57}Co, and ^{58m}Co isotopes production. These are considered to be promising radionuclides in nuclear medicine, with ^{55}Co receiving the most attention as an isotope for diagnostics by positron emission tomography. They determined the yields of nuclear reactions occurring during the irradiation of ^{nat}Ni and ^{60}Ni by bremsstrahlung photons with energy up to 55 MeV and developed a method of fast and simple cobalt isotopes separation from irradiated targets using extraction chromatography. Targets made of ^{nat}Ni and ^{60}Ni were irradiated by bremsstrahlung photons with energy up to 55 MeV. They found that in every case, the activities produced of ^{56,57,58}Co were higher than the activity of ^{55}Co, and therefore enough for preclinical research of radiopharmaceuticals based on cobalt. They also demonstrated that the radionuclide purity of ^{55}Co produced by the photonuclear method at 55 MeV is not sufficient for PET. They demonstrated that the separation of Co(II) is possible in a wide range of HCl concentrations (from 0.01 to 3 M). The separation factor of Ni/Co was $2.8 \cdot \times 10^5$, the yield of Co(II) was close to quantitative, and separation lasted for no longer than 0.5 h.

Another radionuclide of great interest is the PET radionuclide zirconium-89 (^{89}Zr) and the increased interest in immunoPET imaging probes for preclinical and clinical studies has led to a rising demand for this radionuclide. ^{89}Zr emits highly penetrating 511 and 909 keV photons delivering an undesirably high radiation dose, which makes it difficult to produce large amounts manually. Considering also the growing demand for Good Manufacturing Practices (GMP)-grade radionuclides for clinical applications Gaja et al. [22] in their study have adopted the commercially available TRASIS mini AllinOne automated synthesis unit to achieve efficient and reproducible batches of ^{89}Zr. The automated module is used for the target dissolution and separation of ^{89}Zr from the yttrium target material. ^{89}Zr is eluted with a very small volume of oxalic acid (1.5 mL) directly over the sterile filter into the final vial. Using this sophisticated automated purification method, they obtained a satisfactory amount of ^{89}Zr in high radionuclidic and radiochemical purities of over 99.99%. The specific activity of three production batches was calculated and was found to be in the range of 1351–2323 MBq/µmol. ICP-MS analysis of final solutions showed impurity levels always below 1 ppm.

Recently, the technological advancement in the radionuclides cyclotron-based production sector has encouraged the use of novel radioisotopes (mainly radiometals) in medical applications, for implementing the so-called personalized medicine approach. The availability of cyclotron-produced radiometals requires the use of solid targets and a solid target dissolution system and in this context fits the study of Sciacca et al. [23]. They developed a simple and efficient solid target dissolution system compatible with commercial cassette-based synthesis modules. In this way, it would be possible to perform the radiochemical processing, from the dissolution to the labeling, all at once using a single remotely controlled device. Keeping the system compact allows for containing all the processes in a single hot cell, lowering the probability of external and operator contamination. At the same time, this reduces the processing time and maximizes the recovery yield thanks to the absence of wasteful transfers from one system to another. The entire process, starting from dissolution up to radiopharmaceutical formulation, can be applied continuously.

The presented solid target dissolution system concept relies on an open-bottomed vial positioned upon a target coin. In particular, the idea is to use the movement mechanism of a syringe pump to position the vial up and down on the target and to exploit the heater/cooler reactor of the module as a target holder. All the steps can be remotely controlled and are incorporated in the cassette manifold together with the purification and radiolabelling steps.

In their study, Štícha et al. [24] have successfully adapted the Schotten–Baumann (SB) reaction for the derivatization of MS hardly ionizable Re(VII) chlorocomplexes.

Mass spectrometry helps to ensure the chemical identity and desired purity of the prepared complexes before their medical application, and plays an indispensable role in clinical practice. In the context of mass spectrometry, specific problems arise with the low ionization efficiency of particular analytes. Chemical derivatization was used as one of the most effective methods to improve the analyte's response and separation characteristics. They studied the reaction of the Re(VII) bis(catechol) chlorocomplex with the set of halogen and alkyl anilines as derivatization agents and found that the SB reaction products are easily ionizable under common ESI conditions providing structurally characteristic molecular and fragment anions. Based on DFT computation, the effect of Re-N bond shortening in the course of complex deprotonation was simulated and also correlated with the basicity of the aniline derivative used as a derivatization agent. Their conclusions follow the known relation between the basicity of the reaction environment and the yield of the SB reaction. However, an attempt to increase the yield of the derivatization reaction by adding triethylamine (TEA) to the reaction mixture was unsuccessful. Such a conclusion probably refers to the fast reaction providing the dioxorhenium complex as a competition to the SB reaction itself.

A research article reported the development of a solvent extraction and separation process of technetium from molybdenum in a micro-scale in-flow chemistry regime with the aid of a capillary loop and a membrane-based separator, respectively [25]. The developed system can extract and separate quantitatively and selectively ($91.0 \pm 1.8\%$ decay corrected) the [^{99m}Tc]TcO$_4$Na in about 20 min, by using a ZAIPUT separator device. In this work the authors demonstrated for the first time, the high efficiency of a MEK-based solvent extraction process of ^{99m}Tc from a molybdenum-based liquid phased in an in-flow microscale regime. This system allows the extraction time and separation of technetium from the organic phase to be drastically reduced and can be used both to purify technetium from molybdenum metal targets in the direct cyclotron ^{99m}Tc production, as well as in indirect ^{99m}Tc production such as the ^{99}Mo production, by irradiating natural molybdenum using a 14-MeV accelerator-driven neutron source.

Remaining within the scope of SPECT, a specific work has been dedicated to indium-111 [26]. In this study, a CCK-2R-targeting ligand based on the nastorazepide core was syn- thetized and functionalized with a DOTA chelator with the aim to provide a suitable platform for a theragnostic approach with radioactive metals. Avoiding the use of peptide-based sequences in the structure, including the linker, allowed them to obtain a molecule with high stability in physiological media that could be easily labeled with indium-111 as a pivotal radionuclide for future studies. The obtained radiotracer was successfully employed in the imaging of CCK-2R-expressing xenograft tumors in mice, but further structural studies are needed for enhancing receptor affinity and biodistribution. Additionally, the presented results are particularly noteworthy since the ability of a targeting probe to image cancers has been demonstrated using human cells expressing physiological levels of CCK-2R instead of transfected cells like the majority of the studies on the topic so far.

The objective of the research presented by Blumberg et al. [27] was to evaluate and compare different open-chain and bridged p-tert-butylcalix [4] arene derivatives as possible leading compounds that could, upon further modification, yield viable chelators for the selected divalent metal ions Sr^{2+}, Ba^{2+}, and Pb^{2+} in radiopharmaceutical applications and provide information about comparable stability constants. UV titration as a reliable and constant method for the calculation of stability constants was used to determine association constants for the respective ions. Additionally, theoretical calculations involving Ba^{2+} as a surrogate for Ra^{2+} were accomplished to underline the results. They found that the additional proton-ionizable side functions connected with a benzocrown ether structure to the calixarene skeleton led to a considerable improvement of the complex stability resulting in high association constants.

Finally, the only review of the Special Issue covered ^{67}Cu production capabilities [28]. This review reveals the international effort to supply ^{67}Cu, a promising theranostic radionuclide. The increasing availability of intense particle accelerators and the optimization of the associated technologies, targetry, and radiochemical processing, are making ^{67}Cu closer to the clinics. In particular, ^{64}Cu is now widely available for clinical use, promoting the development of innovative Cu-labeled radiopharmaceuticals. The improved availability of ^{67}Cu would speed up further radiopharmaceutical applications for therapy. In addition to a detailed analysis of the possible nuclear reactions to produce ^{67}Cu, the radiochemical procedures to extract and purify Cu from the bulk material were also described in this work. Recent developments in the photoproduction of ^{67}Cu, and in the possibility of having accelerators providing intense 70 MeV proton beams and/or intense 30 MeV deuteron beams, are grounds for a future reliable supply of ^{67}Cu.

Funding: This research received no external funding.

Conflicts of Interest: The authors declare no conflict of interest.

References

1. Fitzsimmons, J.; Foley, B.; Torre, B.; Wilken, M.; Cutler, C.S.; Mausner, L.; Medvedev, D. Optimization of Cation Exchange for the Separation of Actinium-225 from Radioactive Thorium, Radium-223 and Other Metals. *Molecules* **2019**, *24*, 1921. [CrossRef] [PubMed]
2. Larenkov, A.; Bubenschikov, V.; Makichyan, A.; Zhukova, M.; Krasnoperova, A.; Kodina, G. Kodina Preparation of Zirconium-89 Solutions for Radiopharmaceutical Purposes: Interrelation Between Formulation, Radiochemical Purity, Stability and Biodistribution. *Molecules* **2019**, *24*, 1534. [CrossRef] [PubMed]
3. Fitzsimmons, J.; Griswold, J.; Medvedev, D.; Cutler, C.; Mausner, L. Defining Processing Times for Accelerator Produced 225Ac and Other Isotopes from Proton Irradiated Thorium. *Molecules* **2019**, *24*, 1095. [CrossRef] [PubMed]
4. Lymperis, E.; Kaloudi, A.; Kanellopoulos, P.; de Jong, M.; Krenning, E.; Nock, B.; Maina, T. Comparing Gly11/DAla11-Replacement vs. the in-Situ Neprilysin-Inhibition Approach on the Tumor-Targeting Efficacy of the 111In-SB3/111In-SB4 Radiotracer Pair. *Molecules* **2019**, *24*, 1015. [CrossRef]
5. Orteca, G.; Pisaneschi, F.; Rubagotti, S.; Liu, T.; Biagiotti, G.; Piwnica-Worms, D.; Iori, M.; Capponi, P.; Ferrari, E.; Asti, M. Development of a Potential Gallium-68-Labelled Radiotracer Based on DOTA-Curcumin for Colon-Rectal Carcinoma: From Synthesis to In Vivo Studies. *Molecules* **2019**, *24*, 644. [CrossRef]
6. Uccelli, L.; Martini, P.; Cittanti, C.; Carnevale, A.; Missiroli, L.; Giganti, M.; Bartolomei, M.; Boschi, A. Therapeutic Radiometals: Worldwide Scientific Literature Trend Analysis (2008–2018). *Molecules* **2019**, *24*, 640. [CrossRef]
7. Sarnelli, A.; Belli, M.; Di Iorio, V.; Mezzenga, E.; Celli, M.; Severi, S.; Tardelli, E.; Nicolini, S.; Oboldi, D.; Uccelli, L.; et al. Dosimetry of 177Lu-PSMA-617 after Mannitol Infusion and Glutamate Tablet Administration: Preliminary Results of EUDRACT/RSO 2016-002732-32 IRST Protocol. *Molecules* **2019**, *24*, 621. [CrossRef]
8. Skliarova, H.; Cisternino, S.; Cicoria, G.; Marengo, M.; Palmieri, V. Innovative Target for Production of Technetium-99m by Biomedical Cyclotron. *Molecules* **2018**, *24*, 25. [CrossRef]
9. Chen, K.-T.; Nguyen, K.; Ieritano, C.; Gao, F.; Seimbille, Y. A Flexible Synthesis of 68Ga-Labeled Carbonic Anhydrase IX (CAIX)-Targeted Molecules via CBT/1,2-Aminothiol Click Reaction. *Molecules* **2018**, *24*, 23. [CrossRef]
10. Guerreiro, J.; Alves, V.; Abrunhosa, A.; Paulo, A.; Gil, O.; Mendes, F. Radiobiological Characterization of 64CuCl2 as a Simple Tool for Prostate Cancer Theranostics. *Molecules* **2018**, *23*, 2944. [CrossRef]
11. Amor-Coarasa, A.; Kelly, J.; Ponnala, S.; Nikolopoulou, A.; Williams, C.; Babich, J. 66Ga: A Novelty or a Valuable Preclinical Screening Tool for the Design of Targeted Radiopharmaceuticals? *Molecules* **2018**, *23*, 2575. [CrossRef] [PubMed]
12. Borgna, F.; Ballan, M.; Favaretto, C.; Verona, M.; Tosato, M.; Caeran, M.; Corradetti, S.; Andrighetto, A.; Di Marco, V.; Marzaro, G.; et al. Early Evaluation of Copper Radioisotope Production at ISOLPHARM. *Molecules* **2018**, *23*, 2437. [CrossRef] [PubMed]
13. Capogni, M.; Pietropaolo, A.; Quintieri, L.; Angelone, M.; Boschi, A.; Capone, M.; Cherubini, N.; De Felice, P.; Dodaro, A.; Duatti, A.; et al. 14 MeV Neutrons for 99Mo/99mTc Production: Experiments, Simulations and Perspectives. *Molecules* **2018**, *23*, 1872. [CrossRef] [PubMed]
14. Talip, Z.; Favaretto, C.; Geistlich, S.; Meulen, N.P. van der A Step-by-Step Guide for the Novel Radiometal Production for Medical Applications: Case Studies with 68Ga, 44Sc, 177Lu and 161Tb. *Molecules* **2020**, *25*, 966. [CrossRef]
15. Martini, P.; Adamo, A.; Syna, N.; Boschi, A.; Uccelli, L.; Weeranoppanant, N.; Markham, J.; Pascali, G. Perspectives on the Use of Liquid Extraction for Radioisotope Purification. *Molecules* **2019**, *24*, 334. [CrossRef]
16. Esposito, J.; Bettoni, D.; Boschi, A.; Calderolla, M.; Cisternino, S.; Fiorentini, G.; Keppel, G.; Martini, P.; Maggiore, M.; Mou, L.; et al. LARAMED: A Laboratory for Radioisotopes of Medical Interest. *Molecules* **2018**, *24*, 20. [CrossRef]
17. Hernández-Jiménez, T.; Ferro-Flores, G.; Morales-Ávila, E.; Isaac-Olivé, K.; Ocampo-García, B.; Aranda-Lara, L.; Santos-Cuevas, C.; Luna-Gutiérrez, M.; De Nardo, L.; Rosato, A.; et al. 225Ac-RHDL Nanoparticles: A Potential Agent for Targeted Alpha-Particle Therapy of Tumors Overexpressing SR-BI Proteins. *Molecules* **2022**, *27*, 2156. [CrossRef]
18. Da Silva, I.; Johnson, T.R.; Mixdorf, J.C.; Aluicio-Sarduy, E.; Barnhart, T.E.; Nickles, R.J.; Engle, J.W.; Ellison, P.A. A High Separation Factor for 165Er from Ho for Targeted Radionuclide Therapy. *Molecules* **2021**, *26*, 7513. [CrossRef]
19. Helal, M.; Allen, K.J.H.; Burgess, H.; Jiao, R.; Malo, M.E.; Hutcheson, M.; Dadachova, E.; Snead, E. Safety Evaluation of an Alpha-Emitter Bismuth-213 Labeled Antibody to (1→3)-β-Glucan in Healthy Dogs as a Prelude for a Trial in Companion Dogs with Invasive Fungal Infections. *Molecules* **2020**, *25*, 3604. [CrossRef]
20. Summer, D.; Petrik, M.; Mayr, S.; Hermann, M.; Kaeopookum, P.; Pfister, J.; Klingler, M.; Rangger, C.; Haas, H.; Decristoforo, C. Hybrid Imaging Agents for Pretargeting Applications Based on Fusarinine C—Proof of Concept. *Molecules* **2020**, *25*, 2123. [CrossRef]
21. Kazakov, A.G.; Babenya, J.S.; Ekatova, T.Y.; Belyshev, S.S.; Khankin, V.V.; Kuznetsov, A.A.; Vinokurov, S.E.; Myasoedov, B.F. Yields of Photo-Proton Reactions on Nuclei of Nickel and Separation of Cobalt Isotopes from Irradiated Targets. *Molecules* **2022**, *27*, 1524. [CrossRef] [PubMed]
22. Gaja, V.; Cawthray, J.; Geyer, C.R.; Fonge, H. Production and Semi-Automated Processing of 89Zr Using a Commercially Available TRASIS MiniAiO Module. *Molecules* **2020**, *25*, 2626. [CrossRef] [PubMed]
23. Sciacca, G.; Martini, P.; Cisternino, S.; Mou, L.; Amico, J.; Esposito, J.; Gorgoni, G.; Cazzola, E. A Universal Cassette-Based System for the Dissolution of Solid Targets. *Molecules* **2021**, *26*, 6255. [CrossRef] [PubMed]
24. Štícha, M.; Jelínek, I.; Vlk, M. Chemical Conversion of Hardly Ionizable Rhenium Aryl Chlorocomplexes with P-Substituted Anilines. *Molecules* **2021**, *26*, 3427. [CrossRef]

25. Martini, P.; Uccelli, L.; Duatti, A.; Marvelli, L.; Esposito, J.; Boschi, A. Highly Efficient Micro-Scale Liquid-Liquid In-Flow Extraction of 99mTc from Molybdenum. *Molecules* **2021**, *26*, 5699. [CrossRef]
26. Verona, M.; Rubagotti, S.; Croci, S.; Sarpaki, S.; Borgna, F.; Tosato, M.; Vettorato, E.; Marzaro, G.; Mastrotto, F.; Asti, M. Preliminary Study of a 1,5-Benzodiazepine-Derivative Labelled with Indium-111 for CCK-2 Receptor Targeting. *Molecules* **2021**, *26*, 918. [CrossRef] [PubMed]
27. Blumberg, M.; Al-Ameed, K.; Eiselt, E.; Luber, S.; Mamat, C. Synthesis of Ionizable Calix[4]Arenes for Chelation of Selected Divalent Cations. *Molecules* **2022**, *27*, 1478. [CrossRef]
28. Mou, L.; Martini, P.; Pupillo, G.; Cieszykowska, I.; Cutler, C.S.; Mikołajczak, R. 67Cu Production Capabilities: A Mini Review. *Molecules* **2022**, *27*, 1501. [CrossRef]

Article

^{225}Ac-rHDL Nanoparticles: A Potential Agent for Targeted Alpha-Particle Therapy of Tumors Overexpressing SR-BI Proteins

Tania Hernández-Jiménez [1,2], Guillermina Ferro-Flores [1,*], Enrique Morales-Ávila [2,*], Keila Isaac-Olivé [3], Blanca Ocampo-García [1], Liliana Aranda-Lara [3], Clara Santos-Cuevas [1], Myrna Luna-Gutiérrez [1], Laura De Nardo [4], Antonio Rosato [5,6] and Laura Meléndez-Alafort [6]

[1] Department of Radioactive Materials, Instituto Nacional de Investigaciones Nucleares, Ocoyoacac 52750, Mexico; tania.hernandez@inin.gob.mx (T.H.-J.); blanca.ocampo@inin.gob.mx (B.O.-G.); clara.cuevas@inin.gob.mx (C.S.-C.); myrna.luna@inin.gob.mx (M.L.-G.)

[2] Faculty of Chemistry, Universidad Autónoma del Estado de México, Toluca 50180, Mexico

[3] Faculty of Medicine, Universidad Autónoma del Estado de México, Toluca 50180, Mexico; kisaaco@uaemex.mx (K.I.-O.); larandal@uaemex.mx (L.A.-L.)

[4] Department of Physics and Astronomy, University of Padua, 35131 Padova, Italy; laura.denardo@unipd.it

[5] Department of Surgery, Oncology and Gastroenterology, University of Padua, 35138 Padova, Italy; antonio.rosato@unipd.it

[6] Veneto Institute of Oncology IOV-IRCCS, 35138 Padova, Italy; laura.melendezalafort@iov.veneto.it

[*] Correspondence: guillermina.ferro@inin.gob.mx (G.F.-F.); emoralesav@uaemex.mx (E.M.-Á.)

Citation: Hernández-Jiménez, T.; Ferro-Flores, G.; Morales-Ávila, E.; Isaac-Olivé, K.; Ocampo-García, B.; Aranda-Lara, L.; Santos-Cuevas, C.; Luna-Gutiérrez, M.; De Nardo, L.; Rosato, A.; et al. ^{225}Ac-rHDL Nanoparticles: A Potential Agent for Targeted Alpha-Particle Therapy of Tumors Overexpressing SR-BI Proteins. *Molecules* **2022**, *27*, 2156. https://doi.org/10.3390/molecules27072156

Academic Editors: Alessandra Boschi and Petra Martini

Received: 15 February 2022
Accepted: 23 March 2022
Published: 27 March 2022

Publisher's Note: MDPI stays neutral with regard to jurisdictional claims in published maps and institutional affiliations.

Abstract: Actinium-225 and other alpha-particle-emitting radionuclides have shown high potential for cancer treatment. Reconstituted high-density lipoproteins (rHDL) specifically recognize the scavenger receptor B type I (SR-BI) overexpressed in several types of cancer cells. Furthermore, after rHDL-SR-BI recognition, the rHDL content is injected into the cell cytoplasm. This research aimed to prepare a targeted ^{225}Ac-delivering nanosystem by encapsulating the radionuclide into rHDL nanoparticles. The synthesis of rHDL was performed in two steps using the microfluidic synthesis method for the subsequent encapsulation of ^{225}Ac, previously complexed to a lipophilic molecule (^{225}Ac-DOTA-benzene-p-SCN, CLog P = 3.42). The nanosystem (13 nm particle size) showed a radiochemical purity higher than 99% and stability in human serum. In vitro studies in HEP-G2 and PC-3 cancer cells (SR-BI positive) demonstrated that ^{225}Ac was successfully internalized into the cytoplasm of cells, delivering high radiation doses to cell nuclei (107 Gy to PC-3 and 161 Gy to HEP-G2 nuclei at 24 h), resulting in a significant decrease in cell viability down to $3.22 \pm 0.72\%$ for the PC-3 and to $1.79 \pm 0.23\%$ for HEP-G2 at 192 h after ^{225}Ac-rHDL treatment. After intratumoral ^{225}Ac-rHDL administration in mice bearing HEP-G2 tumors, the biokinetic profile showed significant retention of radioactivity in the tumor masses ($90.16 \pm 2.52\%$ of the injected activity), which generated ablative radiation doses (649 Gy/MBq). The results demonstrated adequate properties of rHDL as a stable carrier for selective deposition of ^{225}Ac within cancer cells overexpressing SR-BI. The results obtained in this research justify further preclinical studies, designed to evaluate the therapeutic efficacy of the ^{225}Ac-rHDL system for targeted alpha-particle therapy of tumors that overexpress the SR-BI receptor.

Keywords: scavenger receptor B type I; reconstituted high-density lipoproteins; actinium-225; ^{225}Ac-rHDL; targeted alpha-particle therapy

1. Introduction

High-density lipoproteins (HDLs) are hydrophobic lipid micelles (endogenous nanoparticles; diameter of 5–12 nm) that carry cholesterol esters, free cholesterol, phospholipids, and triglycerides. HDLs transport and release their lipidic content in the liver and into the cytoplasm of different cells by interacting with specific cell membrane proteins. Due to the presence of apolipoprotein AI (Apo AI) as the main HDL peripheral component for targeting the scavenger receptor class B type I (SR-BI), synthetic or reconstituted HDL nanoparticles (rHDL), based on phospholipids and Apo AI, have been used as hydrophobic drug transporters, promoting the research of rHDL as a vehicle for the administration of chemotherapeutic agents [1,2].

The SR-BI is overexpressed on the cell surface of many cancer cells (e.g., ovarian, liver, and prostate cancer) and is a selective and specific receptor for HDL (through Apo AI). The SR-BI–HDL interaction triggers the release of cholesterol from HDL to the cell cytoplasm. Cholesterol is necessary for cell growth and is highly required by cancer cells, because it takes part in the synthesis of new cytoplasmic membranes [2,3].

Radiolabeled rHDLs have been employed for molecular imaging of cancer cells by targeting the SR-BI protein. Perez-Medina et al. attached ^{89}Zr-deferoxamine to the ApoE/phospholipids of rHDL; preclinical PET images showed high ^{89}Zr-rHDL uptake in tumor-associated macrophages of breast cancer [4]. Recently, preclinical studies of ^{99m}Tc-rHDL for SPECT imaging of prostate cancer metastasis have also been reported [5].

Alpha particles are an extraordinary option for targeted radiotherapy due to their capacity to cause damage to cancer cells, because the corresponding deposition of energy at the cellular level is one-hundred times greater than that of β-particles [6]. Targeted alpha-particle therapy (TAT) with ^{225}Ac has demonstrated considerable potential in the treatment of advanced prostate cancer [6].

Despite dozens of previous articles on ^{225}Ac labeling of metallic nanoparticles, liposomes, and micelles, the preparation of ^{225}Ac-HDL nanosystems has not been reported so far. The ^{225}Ac is produced primarily from a ^{229}Th/^{225}Ac generator with subsequent ^{225}Ac purification. The ^{255}Ac is characterized by decay into multiple alpha-emitting daughter radionuclides (four effective alpha emissions from ^{225}Ac, ^{221}Fr, ^{217}At, and ^{213}Po). One possible disadvantage of ^{225}Ac-complexes, however, is the recoil energy (~100 keV) of the nucleus during decay, which could induce the breaking of the bond with the chelator or transporter molecule with the consequent release of the daughter radionuclide to healthy tissues [7].

Therefore, the encapsulation of ^{225}Ac in rHDL nanoparticles (^{225}Ac-rHDL) would prevent the release of daughter radionuclide from the transporter biomolecule due to the recoil energy effect and, at the same time, the specific molecular recognition of ^{225}Ac-rHDL by cancer cells that overexpress the SR-BI protein would allow the cytoplasmic internalization of ^{225}Ac in tumor cells to produce ablative radiation doses.

This research aimed to prepare ^{225}Ac-rHDL and evaluate its preclinical in vitro and in vivo capability as a potential agent for targeted α-therapy of tumors overexpressing SR-BI receptors.

2. Results and Discussion

2.1. rHDL Assembly and Chemical Characterization

rHDL nanoparticles, generated from lipid micelles assembled in a microfluidic system, followed by the incorporation of Apo AI (Figure 1), were obtained with monomodal and monodisperse size distributions, with a hydrodynamic diameter of 13.14 ± 0.20 nm and a polydispersity index of 0.176, as determined by dynamic light scattering (DLS) (Table 1). In addition, transmission electron microscopy (TEM) micrographs demonstrated the presence of nanoparticles with a mean diameter of 11.10 ± 0.17 nm, spherical shape, and a homogeneous and uniform distribution (Table 1 and Figure 2). The protein content analysis of rHDL indicated an Apo AI concentration of 0.131 mg/mL.

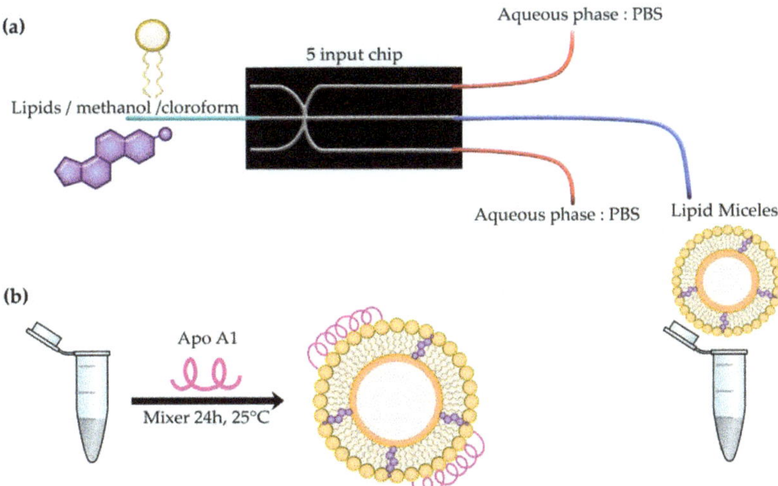

Figure 1. (a) Synthesis of lipid micelles. Preparation of micelles via a microfluidic system with hydrodynamic flow focusing, using a 5-way glass 3D chip that allows the diffusion of lipids in water, as well as water in alcohol, until its concentration decreases below the limit of lipid solubility, triggering the formation of lipid micelles. (b) rHDL synthesis. In the second step, the incorporation of apolipoprotein Apo AI was performed for the formation of rigid rHDL.

Table 1. Physical characteristics of micelles and rHDL nanoparticles.

Parameter	Lipid Micelles	rHDL
Particle diameter (nm), DLS	11.03 ± 0.04	13.14 ± 0.20
Particle diameter (nm), TEM	10.20 ± 0.13	11.10 ± 0.17
Polydispersity index	0.121	0.176
Protein concentration (mg/mL)	NA [1]	0.131

[1] Not applicable.

The FT-IR spectra of the micelles and rHDL are shown in Figure 2. The band at $3400\ \mathrm{cm}^{-1}$, observed in the spectrum of lipid micelles and characteristic of the (OH)υ groups of the cholesterol molecule, was shifted to $3287\ \mathrm{cm}^{-1}$ after Apo AI addition, which is associated with the stretching (OH)υ vibration of water molecules remaining in the interface between the lipoprotein and cholesterol, as well as indicative of the Apo AI incorporation to the lipid micelles due to interactions of the lysine primary amine residues with the lipid groups [8,9]. The band of highest intensity, observed at $2955\ \mathrm{cm}^{-1}$, corresponds to an asymmetric stretch (CH)υ, characteristic of the methyl groups present in the phospholipid chains, while the band at $2854\ \mathrm{cm}^{-1}$ corresponds to the symmetric stretch (CH)υ of the methylene groups that constitute the rHDL. However, in the spectrum obtained from the lipid micelles, these signals may be superimposed by the broad band of the vibration (OH)υ. The region from $1500\ \mathrm{cm}^{-1}$ to $1800\ \mathrm{cm}^{-1}$ is characteristic of (C=O)υ vibrations of the ester bonds present in the phospholipid chains. At $1650\ \mathrm{cm}^{-1}$, there is an important vibration in the rHDL spectrum, which corresponds to Amide I of the α-helix of Apo AI. At $1538\ \mathrm{cm}^{-1}$, a band attributed to amide II corresponding to (N-H)υ bending and (C-N)υ stretching was observed [8,9].

In the case of lipid micelles, a wide band of lower intensity at $1674\ \mathrm{cm}^{-1}$ was attributed to the (C=C)υ vibration characteristic of the second ring of cholesterol. The broad band at $1400\ \mathrm{cm}^{-1}$, in the rHDL spectrum, corresponds to the asymmetric stretching of the (COO-)υ groups of the aspartate and glutamate residues, which are believed to be involved in the coupling of lipoproteins [10].

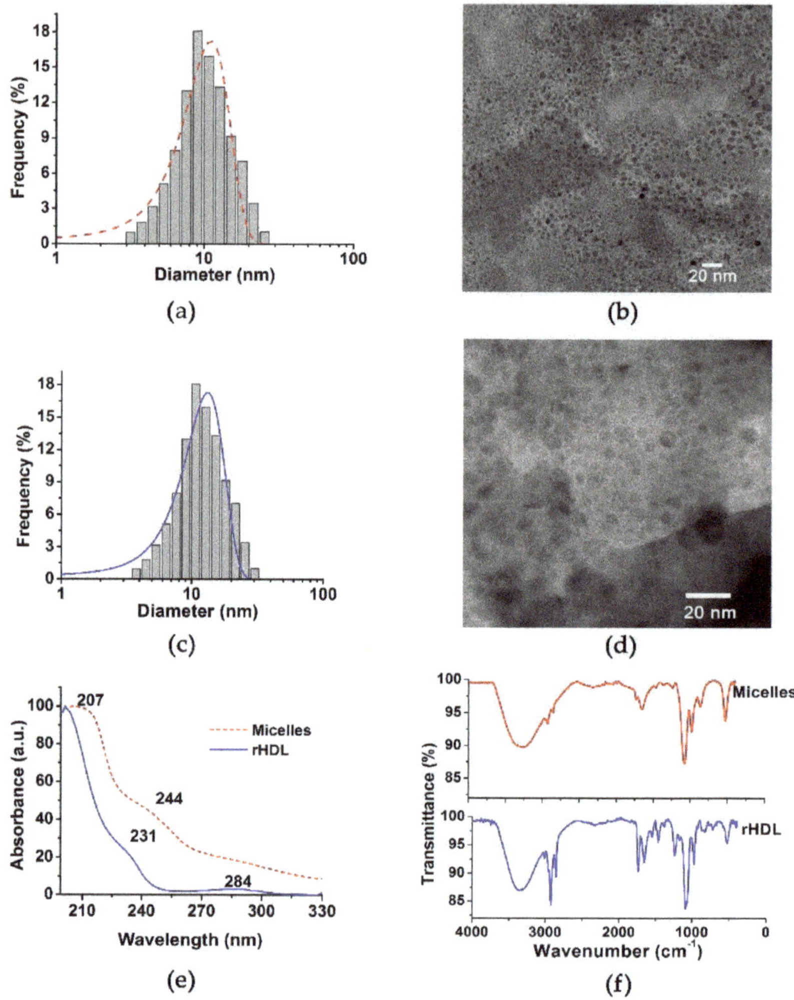

Figure 2. (**a**) Size distribution of lipid micelles by DLS, (**b**) TEM micrographs of lipid micelles, (**c**) size distribution of rHDL by DLS, (**d**) TEM micrographs of rHDL, (**e**) UV-Vis spectra of lipid micelles (red) and rHDL (blue), and (**f**) FT-IR spectra of lipid micelles (red) and rHDL (blue).

In summary, the spectrum of lipid micelles contains characteristic bands attributable to the cholesterol and free cholesterol esters as components of the mixture for the formation of micelles. In the rHDL spectrum, the Apo AI association with acidic lipid membranes, through interactions between lysine residues and negatively-charged lipid groups, can be appreciated. Interactions with the anionic phospholipids increase the Apo AI α-helix content, which is also an important factor in recognizing the specific molecular target [10].

The UV-Vis spectrum of the micelles showed a wide absorption band at 207 nm and a shoulder at 244 nm (Figure 2), in agreement with those reported for cholesterol compounds [8]. The band at 244 nm is associated with the $\pi \rightarrow \pi^*$ transitions of the α and β unsaturated ketones present in the cholesterol structure. After incorporating Apo AI into the micelles, an absorption band at 284 nm was observed, which indicates the coupling of proteins to lipid micelles. A slight band at 231 nm was also observed in the spectrum of rHDL, which is associated with the lipid–protein interaction.

2.2. The 225Ac-rHDL

The incorporation of ^{225}Ac was conducted efficiently with the rHDL vesicles by passive internalization through the encapsulation of ^{225}Ac previously complexed to a lipophilic molecule (^{225}Ac-DOTA-benzene-p-SCN) with a CLog P of 3.42 (Figure 3). As a result, the ^{225}Ac-rHDL system was obtained with a labeling efficiency of 85 ± 3% and a radiochemical purity greater than 99%, as determined by ultrafiltration.

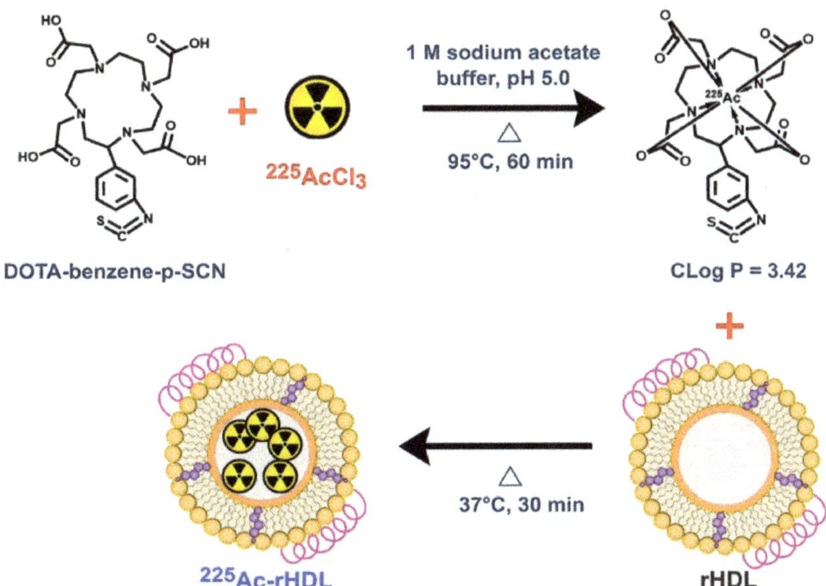

Figure 3. Schematic steps of the incorporation of ^{225}Ac into rHDL nanocapsules (^{225}Ac-rHDL).

2.3. In Vitro Studies

2.3.1. Serum Stability of the ^{225}Ac-rHDL System

The ^{225}Ac-rHDL was incubated in fresh human serum at 37 °C for 10 d. During the analysis of the samples at 24, 72, 144, and 240 h, the radioactivity associated with the rHDL system remained greater than 99%. It is important to mention that, according to what has previously been reported [11,12], the ^{225}Ac daughters (alpha emitters) can escape from the carrier vehicle and irradiate non-target tissues. However, the successful ^{225}Ac-DOTA-benzene-p-SCN encapsulation into the hydrophobic vesicle of rHDL allowed the retention of ^{225}Ac and its progeny within the nanosystem.

2.3.2. Cell Viability Assay and Dose to the Nucleus

Cell viability of PC-3 and HEP-G2 cells (SR-BI positive), and fibroblasts (negative control), was evaluated after treatment with a) ^{225}Ac-DOTA-benzene-p-SCN (control) and b) ^{225}Ac-rHDL at 37 °C for 1 h.

The cell viability of PC-3 and HEP-G2 treated with ^{225}Ac-DOTA-benzene-p-SCN was not significantly different between them nor with regard to that of fibroblasts ($p > 0.05$, two-way ANOVA) (Figure 4a). Furthermore, a low ^{225}Ac-DOTA-benzene-p-SCN internalization was observed in all cell lines (Figure 4b). These results indicate that the ^{225}Ac-DOTA-benzene-p-SCN system, despite being a hydrophobic compound, does not have an interaction mechanism with the surface of the cell membrane that would allow ^{225}Ac to be internalized into the cell cytoplasm. In contrast, the PC-3 and HEP-G2 cells that received treatment with the ^{225}Ac-rHDL nanosystem showed cell viability of 54.67± 3.16% and 53.12 ± 2.93% at 3 h, respectively, with a slight increase at 24 h (62.08 ± 2.44% for

PC-3 and 64.32 ± 3.38% HEP-G2) (Figure 4c). A possible explanation could be that cell repair mechanisms are stimulated upon receiving an initial dose of radiation, promoting cell proliferation in both cell lines [13,14]. Nevertheless, cell viability decreased to 3.22 ± 0.72% for the PC-3 cell line at 192 h after ^{225}Ac-rHDL treatment and to 1.79 ± 0.23% for the HEP-G2 cell line, as a result of the significant internalization of radiation (Figure 4d). In the case of fibroblasts, a cell viability of 81.2 ± 7.21%, and the lowest internalization index regarding the two cell lines (PC-3 and HEP-G2), was observed after ^{225}Ac-rHDL treatment (Figure 4d). As is known, fibroblasts poorly express the SR-BI receptor, and when they interact with rHDL, a different mechanism with slight and unspecific internalization can also occur [15–17].

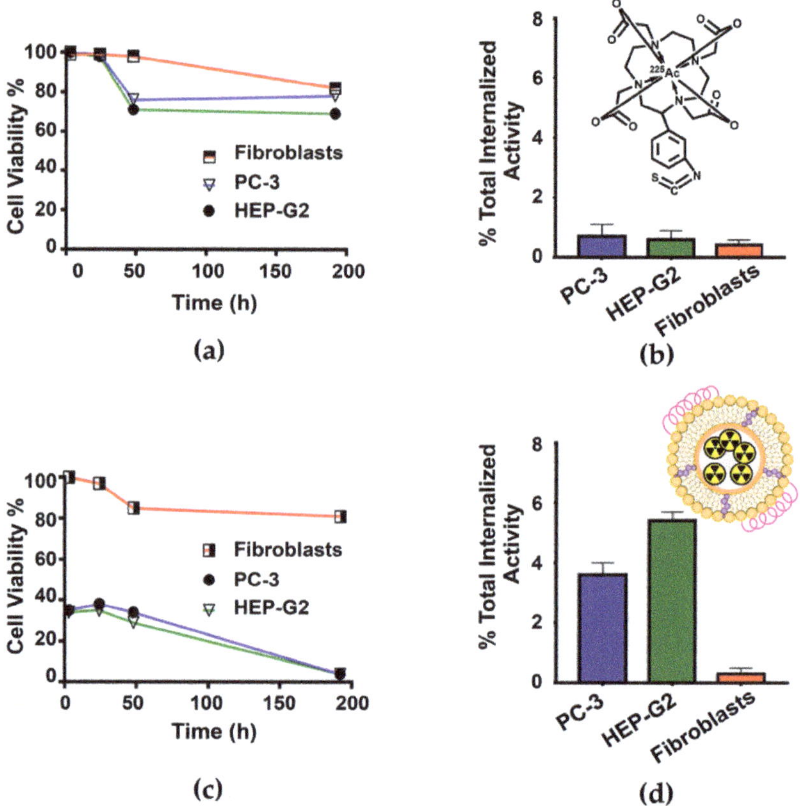

Figure 4. Cell viability assay: (**a**) PC-3, HEP-G2, and fibroblast cell lines treated with the ^{225}Ac-DOTA-benzene-p-SCN system (control); note that viability above 80% was maintained after 192 h for all cell lines. (**b**) Internalization of ^{225}Ac-DOTA-benzene-p-SCN in PC-3, HEP-G2 cells, and fibroblasts; note that the low cellular internalization of the ^{225}Ac-DOTA-benzene-p-SCN complex (without significant difference among cells, $p > 0.05$, two-way ANOVA) correlates with a relatively low effect on cell viability at 192 h. (**c**) PC-3, HEP-G2, and fibroblast cell lines treated with the ^{225}Ac-rHDL system; note a viability of 62–64% after 24 h and less than 3.5% at 192 h for the PC-3 and HEP-G2 cell lines, which overexpress the SR-BI protein. (**d**) Internalization of ^{225}Ac-rHDL in PC-3, HEP-G2 cells, and fibroblasts; note a significant cellular internalization of the ^{225}Ac rHDL complex in PC-3 and HEP-G2 cells, which correlates with a significant effect on cell viability.

The calculation of the absorbed dose, from the cytoplasm to the nucleus of the different cell lines at different times after treatment with the ^{225}Ac-rHDL system, was also evaluated (Table 2). The value used as the dose factor (DF) for the dose calculations was $\left(DF^{\alpha} + DF^{e,ph}\right)_{Ac-225(n \leftarrow Cy)} = 8.96 \times 10^{-2}$ Gy/Bq·s, where the four alpha, electron (e), and photon (ph) emissions of the daughters produced by each ^{225}Ac nuclear transformation were considered (MIRDcellV2.1 software) [18].

Table 2. Mean radiation-absorbed dose (Gy) from the cytoplasm (Cy) to the cell nucleus (n), considering the internalized activity (Bq/cell) with respect to the total initial activity administered as treatment of ^{225}Ac-rHDL (4.0 kBq/well) (1×10^4 cells/well) for each cell line.

Time (h)	Fibroblast	PC-3	HEP-G2
	Radiation Dose ($n \leftarrow Cy$)	Radiation Dose ($n \leftarrow Cy$)	Radiation Dose ($n \leftarrow Cy$)
3	0.5	14.0	20.7
24	3.9	107.3	161.4
48	8.1	208.2	312.2
192	23.8	682.5	1025.5
Internalized Activity (Bq/cell)	0.0005	0.0144	0.0216

The absorbed radiation dose to the cell nucleus at 192 h after ^{225}Ac-rHDL treatment was 1025.5 Gy (256.4 Gy per kBq administered in the well) for HEP-G2 and 682.5 Gy (170.6 Gy per kBq administered in the well) for PC-3, which is 43 and 29 times higher than in the case of fibroblasts, respectively. Therefore, it was demonstrated that the ^{225}Ac-rHDL nanosystem is capable of delivering doses of radiation to cause a significant cytotoxic effect, as a result of the ability to internalize ^{225}Ac into the cell and the multiple alpha-emitting daughter radionuclides generated inside the cell; that is, the SR-BI receptor, expressed on HEP-G2 and PC-3 cells, interacts with the endogenous rHDL lipoproteins, which induces the deposit of the rHDL content (^{225}Ac) directly into the cytoplasm of the cells. An advantage of the alpha radiotherapy system, in addition to its high specificity attributed to rHDL, is that the high number of short-path ionizations, which damage the DNA structure, also inhibit cell repair mechanisms [19,20]. On the other hand, a 20% decrease in cell viability was also observed in fibroblasts with a dose of 8.1 Gy at 48 h after treatment, which can be attributed to the radiosensitivity exhibited by fibroblasts (Figure 4c) [19].

2.4. Biodistribution Assay

The biodistribution profile in healthy mice, after intravenous administration of the ^{225}Ac-rHDL nanosystem, showed an accumulation of activity, mainly in the liver, with a value of 40.4 ± 2.12% ID, but with a relatively rapid clearance, reaching 6.01 ± 1.07% ID at 192 h (Table 3). Radioactivity in the liver is rapidly removed and excreted due to the liver metabolic dynamics of HDL [21]. Given its nanometric size, ^{225}Ac-rHDL elimination occurs both through the hepatobiliary (2.88 ± 0.12% ID at 0.5 h in intestine) and renal (4.85 ± 0.89% ID at 0.5 h in kidney) pathways. Because it is inaccurate to establish the percentage of radioactivity accumulated in the total blood of mice, the % ID in blood was not included in Table 3. However, the activity in blood was 3.91 ± 0.92% ID/g and 1.97 ± 0.81% ID/g at 0.5 h and 24 h, respectively. It should be considered that the elimination time of rHDL in blood plasma is relatively prolonged due to its lipoprotein nature (the half-life in plasma fluctuates between 6 h and 24 h after HDL administration) [21]. Therefore, a prolonged blood circulation of ^{225}Ac-rHDL is expected, which could be an advantage in therapeutic applications given that the nanosystem can produce a significant tumor accumulation of radioactivity before radiopharmaceutical elimination [22]. As expected, rHDL was captured by the liver tissue due to its high expression of SR-BI. These results confirm the specificity and passive targeting of SR-BI receptors, as well as their in vivo affinity for rHDL lipoproteins through a natural process of selective uptake and elimination.

This behavior is of particular importance when demonstrating that it is possible to minimize long retention of ^{225}Ac-rHDL in healthy organs, because, due to the nature of the alpha-emitting daughter radionuclides, high damage to healthy tissues could be produced.

Table 3. Biodistribution of the ^{225}Ac-rHDL nanosystem in healthy mice (Balb-c) after intravenous injection. The percentage of the injected dose per organ (% ID), at various times, is shown (mean $\pm$ SD, $n = 3$).

Tissue	Time (h)						
	0.5	1	3	24	48	96	192
Liver	40.40 ± 2.12	31.87 ± 3.19	23.68 ± 1.13	18.27 ± 2.32	11.12 ± 2.21	8.42 ± 1.32	6.01 ± 1.07
Spleen	1.10 ± 0.20	1.30 ± 0.10	1.74 ± 0.24	1.61 ± 0.21	1.02 ± 0.09	0.92 ± 0.15	0.74 ± 0.08
Lung	1.02 ± 0.32	0.87 ± 0.21	0.54 ± 0.11	0.47 ± 0.10	0.21 ± 0.06	0.11 ± 0.08	0.04 ± 0.01
Kidney	4.85 ± 0.89	3.14 ± 0.47	2.63 ± 0.23	1.98 ± 0.82	1.23 ± 0.45	0.81 ± 0.28	0.32 ± 0.12
Intestine	2.88 ± 0.12	1.77 ± 0.92	1.25 ± 0.15	0.84 ± 0.31	0.81 ± 0.12	0.68 ± 0.10	0.42 ± 0.06
Heart	0.50 ± 0.14	0.34 ± 0.09	0.21 ± 0.05	0.17 ± 0.04	0.09 ± 0.02	0.05 ± 0.02	0.01 ± 0.01

Biodistribution in mice with induced HEP-G2 tumors was performed in order to assess tumor uptake activity as a function of time. The biodistribution profile of ^{225}Ac-rHDL, after intratumoral administration, shows that the radioactivity of the system is retained within the tumor masses (Figure 5a), with an uptake of 90.16 ± 2.52% at 0.5 h, with regard to the initially-administered activity. These results confirm the potential of ^{225}Ac-rHDL to produce a localized cytotoxic effect. Low ^{225}Ac-rHDL accumulation was observed in liver (0.65% ID at 24 h), kidney (0.51% ID at 24 h), and spleen (0.34% ID at 24 h), which indicates that ^{225}Ac does not leak in vivo from the rHDL nanocapsule, because unchelated actinium (^{225}Ac^{3+}) accumulates significantly in the liver [23].

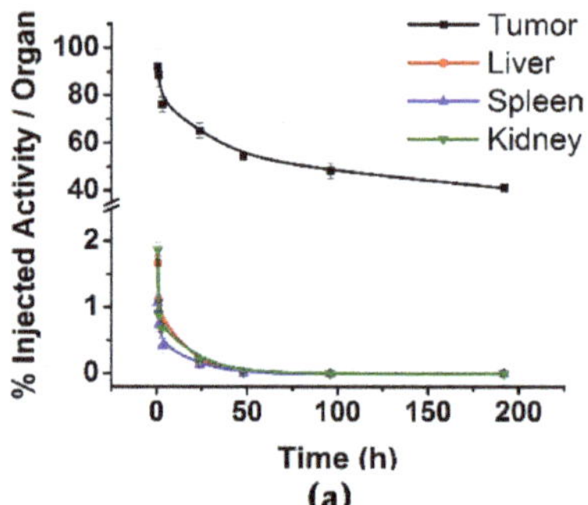

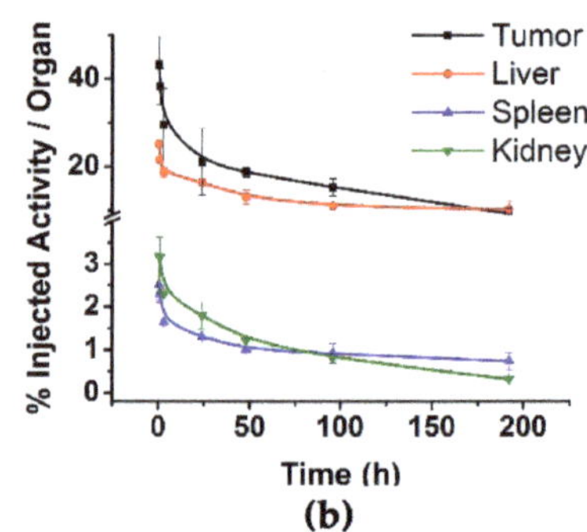

Figure 5. Comparison of the biokinetic profile between the (**a**) ^{225}Ac-rHDL nanosystem and (**b**) ^{225}Ac-DOTA-benzene-p-SCN in nude mice bearing HEP-G2 tumors after intratumoral administration.

When ^{225}Ac-DOTA-benzene-p-SCN was injected intratumorally in nude mice bearing HEP-G2 tumors, a very rapid clearance of radioactivity from the tumor was observed (Figure 5b); and almost 10-fold higher uptake in liver and spleen was seen with regard to the ^{225}Ac-rHDL biodistribution pattern (Figure 5b). These findings suggest that although the DOTA chelating agent is highly stable for positive trivalent radiometals [12,24,25], ^{225}Ac-DOTA-benzene-p-SCN does not remain in the tumor, because the recoil energy of the ^{225}Ac daughters (1000 times greater than the binding energy of any chemical compound) is possibly causing the breaking of chemical bonds, with the consequent retention of ^{225}Ac^{3+} and its progeny in the liver as ionic forms [23].

Based on the biodistribution results in healthy and tumor-bearing mice, it is possible to propose ^{225}Ac-rHDL as a convenient natural nanocarrier for the use of ^{225}Ac in targeted radiotherapy, in a safe and efficient manner.

Table 4 shows the biokinetic models of the ^{225}Ac-rHDL nanosystem in tumor and radiation source organs. The results show both the biological or pharmacokinetic model ($q_h(t)$) and the radiopharmacokinetic model ($A_h(t)$), the latter associated with the number of nuclear transformations that occurred in each tissue (N) for the radiation-absorbed dose calculation.

Table 4. Biokinetic models and average radiation-absorbed doses in different tissues (liver, kidney, spleen, and HEP-G2 tumor) of mice intratumorally injected with ^{225}Ac-rHDL (1 MBq).

Organ	Biokinetic Model	$\int_0^\infty q_h(t)dt$ Biological Residence Time (h)	$N=\int_0^\infty A_h(t)dt$ Total Nuclear Transformations	Absorbed Dose (Gy)
Tumor	$q_h(t) = 26.7e^{-0.542t} + 46.4e^{-0.0088t} + 26.2e^{-0.000039t}$ $A_h(t) = 26.7e^{-0.5449t} + 46.4e^{-0.0117t} + 26.2e^{-0.0029t}$	6772	130	649.20
Liver	$q_h(t) = 0.979e^{-0.0784t} + 0.929e^{-1.66t} + 3.71e^{-4.88t}$ $A_h(t) = 0.979e^{-0.0813t} + 0.929e^{-1.6629t} + 3.71e^{-4.8829t}$	0.138	0.928	2.17
Kidneys	$q_h(t) = 0.619e^{-0.0583t} + 0.193e^{-0.0582t} + 8.68e^{-4.16t}$ $A_h(t) = 0.619e^{-0.0612t} + 0.193e^{-0.0611t} + 8.68e^{-4.1629t}$	0.160	0.153	2.06
Spleen	$q_h(t) = 0.484e^{-0.0588t} + 1.27e^{-1.51t} + 0.884e^{-13.6t}$ $A_h(t) = 0.484e^{-0.0617t} + 1.27e^{-1.5129t} + 0.884e^{-13.6029t}$	0.091	0.087	3.18

The highest dose occurred in the tumor (649 Gy), while doses were low to the other organs (Table 4), indicating that the energy deposited by ^{225}Ac produces ablative radiation doses to the target malignant lesions, avoiding cytotoxic effects on healthy tissues.

As is known, some of the key properties of ^{225}Ac as a radionuclide for targeted alpha radiotherapy of micrometastases are: (1) range in tissue of a few cell diameters, (2) high linear energy transfer, which leads to direct damage of the DNA structure (LET = 100 KeV/μm), (3) half-life of 10 days, which allows sufficient time for the administration of the dose and its binding and retention in tumor masses, (4) emission of four alpha particles per nuclear transformation [26–29]. Another important aspect to consider is that ^{225}Ac requires a much lower activity to produce cytotoxic effects at the cellular level with regard to beta emitters, because its energy is deposited in an extremely localized manner. Furthermore, ^{225}Ac-rHDL have the appropriate physicochemical properties and size to reach the tumor tissue and locally deposit their content (^{225}Ac) within the cytoplasm of tumor cells, being a convenient vehicle for targeted alpha-particle radiotherapy.

Although the mechanism by which the interaction of the SR-BI receptor with HDL occurs has not yet been fully elucidated, it is possible that there is a non-aqueous "channel" in the SR-BI receptor that can accommodate cholesterol esters in such a way that it can couple with rHDL and capture its content through an internal tunnel [30] (Figure 6).

The biokinetic profile of the ^{225}Ac-rHDL system is comparable to those reported for ^{99m}Tc-HYNIC-DA-rHDL, with regard to tumor uptake and clearance because the delivery mechanism is the same [5]. In addition, the accumulation of activity in organs that express SR-BI (mainly liver, spleen, and kidneys) is comparable.

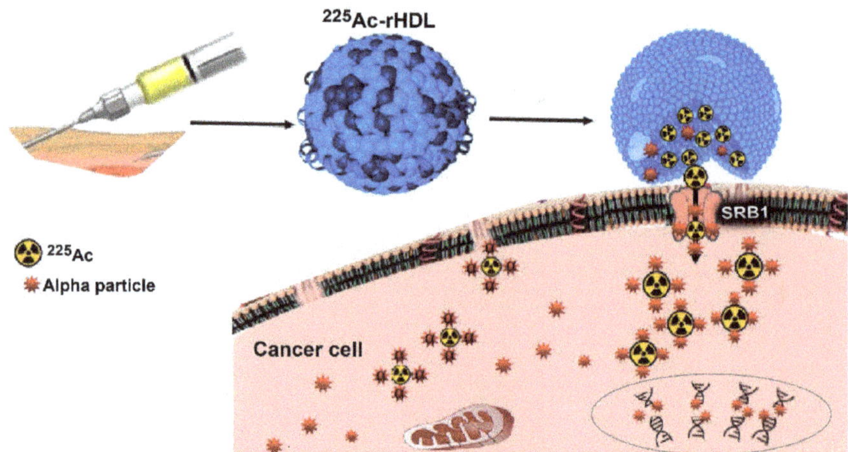

Figure 6. Cellular interaction mechanism of rHDL containing ^{225}Ac. Once endogenous rHDL interacts with the SR-BI receptor found on the surface of the cell membrane, its content is released directly into the cell's cytoplasm. Therefore, ^{225}Ac-rHDL is an effective nanosystem for depositing within the malignant cells, an in vivo generator for alpha particle radiotherapy, which delivers lethal radiation doses to cells due to the four alpha particles emitted by ^{225}Ac and its progeny in each nuclear transformation. The ionizations produced by ^{225}Ac and its daughters produce direct damage to the DNA structure, preventing tumor proliferation.

3. Materials and Methods

Free cholesterol (FC), egg yolk phosphatidylcholine (EYPC), cholesterol oleate (CE), and sodium cholate reagents were supplied by Landsteiner Scientific, and apolipoprotein Apo AI and human plasma by Alfa Aesar (Thermo Fisher, Tewksbury, MA, USA). Buffer Tris-EDTA was prepared in the laboratory. Dialysis tubbing (14000 Daltons) was obtained from MEMBRA CEL® (Thermo Fisher, Tewksbury, MA, USA). Macrocycle S-2-(4-isothiocyanatobenzyl)-1,4,7,10-tetraazacyclododecane acid, DOTA-benzene-p-SCN (p-SCN-Bn-DOTA) was obtained from Macrocyclics (Dallas, TX, USA). Actinium-225 (^{225}Ac), as ^{225}AcCl$_3$, was supplied from ITM, Germany. The bladder fibroblast cell human cell line, PC-3, and HEP-G2 cell lines (SR-BI positive) were obtained from American Type Culture Collection (ATCC®, Manassas, VA, USA).

3.1. Preparation of Lipid Micelles and rHDL

The synthesis of the lipid micelles was performed via the microfluidic method with hydrodynamic flow focusing (MFH), using a Dolomite Microfluidics System (Dolomite) equipped with a 150-μm 5-way glass 3D chip, with dimensions of 22.5 mm long, 15 mm wide, and 4 mm high. A 4-way H interface with two 4-way linear connectors and two connector seals were used. Flow patterns in the mixing zone of the microfluidic device were visualized using a high-speed digital microscope (Meros, Dolomite). Fluids were administered by two pressure pumps. The first pump for lipid, ethanol, and chloroform delivery. The second pump for the administration of PBS, with the corresponding flow sensors. The chip used presents a hydrodynamic flow approach that allows the formation of a stable laminar flow confined by two lateral flows.

rHDL was prepared in two steps. In the first, the formation of lipid micelles was performed via the microfluidic method, for which an organic solution consisting of 2 mL of methanol: chloroform (1:0.02 v/v), containing a mixture of the following lipids, was used: 300 μL of egg yolk phosphatidylcholine (10 mg/mL in methanol), 7 μL of free cholesterol (10 mg/mL in methanol), and 7.5 μL of cholesterol ester (4 mg/mL in chloroform). PBS (pH 7.4) was used as aqueous phase. Both solutions were filtered through a 0.22-μm PVDF membrane before being introduced into the microfluidic device. The organic phase was placed in the central channel and the aqueous solution in the coaxial channels, adjusting the organic phase to a flow of 50 μL/min and the aqueous phase to 500 μL/min, with a total flow ratio (TRF) of 550 μL/min and a flow rate ratio (FRR) of 47. Once the micelles were obtained, the size was measured by dynamic light scattering (DLS) (Nanotrac Wave, Model MN401, Microtract, Montgomeryville, PA, USA). Dialysis for purification (5 °C; 24 h) was performed using a 14000-kDa membrane (MEBRA CEL®, Thermo Fisher, Tewksbury, MA, USA) with PBS pH 7.4 to remove any residual solvent. Then, 4 mg/mL of Apo AI and 140 μL of sodium cholate (20 mg/mL) were added. Dialysis was performed again with stirring at 5 °C for 24 h to remove excess surfactant. The obtained solution was filtered using a 0.45-μm Millipore filter and stored at 4 °C.

3.2. Physicochemical Characterization of rHDL

3.2.1. Nanoparticle Size

The size of lipid micelles and rHDL was determined via dynamic light scattering (DLS) (Nanotrac Wave, Model MN401, Microtract, Montgomeryville, PA, USA). The size was also determined by transmission electron microscopy (TEM) (JEOL JEM2010 HT microscope operated at 200 kV), with a magnification of $50,000\times g$ for micelles and $200,000\times g$ for rHDL. Samples for TEM were stained with 2% sodium phosphotungstate solution (pH 7.2) and placed in carbon-lined 200-mesh copper support cells.

3.2.2. Protein Content

The rHDL protein content was determined using a colorimetric assay based on the formation of copper complexes using the BCA (bicinchoninic acid assay).

3.2.3. UV-Vis Spectra

The UV-Vis spectra (Perkin Elmer Lambda Bio spectrometer) of the lipid micelles and rHDL (1 mg/mL) were obtained, in the range of 200 to 500 nm, using a low-volume quartz cell (0.5-mL capacity) and an optical path of 1 cm.

3.2.4. Fourier Transform Infrared Spectroscopy (FT-IR)

The IR spectra of lipid micelles and rHDL were acquired on a PerkinElmer 2000 spectrophotometer with an attenuated total reflection platform (Pike Technologies; Madison, WI, USA). Spectra were acquired from 50 scans at $0.4\ cm^{-1}$, from 400 to $4000\ cm^{-1}$.

3.3. Preparation of ^{225}Ac-DOTA-Benzene-p-SCN

DOTA-benzene-p-SCN (pDOTA-Bz-SCN) (1 mg) was dissolved in 50 μL of 0.01 M NaOH, adjusting to a final volume of 2 mL with 1 M acetate buffer (pH 5.0). The ^{225}Ac progeny in the decay chain have γ emissions. Therefore, to quantify the ^{225}Ac activity, under secular equilibrium, a CRC-55tR radioisotope calibrator setting was used (Capintec Inc., Mirion Technologies, Florham Park, NJ, USA) (calibration # 775 with a $5\times$ multiplier), which mainly considers the 213Bi 440 keV γ emission [12,31]. To 200 μL of the DOTA-benzene-p-SCN, 50 μL of ^{225}AcCl$_3$ (18 MBq in 0.01 M HCl) were added. Finally, the mixture was incubated at 95 °C for 30 min. The radiochemical purity of ^{225}Ac-DOTA-benzene-p-SCN was determined using an HPLC reverse phase chromatography system (Waters Corporation, Milford, MA, USA). The separation of the samples was performed with a Waters μBondapak-C18 column at a flow rate of 1 mL/min. A linear gradient of H_2O with 0.1% TFA (A)/CH_3CN with 0.1% TFA (B), from 100% to 10% of A in 20 min, was used. Fractions of 0.5 mL (40 fractions) were collected and the activity measured in a well-type scintillation NaI(Tl) detector (Auto In-v-tron 4010; NML Inc., Houston, TX, USA). The retention times of ^{225}AcCl$_3$ and ^{225}Ac-DOTA-benzene-p-SCN were 3 min and 12.5 min, respectively. The ^{225}Ac-DOTA-benzene-p-SCN was obtained with radiochemical purity greater than 99%.

3.4. Preparation of ^{225}Ac-rHDL

To 1 mL of the previously prepared rHDL, 200 μL of ^{225}Ac-DOTA-benzene-p-SCN (12 MBq) were added, and the mixture was incubated at 37 °C for 1 h to obtain the ^{225}Ac-rHDL nanosystem.

The labeling efficiency of ^{225}Ac-rHDL was evaluated by ultracentrifugation (2500 g for 0.5 h; MWCO 30-kDa filter units, Amicon Ultra, Millipore, MilliporeSigma, Burlington, MA, USA). The fraction that is not retained in the membrane was considered as the radioactivity associated with ^{225}Ac-DOTA-benzene-p-SCN and ^{225}Ac^{3+} that was not bound to rHDL, while the fraction that remained in the membrane represents the ^{225}Ac-rHDL system. Both fractions were counted in a well-type scintillation NaI(Tl) detector to evaluate the labeling efficiency. The radioactive sample that remained in the filter membrane was resuspended in 10 mL of acetate buffer (pH 5.0)/0.1% ascorbic acid. Finally, a sample of the final solution (100 μL; activity measured in a well-type scintillation NaI(Tl) detector) was centrifuged again under the same conditions (2500× g for 0.5 h; MWCO 30-kDa filter units) to verify the radiochemical purity (RP) (RP = (activity remained in the filter membrane/total activity) × 100).

3.5. Serum Stability of the ^{225}Ac-rHDL System

The ^{225}Ac-rHDL system (150 μL) was added to 5 mL of diluted human serum (5×). The solutions (n = 3) were incubated at 37 °C for 10 days. A total of 1 mL of sample was taken at 24, 72, 144, and 240 h to evaluate the stability of the nanosystem. To each sample, 300 μL of TFA was added for protein precipitation. Samples were centrifuged at 1200× g for 5 min. The activity (measured in a in a well-type scintillation NaI(Tl) detector) obtained in the sediment corresponded to ^{225}Ac-rHDL, and the activity in the supernatant, to the free fraction of ^{225}Ac^{3+} leaked from the rHDL nanosystem.

3.6. Cell Internalization

The HEP-G2 (human hepatocellular carcinoma), PC-3 (human prostate cancer), and fibroblasts cell lines, used for the internalization assays, were cultured in RPMI-1640 medium containing antibiotics (penicillin and streptomycin; 100 μg/mL) and fetal bovine serum at a concentration of 15% in an atmosphere of 5% carbon dioxide and 37 °C.

HEP-G2, PC-3, and fibroblast cells were harvested and diluted in PBS (pH 7.4). Each cell line (1 × 10^5 cells/tube) received two different treatments: (a) ^{225}Ac-rHDL (4 kBq/200 μL PBS at pH 7.4) (n = 3), and (b) ^{225}Ac-DOTA-benzene-p-SCN (4 kBq/200 μL PBS at pH 7.4) (n = 3). Cells were incubated with each treatment at 37 °C for 1 h. After the incubation time, tubes were measured in a well-type NaI(Tl) scintillation detector to determine the initial activity (100%). The tubes were centrifuged at 500× g for 10 min. The button was washed 2 times with PBS, pH 7.4. Then, a mixture of acetic acid/0.5 M NaCl was added, and the tubes were centrifuged again at 500× g for 10 min. The supernatant was removed, and activity of the button was measured, which corresponded to the percentage of activity internalized in the cells with regard to the initial activity.

3.7. Cell Viability Assay

The cytotoxic effect on HEP-G2, PC-3, and fibroblast cells after treatment with (a) ^{225}Ac-rHDL and (b) ^{225}Ac-DOTA-benzene-p-SCN was evaluated by performing an assay of mitochondrial dehydrogenase activity using the XTT kit (Roche Holding AG, Rotkreuz, Switzerland). Briefly, HEP-G2, PC-3, and fibroblast cells (1 × 10^4 cells/well) were seeded in 96-well microtiter plates. The medium was removed after overnight incubation and the cells were incubated for 1 h with each treatment (4 kBq/200 μL). After removing the treatments, the cells were kept at 37 °C, 5% carbon dioxide, and 85% relative humidity. Cell viability was assessed at 3, 24, 48, and 192 h by spectrophotometric measurements (microplate absorbance reader, EpochTM, BioTek) at 450 nm. Fibroblast cells that did not receive treatment were considered as the control group with 100% cell viability at the different time points.

3.8. Radiation-Absorbed Dose Calculation

For the calculation of the total number of nuclear transformations (N)(Bq·s) that occurred in the cytoplasm of the PC-3, HEP-G2, and fibroblast cells at different times, the mathematical integration of the activity (^{225}Ac) in the cytoplasm (A_h), as a function of time ($N = \int_{t=1}^{t=2} A_h(t)dt$), was performed. The absorbed dose was calculated by multiplying N by the value of the dose factor (DF)(Gy/Bq·s). The DF value (from the cytoplasm to the nucleus) was calculated by using the MIRDcellV2.1 software, considering cells with a diameter of 20 μm, a nucleus radius of 3 μm, and a density of 1 mg/mL. For the dose calculation, the progeny of ^{225}Ac with significant decay yield (^{225}Ac, ^{221}Fr, ^{217}At, ^{213}Bi, ^{213}Po, ^{209}Tl, ^{209}Pb) was considered, as expressed in Equation (1):

$$\overline{D}_{225Ac-rHDL\ (n\leftarrow Cy)} = \sum_{j=1} N_{(Cy)} \sum_{k} b_k DF^{\alpha}_{k\ (n\leftarrow Cy)} + \sum_{j=1} N_{(Cy)} \sum_{k} b_k DF^{e,\ ph}_{k\ (n\leftarrow Cy)} \qquad (1)$$

$\overline{D}_{225Ac-rHDL\ (n\leftarrow Cy)}$ = radiation-absorbed dose (Gy) to the cell nucleus.
b_k = branching fraction for daughter k.
$DF^{\alpha}_{k\ (n\leftarrow Cy)}$ = absorbed dose to the cell nucleus from α emission per nuclear transformation (Gy/Bq·s) from decay of daughter radionuclide "k" originated in the cytoplasm.
$DF^{e,\ ph}_{k}$ = absorbed dose to the cell nucleus from electron (e) and photon (ph) emission per nuclear transformation (Gy/Bq·s) from decay of daughter radionuclide "k" originated in the cytoplasm.
$N_{(Cy)}$ = total number of nuclear transformations from each daughter radionuclide "k" in the cytoplasm.

3.9. Biodistribution Studies

A total of sixty-three animals were injected in three groups: one group of healthy mice (n = 21), and two groups of mice bearing HEP-G2 tumors (n = 42). In vivo studies with Balb-C mice (7-week-old; weight of 18–20 g) were performed in accordance with the corresponding ethical regulations for handling laboratory animals ("Official Mexican Standard" NOM-062-ZOO-1999). Mice were injected with the ^{225}Ac-rHDL system (50 μL, 0.9 MBq) in the tail vein and sacrificed at 0.5, 1, 3, 24, 48, 96, and 192 h (n = 3). The spleen, liver, kidneys, lungs, intestine, and heart were removed to measure their activity in a (NaI(Tl)) radioactivity counter, to determine the percentage of injected dose (% ID) per organ, with regard to the total injected activity.

Nude mice bearing HEP-G2 tumors (0.22 ± 0.03 g) were injected intratumorally with the ^{225}Ac-rHDL system (50 μL, 1.0 MBq) and sacrificed at 0.5, 1, 3, 24, 48, 96, and 192 h (n = 3). As a control, the ^{225}Ac-DOTA-benzene-p-SCN system (50 μL, 1.0 MBq) (n = 3) was administered and mice were sacrificed at the same time. In all mice, the tumor, liver, spleen, and kidney were removed for activity measurement in a (NaI(Tl)) radioactivity counter, to determine the percentage of injected dose (% ID) per organ, with regard to the total injected activity.

The values obtained from % ID of each organ or tumor were adjusted to exponential functions to obtain the biokinetic models ($q_h(t)$). The $A_h(t)$ functions were obtained by decay-correcting the biokinetic models; that is, by adding to the biological constant (λ_B) the radioactive constant (λ_R), as follows (Equation (2)):

$$A(t) = Be^{-(\lambda_R+\lambda_B)t} + Ce^{-(\lambda_R+\lambda_B)t} + De^{-(\lambda_R+\lambda_B)t} \qquad (2)$$

To obtain the biokinetic model, each of the radionuclides generated in the ^{225}Ac decay chain was considered as described above. The total number of nuclear transformations (N) in each murine organ were obtained by the mathematical integration (from t = 0 to t = ∞) of the biokinetic models. The DF values were obtained via the OLINDA 2.0 software, considering a normalized tumor mass of 1.0 g and all the progeny of ^{225}Ac.

4. Conclusions

In this study, the ^{225}Ac-rHDL system was prepared and evaluated as a potential targeted radiotherapeutic agent. The results showed adequate physicochemical properties of the rHDL nanocarrier to specifically deposit ^{225}Ac into the cytoplasm of HEP-G2 and PC-3 cancer cells and produce a significant cell cytotoxic effect. Biodistribution studies of ^{225}Ac-rHDL in healthy mice showed mainly liver uptake with hepatobiliary and renal excretion without appreciable accumulation in other tissues, while, in tumor-bearing mice, the ^{225}Ac-rHDL nanosystem remained stable in the tumors and generated ablative radiation doses. The results obtained in this research justify further preclinical studies designed to evaluate the therapeutic efficacy of ^{225}Ac-rHDL for targeted alpha-particle therapy of tumors that overexpress the SR-BI receptor.

Author Contributions: Conceptualization, G.F.-F., K.I.-O. and E.M.-Á.; methodology, T.H.-J., B.O.-G., C.S.-C., L.A.-L., M.L.-G. and L.D.N.; formal analysis, E.M.-Á., C.S.-C. and G.F.-F.; writing—original draft preparation, T.H.-J.; writing—review and editing, G.F.-F. and L.M.-A.; funding acquisition, G.F.-F. and K.I.-O.; writing—review, A.R. All authors have read and agreed to the published version of the manuscript.

Funding: This research was funded by the Mexican National Council of Science and Technology (CONACyT), grant CB2017-2018-A1-S-36841.

Institutional Review Board Statement: This research was approved by the Ethics Internal Committee of Use and Care of Laboratory Animals (CICUAL-ININ) of the National Institute of Nuclear Research, Approval No. 04-2018-2021.

Informed Consent Statement: Not applicable.

Data Availability Statement: Not applicable.

Acknowledgments: This study was conducted as a component of the activities of the "Laboratorio Nacional Investigación y Desarrollo de Radiofármacos, CONACyT" (LANIDER).

Conflicts of Interest: The authors declare no conflict of interest.

Sample Availability: Samples of the compounds not available are available from the authors.

References

1. Simonsen, J.B. Schwendeman, Evaluation of reconstituted high-density lipoprotein (rHDL) as a drug delivery platform—A detailed survey of rHDL particles ranging from biophysical properties to clinical implications. *Nanomedicine* **2016**, *12*, 2161–2179. [CrossRef]
2. Mooberry, L.K.; Sabnis, N.A.; Panchoo, M.; Nagarajan, B.; Lacko, A.G. Targeting the SR-B1 Receptor as a Gateway for Cancer Therapy and Imaging. *Front. Pharmacol.* **2016**, *7*, 466. [CrossRef]
3. Shah, S.; Chib, R.; Raut, S.; Bermudez, J.; Sabnis, N.; Duggal, D.; Kimball, J.; Lacko, A.G.; Gryczynski, Z.; Gryczynski, I. Photophysical Characterization of Anticancer Drug Valrubicin in rHDL Nanoparticles and Its Use as an Imaging Agent. *J. Photochem. Photobiol. B Biol.* **2016**, *155*, 60–65. [CrossRef]
4. Pérez-Medina, C.; Tang, J.; Abdel-Atti, D.; Hogstad, B.; Merad, M.; Fisher, E.A.; Fayad, Z.A.; Lewis, J.; Mulder, W.J.; Reiner, T. Pet Imaging of Tumor-Associated Macrophages with 89Zr-Labeled High-Density Lipoprotein Nanoparticles. *J. Nucl. Med.* **2015**, *56*, 1272–1277. [CrossRef]
5. Isaac-Olive, K.; Ocampo-García, B.E.; Aranda-Lara, L.; Santos-Cuevas, C.L.; Jiménez-Mancilla, N.P.; Luna-Gutiérrez, M.A.; Medina, L.A.; Nagarajan, B.; Sabnis, N.; Raut, S.; et al. [99mTc-Hynic-N-Dodecylamide]: A New Hydrophobic Tracer for Labelling Reconstituted High-Density Lipoproteins (Rhdl) for Radioimaging. *Nanoscale* **2018**, *11*, 541–551. [CrossRef]
6. Allen, B. Systemic Targeted Alpha Radiotherapy for Cancer. *J. Biomed. Phys. Eng.* **2013**, *3*, 67–80. [CrossRef]
7. Lacoeuille, F.; Arlicot, N.; Faivre-Chauvet, A. Targeted Alpha and Beta Radiotherapy: An Overview of Radiopharmaceutical and Clinical Aspects. *Méd. Nucl.* **2018**, *42*, 32–44. [CrossRef]
8. Gupta, U.; Singh, V.K.; Kumar, V.; Khajuria, Y. Spectroscopic Studies of Cholesterol: Fourier Transform Infra-Red and Vibrational Frequency Analysis. *Mater. Focus* **2014**, *3*, 211–217. [CrossRef]
9. Beć, K.B.; Grabska, J.; Huck, C.W. Near-Infrared Spectroscopy in Bio-Applications. *Molecules* **2020**, *25*, 2948. [CrossRef]
10. Peng, Y.; Akmentin, W.; Connelly, M.A.; Lund-Katz, S.; Phillips, M.C.; Williams, D.L. Scavenger Receptor BI (SR-BI) Clustered on Microvillar Extensions Suggests that This Plasma Membrane Domain Is a Way Station for Cholesterol Trafficking between Cells and High-Density Lipoprotein. *Mol. Biol. Cell* **2004**, *15*, 384–396. [CrossRef]

11. Chang, M.-Y.; Seideman, J.; Sofou, S. Enhanced Loading Efficiency and Retention of 225Ac in Rigid Liposomes for Potential Targeted Therapy of Micrometastases. *Bioconjug. Chem.* **2008**, *19*, 1274–1282. [CrossRef]
12. McDevitt, M.R.; Ma, D.; Simon, J.; Frank, R.; Scheinberg, D.A. Design and synthesis of 225Ac radioimmunopharmaceuticals. *Appl. Radiat. Isot.* **2002**, *57*, 841–847. [CrossRef]
13. Bandekar, A.; Zhu, C.; Jindal, R.; Bruchertseifer, F.; Morgenstern, A.; Sofou, S. Anti–Prostate-Specific Membrane Antigen Liposomes Loaded with 225Ac for Potential Targeted Antivascular α-Particle Therapy of Cancer. *J. Nucl. Med.* **2014**, *55*, 107–114. [CrossRef]
14. Wang, G.; de Kruijff, R.; Rol, A.; Thijssen, L.; Mendes, E.; Morgenstern, A.; Bruchertseifer, F.; Stuart, M.; Wolterbeek, H.; Denkova, A. Retention Studies of Recoiling Daughter Nuclides of 225Ac in Polymer Vesicles. *Appl. Radiat. Isot.* **2014**, *85*, 45–53. [CrossRef]
15. Biesbroeck, R.; Oram, J.F.; Albers, J.J.; Bierman, E.L. Specific High-Affinity Binding of High-Density Lipoproteins to Cultured Human Skin Fibroblasts and Arterial Smooth Muscle Cells. *J. Clin. Investig.* **1983**, *71*, 525–539. [CrossRef]
16. Sticozzi, C.; Belmonte, G.; Pecorelli, A.; Cervellati, F.; Leoncini, S.; Signorini, C.; Ciccoli, L.; De Felice, C.; Hayek, J.; Valacchi, G. Scavenger Receptor B1 Post-Translational Modifications in Rett Syndrome. *FEBS Lett.* **2013**, *587*, 2199–2204. [CrossRef]
17. Connelly, A.M.; Williams, D.L. Scavenger Receptor B1: A Scavenger Receptor with a Mission to Transport High Density Lipoprotein Lipids. *Curr. Opin. Lipidol.* **2004**, *15*, 287–295. [CrossRef]
18. Tiggemann, M. MIRD Pamphlet No. 25: MIRDcell V2.0 Software Tool for Dosimetric Analysis of Biologic Response of Multicellular Populations. *J. Nucl. Med.* **2014**, *55*, 1557–1564. [CrossRef]
19. Bannik, K.; Madas, B.; Jarzombek, M.; Sutter, A.; Siemeister, G.; Mumberg, D.; Zitzmann-Kolbe, S. Radiobiological Effects of the Alpha Emitter Ra-223 on Tumor Cells. *Sci. Rep.* **2019**, *9*, 1–11. [CrossRef]
20. Hartman, T.; Lundqvist, H.; Westlin, J.-E.; Carlsson, J. Radiation Doses to the Cell Nucleus in Single Cells and Cells in Micrometastases in Targeted Therapy with 131I Labeled Ligands or Antibodies. *Int. J. Radiat. Oncol.* **2000**, *46*, 1025–1036. [CrossRef]
21. Kuai, R.; Li, D.; Chen, Y.E.; Moon, J.J.; Schwendeman, A. High-Density Lipoproteins: Nature's Multifunctional Nanoparticles. *ACS Nano* **2016**, *10*, 3015–3041. [CrossRef]
22. Liu, T.; Liu, C.; Ren, Y.; Guo, X.; Jiang, J.; Xie, Q.; Xia, L.; Wang, F.; Zhu, H.; Yang, Z. Development of an Albumin-Based PSMA Probe with Prolonged Half-Life. *Front. Mol. Biosci.* **2020**, *7*, 585024. [CrossRef]
23. Beyer, G.J.; Bergmann, R.; Schomäcker, K.; Rösch, F.; Schäfer, G.; Kulikov, E.V.; Novgorodov, A.F. Comparison of the Biodistribution of 225Ac and Radio-Lanthanides as Citrate Complexes. *Isot. Isot. Environ. Health Stud.* **1990**, *26*, 111–114. [CrossRef]
24. Gao, Y.; Grover, P.; Schreckenbach, G. Stabilization of Hydrated AcIII Cation: The Role of Superatom States in Actinium-Water Bonding. *Chem. Sci.* **2021**, *12*, 2655–2666. [CrossRef]
25. Thiele, N.A.; Wilson, J.J. Actinium-225 for Targeted α Therapy: Coordination Chemistry and Current Chelation Approaches. *Cancer Biother. Radiopharm.* **2018**, *33*, 336–348. [CrossRef]
26. Makvandi, M.; Dupis, E.; Engle, J.W.; Nortier, F.M.; Fassbender, M.; Simon, S.; Birnbaum, E.R.; Atcher, R.W.; John, K.D.; Rixe, O.; et al. Alpha-Emitters and Targeted Alpha Therapy in Oncology: From Basic Science to Clinical Investigations. *Target. Oncol.* **2018**, *13*, 189–203. [CrossRef]
27. Miederer, M.; Scheinberg, D.A.; McDevitt, M.R. Realizing the Potential of the Actinium-225 Radionuclide Generator in Targeted Alpha Particle Therapy Applications. *Adv. Drug Deliv. Rev.* **2008**, *60*, 1371–1382. [CrossRef]
28. Tafreshi, N.K.; Doligalski, M.L.; Tichacek, C.J.; Pandya, D.N.; Budzevich, M.M.; El-Haddad, G.; Khushalani, N.I.; Moros, E.G.; McLaughlin, M.L.; Wadas, T.J.; et al. Development of Targeted Alpha Particle Therapy for Solid Tumors. *Molecules* **2019**, *24*, 4314. [CrossRef]
29. Shahzad, M.M.; Mangala, L.S.; Han, H.D.; Lu, C.; Bottsford-Miller, J.; Nishimura, M.; Mora, E.M.; Lee, J.-W.; Stone, R.L.; Pecot, C.V.; et al. Targeted Delivery of Small Interfering RNA Using Reconstituted High-Density Lipoprotein Nanoparticles. *Neoplasia* **2011**, *13*, 309–319. [CrossRef]
30. Peltek, O.O.; Muslimov, A.R.; Zyuzin, M.V.; Timin, A.S. Current Outlook on Radionuclide Delivery Systems: From Design Consideration to Translation into Clinics. *J. Nanobiotechnol.* **2019**, *17*, 1–34. [CrossRef]
31. Beattie, B.J.; Thorek, D.L.J.; Schmidtlein, C.R.; Pentlow, K.S.; Humm, J.L.; Hielscher, A.H. Quantitative Modeling of Cerenkov Light Production Efficiency from Medical Radionuclides. *PLoS ONE* **2012**, *7*, e31402. [CrossRef]

molecules

Article

A High Separation Factor for ^{165}Er from Ho for Targeted Radionuclide Therapy

Isidro Da Silva [1,2], Taylor R. Johnson [1], Jason C. Mixdorf [1], Eduardo Aluicio-Sarduy [1], Todd E. Barnhart [1], R. Jerome Nickles [1], Jonathan W. Engle [1,3] and Paul A. Ellison [1,*]

1 Department of Medical Physics, University of Wisconsin School of Medicine and Public Health, 1111 Highland Avenue, Madison, WI 53705, USA; isidro.dasilva@cnrs-orleans.fr (I.D.S.); trjohnson32@wisc.edu (T.R.J.); jmixdorf@wisc.edu (J.C.M.); aluiciosardu@wisc.edu (E.A.-S.); tebarnhart@wisc.edu (T.E.B.); rnickles@wisc.edu (R.J.N.); jwengle@wisc.edu (J.W.E.)

2 Conditions Extrêmes et Matériaux: Haute Température et Irradiation, Centre National de la Recherche Scientifique, UPR3079, Energy, Materials Earth and Universe Science Doctoral School, Université d'Orléans, F-45071 Orléans, France

3 Department of Radiology, University of Wisconsin School of Medicine and Public Health, 1111 Highland Avenue, Madison, WI 53705, USA

* Correspondence: paellison@wisc.edu

Citation: Da Silva, I.; Johnson, T.R.; Mixdorf, J.C.; Aluicio-Sarduy, E.; Barnhart, T.E.; Nickles, R.J.; Engle, J.W.; Ellison, P.A. A High Separation Factor for ^{165}Er from Ho for Targeted Radionuclide Therapy. *Molecules* **2021**, *26*, 7513. https://doi.org/10.3390/molecules26247513

Academic Editors: Alessandra Boschi and Petra Martini

Received: 8 November 2021
Accepted: 7 December 2021
Published: 11 December 2021

Publisher's Note: MDPI stays neutral with regard to jurisdictional claims in published maps and institutional affiliations.

Abstract: Background: Radionuclides emitting Auger electrons (AEs) with low (0.02–50 keV) energy, short (0.0007–40 μm) range, and high (1–10 keV/μm) linear energy transfer may have an important role in the targeted radionuclide therapy of metastatic and disseminated disease. Erbium-165 is a pure AE-emitting radionuclide that is chemically matched to clinical therapeutic radionuclide ^{177}Lu, making it a useful tool for fundamental studies on the biological effects of AEs. This work develops new biomedical cyclotron irradiation and radiochemical isolation methods to produce ^{165}Er suitable for targeted radionuclide therapeutic studies and characterizes a new such agent targeting prostate-specific membrane antigen. **Methods:** Biomedical cyclotrons proton-irradiated spot-welded Ho$_{(m)}$ targets to produce ^{165}Er, which was isolated via cation exchange chromatography (AG 50W-X8, 200–400 mesh, 20 mL) using alpha-hydroxyisobutyrate (70 mM, pH 4.7) followed by LN2 (20–50 μm, 1.3 mL) and bDGA (50–100 μm, 0.2 mL) extraction chromatography. The purified ^{165}Er was radiolabeled with standard radiometal chelators and used to produce and characterize a new AE-emitting radiopharmaceutical, [^{165}Er]PSMA-617. **Results:** Irradiation of 80–180 mg ^{nat}Ho targets with 40 μA of 11–12.5 MeV protons produced ^{165}Er at 20–30 MBq·μA^{-1}·h^{-1}. The 4.9 ± 0.7 h radiochemical isolation yielded ^{165}Er in 0.01 M HCl (400 μL) with decay-corrected (DC) yield of 64 ± 2% and a Ho/^{165}Er separation factor of (2.8 ± 1.1) · 10^5. Radiolabeling experiments synthesized [^{165}Er]PSMA-617 at DC molar activities of 37–130 GBq·μmol^{-1}. **Conclusions:** A 2 h biomedical cyclotron irradiation and 5 h radiochemical separation produced GBq-scale ^{165}Er suitable for producing radiopharmaceuticals at molar activities satisfactory for investigations of targeted radionuclide therapeutics. This will enable fundamental radiation biology experiments of pure AE-emitting therapeutic radiopharmaceuticals such as [^{165}Er]PSMA-617, which will be used to understand the impact of AEs in PSMA-targeted radionuclide therapy of prostate cancer.

Keywords: targeted radionuclide therapy; auger emission; radionuclide production; lanthanide separation; erbium-165; ^{165}Er

1. Introduction

The recent phase III clinical trial of Lutathera® ([^{177}Lu]DOTATATE) for neuroendocrine tumors [1] and phase II clinical trial of [^{177}Lu]PSMA-617 for prostate cancer [2] show receptor targeted, medium energy electron-emitting radiopharmaceuticals are effective in treating these solid tumors. However, for the treatment of micrometastatic or disseminated cancers, radiopharmaceuticals emitting shorter range, higher linear energy transfer (LET) radiations, such as Auger electrons (AEs, also known as Auger–Meitner or Meitner–Auger electrons [3,4]), show potential in preclinical studies [5–7]. Following decay, AE-emitting radionuclides release a cascade of 0.02–50 keV electrons that stop with high LET over subcellular, nano- to micrometer ranges. These properties give AE-based radiopharmaceuticals reduced crossfire irradiation of healthy tissues and off-target dose burden compared with β^--emitting radionuclide therapeutics. Furthermore, when AE-emitting radiopharmaceuticals are specifically localized to radiation-sensitive subcellular structures such as DNA [8], this may result in higher maximum tolerated doses and expanded therapeutic windows. The biological impact of these AEs is also evident in comparison studies of ^{177}Lu- and ^{161}Tb-based radiopharmaceuticals [9–11], which are biologically, chemically, and physically matched, with the exception that ^{161}Tb has ten times larger AE emission yields (Table 1) [12]. Fundamental radiation biology studies of the effects of AEs require a pure AE-emitting radionuclide, with minimal concomitant medium/high energy electron or photon emissions. Erbium-165 decays by electron capture with only low energy X-ray and AE emissions (Table 1) [12]. As a heavy lanthanide, ^{165}Er can radiolabel the same biological targeting vectors used to deliver ^{177}Lu and ^{161}Tb [13], allowing for comparative studies of pure AE-emitting (^{165}Er), β^--emitting (^{177}Lu), and mixed AE- and β^--emitting (^{161}Tb) radiopharmaceuticals. Thus, ^{165}Er is a useful tool for fundamental studies on the biological effects of AEs and, if incorporated into an appropriate biological targeting vector, for the targeted radionuclide therapy of metastatic and disseminated disease.

Table 1. Comparison of AEs and β^--particles emitted by ^{165}Er, ^{161}Tb, and ^{177}Lu.

R.	Half-Life (d)	Avg. AEs per Decay	Avg. Energy per AE (keV)	Avg. β^- per Decay	Avg. Energy per β^- (keV)
^{177}Lu	6.64	1.1	1	1	133
^{161}Tb	6.89	11	5.7	1	154
^{165}Er	0.43	7.3	11	0	0

Proton, deuteron, or alpha particle irradiation of erbium or holmium targets produces no-carrier-added ^{165}Er through a variety of nuclear reaction routes [14]. Proton or deuteron irradiation of erbium produces ^{165}Tm ($t_{1/2}$ = 30.06 h) via ^{nat}Er$(p,xn)^{165}$Tm [15,16] or ^{nat}Er$(d,xn)^{165}$Tm [17], respectively, which can be chemically isolated prior to its β^- decay to ^{165}Er. While these routes offer high ^{165}Er yields, they require a medium energy, multiparticle research cyclotron and expensive, isotopically enriched targets for highest yields. Irradiation of naturally monoisotopic holmium targets produces ^{165}Er using 35–70 MeV alpha particles via ^{165}Ho$(\alpha,4n)^{165}$Tm$(\beta^-$ decay$)^{165}$Er [18,19], 10–20 MeV deuterons via ^{165}Ho$(d,2n)^{165}$Er [20,21], or 6–16 MeV protons via ^{165}Ho$(p,n)^{165}$Er [22–24]. The existence of more than 500 biomedical cyclotrons capable of accelerating low energy protons makes the latter route particularly accessible to the research community worldwide [25].

Receptor-targeted therapeutic radiopharmaceuticals for human use are radiolabeled using 10–40 GBq ^{177}LuCl$_3$ with ~90% radiolabeling yield at ~60 MBq/nmol molar activity [26–28]. To accomplish these molar activities for biomedical cyclotron-produced ^{165}Er, the radionuclide must be chemically isolated from the holmium target material with a high ($>10^5$) separation factor (SF) due to residual target material competition with ^{165}Er during radiolabeling chemistry. Radiochemical separation of ^{165}Er from a Ho cyclotron target is challenging as adjacent lanthanides have identical oxidation states, similar coordination chemistry, and unfavorable ^{165}Er (~10 ng) to Ho (~100 mg) mass ratio. Radiochemical separations of adjacent lanthanides have traditionally been performed using cation exchange (CX) chromatography with the complexing agent α-hydroxy isobutyrate (αHIB) [29,30], with recent reports effectively separating Gd/Tb [31,32], Dy/Ho [33], Ho/Er [22,24], and Er/Tm [34,35]. Additionally, extraction chromatography (EXC) resins [36,37] impregnated with acidic organophosphorous extractants including bis(2-ethylhexyl) phosphoric acid (HDEHP, LN resin) [38,39], 2-ethylhexyl phosphonic acid mono-2-ethylhexyl ester

(HEHEHP, LN2 resin) [40–43], and bis(2,4,4-trimethyl-1-pentyl)phosphinic acid (H[TMPeP], LN3 resin) [24,43] have been used to separate adjacent heavy lanthanides.

A recent publication of the biomedical cyclotron production and radiochemical isolation of ^{165}Er from holmium reported the production of 1.6 GBq ^{165}Er in a 10 h, 10 μA proton irradiation [24]. The ^{165}Er was isolated in a 10 h process through a CX/αHIB column using a proprietary resin with 12–22 μm particle size followed by an EXC column using LN3 resin to concentrate the product. While the isolation of ^{165}Er from macroscopic Ho was achieved, neither the residual mass of Ho in the ^{165}Er product, the Ho/Er SF, nor radiopharmaceutical labeling results were reported [24]. The present work aims to improve the irradiation intensity tolerance of holmium targets to allow for the production of GBq-scale ^{165}Er in shorter irradiations, develop a radiochemical isolation process that utilizes commercially available resins to achieve a high Ho/Er SF in shorter times, and demonstrate the radiopharmaceutical quality of the produced ^{165}Er through chelator-based titration apparent molar activity (AMA) measurements and labeling a clinically relevant DOTA-based radiopharmaceutical (PSMA-617).

2. Materials and Methods

2.1. Ho Target Preparation and Irradiation

Cyclotron targets were prepared from holmium metal foils. Initially, 300–640 μm thick holmium foils (99.9%, Alfa Aesar, Haverhill, MA, USA) were used. Based on certificates of analysis, two differing foil lots contained 0.06% (600 ppm) or <0.01% (<100 ppm) erbium. High-purity 0.5 mm-thick, 10 mm diameter holmium metal discs with 0.5 ppm Er were purchased from the U.S. Department of Energy (DOE) Ames Laboratory Materials Preparation Center (MPC), Ames, IA, USA. Based on proton energy loss calculations [44], a 300 μm holmium degrades 12.5 MeV protons to 7.1 MeV, below which the ^{165}Ho$(p,n)^{165}$Er nuclear reaction cross-section vanishes [24]. Holmium foils were wrapped in 25 μm stainless steel foil and rolled to desired thickness (190–400 μm) using a commercial rolling mill. A commercial disc cutter (Pepetools) was used to punch holmium discs of desired diameter (3–9.5 mm). The holmium target disc was centered and spot-welded to a 0.5 mm-thick, 19 mm diameter tantalum disc using a variable transformer-controlled (60–75% power) commercial 115 V spot welder fitted with a copper or silver electrode as previously described [45]. Roughly 6–10 individual spot welds uniformly cover the 7 to 70 mm^2 holmium target.

Spot-welded holmium metal targets were proton irradiated at the University of Wisconsin using two biomedical cyclotrons—an RDS-112 (CTI Cyclotron Systems, Knoxville, TN, USA) and a PETtrace (GE Healthcare, Uppsala, Sweden). For RDS-112 irradiations, the target disc was clamped to a water-jet-cooled target support fixture with a 12.7 mm apertured aluminum ring and irradiated with 10–20 μA of undegraded 11 MeV protons. For PETtrace irradiations, a commercial solid target irradiation and transfer system (ARTMS QIS, Vancouver, BC, Canada) was used. Holmium targets discs were assembled into the water-jet cooled transfer capsule, positioned 3.6 cm down-beam from a water-cooled 500 μm thick aluminum degrader, and irradiated with 20–40 μA of 12.5 MeV protons.

For low intensity irradiations, ^{165}Er was quantified by high-purity germanium (HPGe) gamma spectrometry (full width at half maximum at 1333 keV = 1.6 keV, Canberra Inc., Meriden, CT, USA) of the irradiated target disc, while correcting for the self-attenuation of the 46–55 keV x-rays. The HPGe low energy range (30–300 keV) was efficiency calibrated using ^{133}Ba and ^{241}Am calibration standards (Amersham, United Kingdom).

2.2. ^{165}Er Radiochemical Isolation

After irradiation, holmium cyclotron targets were dissolved in 11 M HCl (2 mL, Trace Select Ultra, Fluka Analytical, Buchs, Switzerland), followed by evaporation to dryness at 120 °C under Ar flow. The resulting yellow/pink salts were dissolved in 70 mM αHIB (2 mL, pH = 4.7). The αHIB solution was freshly prepared by dissolving α-hydroxyisobutyric acid (99%, Sigma Aldrich, St. Louis, MO, USA or 98%, Acros Organics, Geel, Belgium) with 18 MΩ·cm ultrapure water (Milli-Q) and pH adjusted with 25% ammonia solution (Fisher Scientific, Waltham, MA, USA) using a pH probe (Checker® pH tester, Hanna Instruments, Woonsocket, RI, USA). A 30–60 μL aliquot of the dissolved, reconstituted target was assayed for ^{165}Er radioactivity by ionization chamber dose calibrator (CRC-15R, setting #260, Capintec Inc., Florham Park, NJ, USA). The dose calibrator setting #260 was experimentally determined by cross-calibration with HPGe gamma spectrometry. The ^{165}Er activity in highly concentrated (50–90 mg/mL) holmium solutions was corrected for self-attenuation of the 46–55 keV X-rays.

Preparation of radiopharmaceutical quality ^{165}Er from bulk holmium was accomplished through a three-step radiochemical isolation process. First, a 1 cm diameter, 25 cm-long CX column (AG50W-X8, 230–400 mesh, 63–150 μm, 1.7 meq/mL, NH_4^+ form, Bio-Rad, Hercules, CA, USA) was equilibrated with water (~100 mL), then 70 mM αHIB (pH = 4.7, ~100 mL), followed by injection of the ^{165}Er/Ho/αHIB solution and elution with 5 mL/min 70 mM αHIB (pH = 4.7, 440–820 mL). Under these conditions, ^{165}Er elutes before holmium and the first ~90% of eluted ^{165}Er was collected (150–350 mL) for secondary purification, as determined by monitoring the column effluent using a shielded inline radiation detection system consisting of a CsI(Tl) scintillator (1 × 1 cm, Hilger Crystals Ltd., Kent, UK) coupled to a photomultiplier tube (E849-35, Hamamatsu Photonics, Hamamatsu City, Japan) powered and processed with bench-top electronics (925-SCINT, Ortec, Oak Ridge, TN, USA) and logged using a digital counter (USB-6008, National Instruments Corp., Austin, TX, USA). Then, the mobile phase was changed to 0.5 M αHIB (pH = 4.7, 250 mL) for stripping the remaining bulk holmium from the cation exchange column followed by water (200–500 mL) for column storage. The ^{165}Er radioactivity eluting in each collected CX fraction was quantified by a dose calibrator immediately after elution.

The second step utilized EXC with a commercially available HEHEHP-impregnated resin (LN2, 20–50 μm, Triskem Int., Bruz, France) [37], based on previous literature work on its use for Ho/Er separations [41]. Fritted polypropylene columns (5.5 mm diameter, 1 mL, Supelco Inc., Bellefonte, PA, USA) were dry-packed with resin (500 mg) and preconditioned with 1 M HNO_3 (5 mL) followed by 0.1 M HNO_3 (25 mL). All nitric acid solutions were prepared using 16 M HNO_3 (TraceSelect, Fluka Analytical, Buchs, Switzerland) and 18 MΩ·cm ultrapure water (Milli-Q). The ^{165}Er/Ho/αHIB eluate from separation step 1 was acidified to 0.1 M HNO_3 using 16 M HNO_3 and passed through the EXC column at 5.8 ± 0.9 mL/min using a peristaltic pump (*n* = 10, WPM1-P1CA-WP Welco Co. Ltd., Fuchū, Japan). With a lower flow rate of 1.2 ± 0.1 mL/min (*n* = 10, WPM1-P1BB-BP, Welco Co. Ltd., Fuchū, Japan), holmium was eluted with 0.4 M HNO_3 (40–60 mL), followed by ^{165}Er eluted with 1 M HNO_3 (4–6 mL).

In the third separation step, the ^{165}Er-rich fractions from separation step 2 were acidified from 1 M to 5 M HNO_3, loaded onto a N,N,N′,N′-tetra-2-ethylhexyldiglycolamide (100 mg, bDGA, 50–100 μm, Triskem Int., Bruz, France) EXC column. The column was then rinsed with 3 M HNO_3 (15 mL) and 0.5 M HNO_3 (2 mL). The ^{165}Er was subsequently eluted with 0.01 M HCl (1–1.5 mL).

2.3. ^{165}Er Quality Control

Holmium mass across the separation procedure was quantified using microwave plasma atomic emission spectrometry (MP-AES, MP4200, Agilent Technologies, Santa Clara, CA, USA). Holmium standard solutions of 0.1–50 ppm were made by dissolving holmium chloride hydrate (REaction® 99.99% (REO), Alfa Aesar, Haverhill, MA, USA) in water or holmium metal (US DOE Ames Laboratory MPC, Ames, IA, USA) in 11 M HCl, followed by dilution in 0.1 M HCl. The limit of detection for holmium was determined to be ~0.1 ppm in an undiluted sample solution. The MP-AES-analyzed samples were typically diluted by a factor of 2–10 in 0.1 M HCl. The Ho/Er SFs of the various separation steps and the overall separation procedure were calculated according to Equation (1) with $m_{Ho,before/after}$ being the MP-AES-quantified holmium mass before/after the separation step and $A_{165Er,\,before/after}$ being the dose-calibrator-quantified ^{165}Er activity before/after the separation step.

$$SF_{Ho/Er} = \frac{\dfrac{m_{Ho,\,before}}{m_{Ho,after}}}{\dfrac{A_{165Er,before}}{A_{165Er,after}}} \tag{1}$$

The apparent molar activity (AMA) of the final ^{165}Er solution was determined by titration using tetraazacyclododecane-1,4,7,10-tetraacetic acid (DOTA, Macrocyclics Inc, Plano, TX, USA) and diethylenetriaminepentaacetic acid (DTPA, Acros Organics, Geel, Belgium) as previously described [46]. To polypropylene tubes, 0.5–3 MBq (20 µL) of ^{165}Er in 0.01 M HCl, 1 M NaOAc (100 µL, pH 4.7, 99.995% trace metals basis, Aldrich, St. Louis, MO), and DOTA or DTPA (100 µL, 0.03–3 µg/mL) in 18 MΩ·cm water were added. Following incubation at 85 °C for DOTA and 21 °C for DTPA for 1 h, each tube was assayed by thin layer chromatography (TLC) using silica-based stationary phase (J.T. Baker, Phillipsburg, NJ, USA) and 0.05 M disodium ethylenediaminetetraacetic acid (EDTA, Fisher Scientific Co., Pittsburgh, PA, USA) as mobile phase. Radioactivity distribution on the TLC plates was visualized using a Cyclone Plus phosphor plate reader (Perkin Elmer, Waltham, MA, USA). Free ^{165}Er had a retention factor (R_f) of ~1, chelated ^{165}Er-DOTA had an R_f of ~0.2, and ^{165}Er-DTPA had an R_f of ~0.7. To compute the AMA, ^{165}Er activity in MBq was divided by twice the number of moles of DOTA/DTPA required to complex 50% of the radioactivity, and the value was reported as MBq of ^{165}Er per nmol of ligand, or MBq/nmol (mean $\pm$ standard deviation, SD).

2.4. Radiosynthesis and Characterization of $[^{165}Er]PSMA$-617

PSMA-617 (MedChemExpress, Monmouth Junction, NJ, USA) was dissolved in 18 MΩ·cm water to a concentration of 0.1 µg/µL and distributed into 10 aliquots that were stored at –20 °C. A sodium acetate solution (1 M, 99.995% trace metals basis, Aldrich) was buffered with hydrochloric acid until a pH of 5.7 was obtained. Er-165 (41–52 MBq in 70–80 µL 0.01 M HCl) was added to a solution of PSMA-617 in water (1 nmol in 10 µL, 2 nmol in 10 µL, or 5 nmol in 25 µL), NaOAc aq. (1M, pH 5.7, 50 µL), and L-ascorbic acid (0.3–0.4 mg, 25 µL, $\geq$99.9998 trace metals basis, Honeywell-Fluka), which were reacted at 80 °C and 1000 rpm for 30 min. The reaction was diluted in 18 MΩ·cm water (10 mL) and loaded onto a pre-equilibrated C-18 Plus Light cartridge; after rinsing with water (10 mL), elution was performed with pure ethanol (700 µL). An aliquot (25 µL) was diluted with 18 MΩ·cm water (25 µL) and set aside for quality control analysis. Analytical HPLC was performed on $[^{165}$Er]PSMA-617 using an Agilent 1260 Infinity II module coupled with an inline radioactivity detection system similar to that described in Section 2.2 but with electronic analog pulse converted to voltage (Model 106, Lawson Labs Inc., Malvern, PA, USA) and logged using an Agilent 1200 universal interface box. The separation was performed on reverse-phase C18 column (InfinityLab Poroshell 120 EC-C18, 4.6 × 100 mm, 2.7 µm, Agilent Technologies, Santa Clara, CA, USA) using a linear gradient (95 to 20% over 15 min) of 0.1% trifluoroacetic acid (Thermo Scientific in 18 MΩ·cm water) in acetonitrile (HPLC-grade, Fisher Scientific, Pittsburgh, PA, USA) at 1 mL/min.

In vitro stability of $[^{165}$Er]PSMA-617 was investigated in the presence of L-ascorbic acid and freshly prepared normal human serum prepared from lyophilized powder (009-000-001, Jackson ImmunoResearch Laboratories, Inc., West Grove, PA, USA) reconstituted using phosphate buffered saline (2.0 mL, PBS, Lonza Bioscience). Dry $[^{165}$Er]PSMA-617 (3.1 MBq) and L-ascorbic acid (0.6 mg) were dissolved in serum (300 µL) and incubated at 37 °C for 12 h. Aliquots of the $[^{165}$Er]PSMA-617 complex were assessed by analytical HPLC at t = 1 and 12 h. The chromatograms were analyzed by integration of the $[^{165}$Er]PSMA-617 peak compared to all radioactive peaks (free erbium or decomposition products) in the chromatogram.

The distribution coefficient (logD value) of $[^{165}$Er]PSMA-617 was determined using a 1:1 (v/v) solution of n-octanol (Alfa Aesar, Haverhill, MA, USA) and PBS according to previously reported methods [11]. A sample of $[^{165}$Er]PSMA-617 (8.2–51 MBq,

2.8–8.4 MBq/nmol) was dried and diluted with PBS and n-octanol (700 μL each). The solution was vigorously agitated for 5 min before being centrifuged at 1000 rpm for 5 min. An aliquot of PBS and n-octanol was analyzed by HPGe gamma spectrometry under identical geometries and the ratio of decay-corrected net counts per minute was used to determine the distribution coefficient. The PBS aliquot required 4–5 days of decay before acquisition of an HPGe spectrum with acceptable (<5%) dead-time. Uncertainty in the LogD value was calculated by propagation of error associated with HPGe counting statistics and in the half-life of ^{165}Er (10.36 ± 0.04 h [47]). The logD experiment was repeated for five independent preparations of [^{165}Er]PSMA-617 and logD reported as average and standard deviation of results.

3. Results

3.1. Ho Target Preparation and Irradiation

Cold rolling, disc cutting, and spot-welding methods were well suited for the fabrication of cyclotron irradiation targets of tightly controlled dimensions and mass from a variety of commercial holmium metal sources. The malleability of holmium allowed for the dramatic thinning of metal foils through rolling. While thickness reduction by factors of two or three was readily achieved, when an eight-fold change in thickness was attempted, cracking around the edges was observed, as shown in supplementary Figure S1. Spot-welded holmium was well adhered to the tantalum backing and withstood proton irradiation at all investigated intensities with only minor discoloration, as shown in Figure 1.

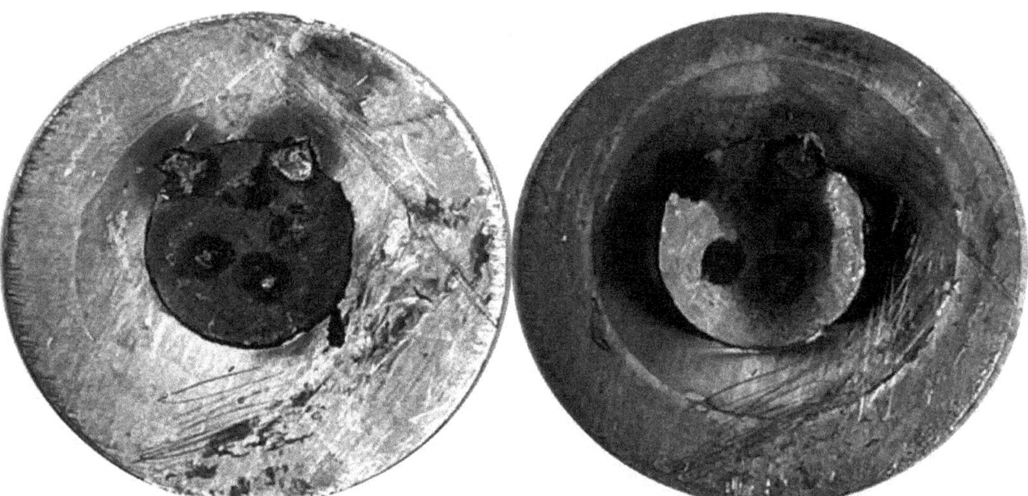

Figure 1. Representative image of a 7.9 mm ø, 270 μm-thick, 106 mg holmium disc spot-welded to tantalum, before and after 68 min, 40 μA·h PETtrace irradiation.

For the PETtrace cyclotron, a proton irradiation energy of 12.5 MeV centers the ^{165}Ho$(p,n)^{165}$Er excitation function peak (see supplementary Figure S2 [22–24]) within the energy loss window of the protons traversing a 200–300 μm-thick holmium target. Experimental end-of-bombardment (EoB) ^{165}Er physical yields were measured via attenuation-corrected HPGe of target discs or dose calibrator measurements of dissolved target aliquots, or CX elution fractions (Table 2). The yields show a significant dependence on the holmium target dimensions and the irradiating cyclotron. Calculated using literature cross-sections [23,24], a 1–2 h proton irradiation of 275 μm-thick holmium results in physical ^{165}Er yields of 50 MBq·μA^{-1}·h^{-1} at 12.5 MeV and 39 MBq·μA^{-1}·h^{-1} at 11 MeV.

Experimental ^{165}Er physical yields were significantly lower than these theoretical maxima, likely because the cyclotron-integrated charge includes protons impinging on the target outside the holmium diameter. This is especially problematic for the PETtrace cyclotron, which has an oblong beam spot with full width at half maxima of 11 and 8.7 mm, as measured by autoradiography of irradiated aluminum discs. The RDS-112 cyclotron provides a significantly smaller beam spot, resulting in higher overall ^{165}Er physical yield, despite the lower irradiation energy and smaller target diameter. However, the RDS-112 cyclotron is limited to maximum irradiation current of 20 µA, a factor of 2 below that routinely used with the PETtrace cyclotron.

Table 2. Erbium-165 physical yields for different Ho targets and irradiation configurations.

| Cyclotron | Ho Dimensions | | | E_{in} (MeV) | E_{out} (MeV) | ^{165}Er Physical Yield (MBq·µA^{-1}·h^{-1}) | n |
	Diam (mm)	Thick. (mm)	Mass (mg)				
PETtrace	9.5	280–300	174 ± 8	12.5	7.5	24.1 ± 0.5	5
PETtrace	9.5	200–240	125 ± 6	12.5	8.4–9.1	19.1 ± 1.1	3
PETtrace	7.9	270–280	108 ± 4	12.5	7.8	14.1 ± 1.4	3
PETtrace	7.9	190	69 ± 1	12.5	9.3	12.0 ± 0.9	4
RDS-112	7.9	190–280	84 ± 22	11	5.3–7.5	28.0 ± 1.8	5
RDS-112	6.4	320–620	121 ± 75	11	<4.8	30.0 ± 6.1	2
RDS-112	4.8	320	48	11	4.2	23	1
RDS-112	4.8	180	23	11	7.7	13	1
RDS-112	3	320	22	11	4.2	16	1
RDS-112	3	180	10	11	7.7	9	1

3.2. ^{165}Er Radiochemical Isolation

The holmium target was dissolved over 5 min at room temperature and evaporated to dryness in 30 min. The time of dissolution/evaporation/reconstitution/CX injection was 50 ± 20 min (n = 13).

Step 1: CX/αHIB

The first step of the ^{165}Er isolation procedure accomplishes a bulk holmium/erbium separation while accommodating 180 mg holmium loading masses through CX/αHIB column chromatography with commercially available CX resin in a standard stainless steel semipreparative high pressure chromatography column housing. This 19.6 mL column has a theoretical capacity of 1.8 g of trivalent Ho^{3+}, ten times larger than the intended holmium loading masses. Based on the Dy/Ho separation process of Mocko et al. [33], 70 mM αHIB (pH = 4.7) was used as mobile phase. When mobile phase was freshly prepared and carefully pH adjusted to within 0.05 pH units, consistent retention times were observed with 90% of the total ^{165}Er radioactivity eluting in a Gaussian-shaped peak 40–90 min after injection, as shown in the representative radiochromatogram (Figure 2). Holmium elutes after ^{165}Er with its leading edge beginning at ~350 mL as seen through the diminishing Ho/Er SF with increasing ^{165}Er fraction collection volume. Use of mobile phase that was not freshly prepared or incorrectly adjusted to too low of a pH resulted in significantly longer retention times and diminished separation of ^{165}Er and Ho. With 5 mL/min flow, the column pressure was routinely 17–19 MPa. Following ^{165}Er elution, the column was stripped with freshly prepared 0.5 M αHIB (100 mL, pH = 4.7), followed by water (250 mL). The column was re-used until the flow pressure significantly increased (>22 MPa), upon which it was disassembled and repacked with fresh resin slurry, every ~10 uses.

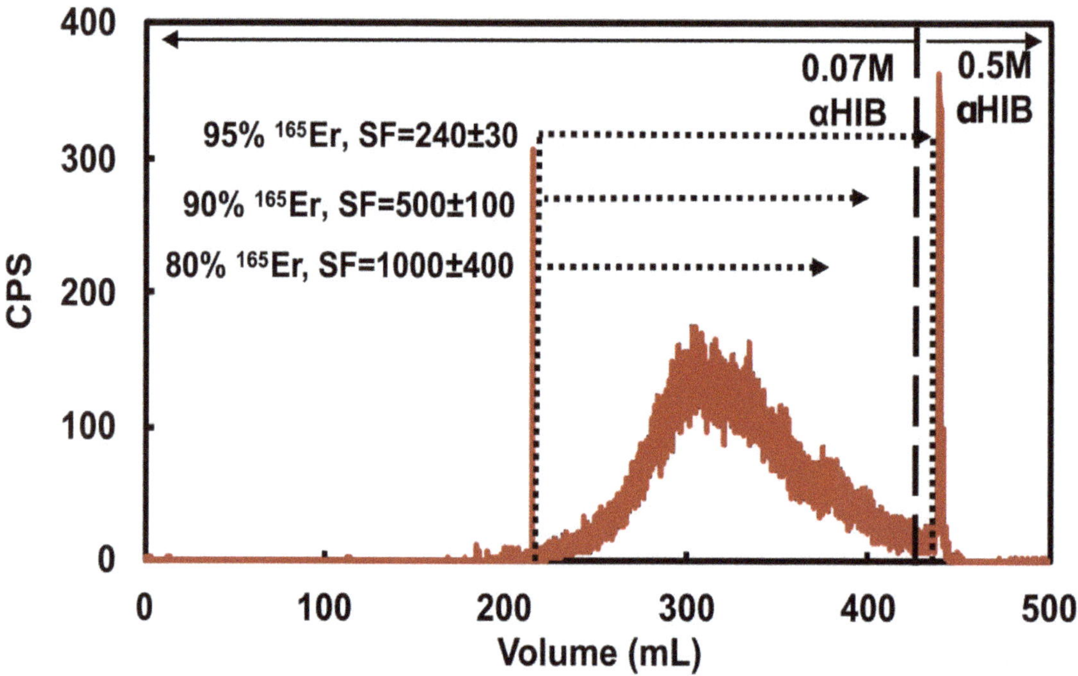

Figure 2. Representative ^{165}Er radioactivity elution profile from a cation exchange column loaded with 178 mg holmium and eluted with 5 mL/min 0.07 M αHIB (pH = 4.7). Dotted lines and corresponding text highlight three possible ^{165}Er-rich fraction collection volumes demonstrating the balance between ^{165}Er recovery and Ho/Er SF. The 0.5 M αHIB (pH = 4.7) rapidly elutes remaining ^{165}Er along with bulk holmium.

As determined by dose calibrator measurements of ^{165}Er and MP-AES measurements of Ho in the eluted CX fractions, the CX/αHIB column accommodated 180 mg Ho loading mass and effectively removed bulk Ho with acceptable ^{165}Er yield. Loading 111 ± 17 mg Ho and recovering 94.7 ± 2.5 % of ^{165}Er resulted in a SF of 320 ± 210 ($n = 6$). Loading 174 ± 8 mg Ho and recovering 95.3 ± 1.8 % of ^{165}Er resulted in a SF of 130 ± 60 ($n = 5$). For the larger loading masses, decreasing ^{165}Er recovery to 90.5 ± 1.4 % resulted in an SF of 250 ± 150 ($n = 5$) and further decreasing ^{165}Er recovery to 80% resulted in a SF of 1000 ± 400 ($n = 1$). These results indicate the sensitivity of the CX/αHIB separation to Ho loading mass and demonstrate the challenging balance between ^{165}Er recovery and Ho/Er SF. To ensure a reproducible, optimal balance between yield and SF for this step, an inline radiation detector was used to determine when to stop collecting the ^{165}Er fraction. Following loading ≤ 120 mg Ho, ^{165}Er fraction collection was ended when the radioactivity signal was ~1/10th maximum value, resulting in ~95% ^{165}Er recovery. Following loading ~180 mg Ho, ^{165}Er fraction collection was ended when the radioactivity signal was ~1/4th maximum value, resulting in ~90% ^{165}Er recovery.

Step 2: LN2 EXC

The second step of the ^{165}Er isolation procedure accomplishes a high Ho/Er SF while accommodating milligram quantity holmium masses through EXC using commercially available LN2 resin in a polypropylene column. When filled to maximum capacity (500 mg), the 1.3 mL column has a theoretical capacity of 36 mg of trivalent lanthanides according to the Triskem product sheet. However, a significant decrease in chromatographic performance was observed when loading more than 5% theoretical capacity, limiting the LN2 column capacity to 1–2 milligrams of holmium (see Supplementary Material Section S1).

Following the CX/αHIB column, the acidified Ho/^{165}Er solution was loaded onto the LN2 column, trapping both ^{165}Er and Ho. Based on previously published studies [41], the column was rinsed with 0.4 M HNO$_3$ to affect the differential elution of holmium before erbium. As shown in Figure 3, elution with 0.4 M HNO$_3$ (50 mL) removed >99% of the holmium, along with a cumulated ~20% of the ^{165}Er. The remaining ^{165}Er was rapidly eluted with 1 M HNO$_3$ (~5 mL). To avoid moderate to severe decrease in chromatographic performance, care was taken to prevent the resin bed from going dry during use and columns were freshly packed and conditioned prior to each experiment (see Supplementary Material Sections S2 and S3).

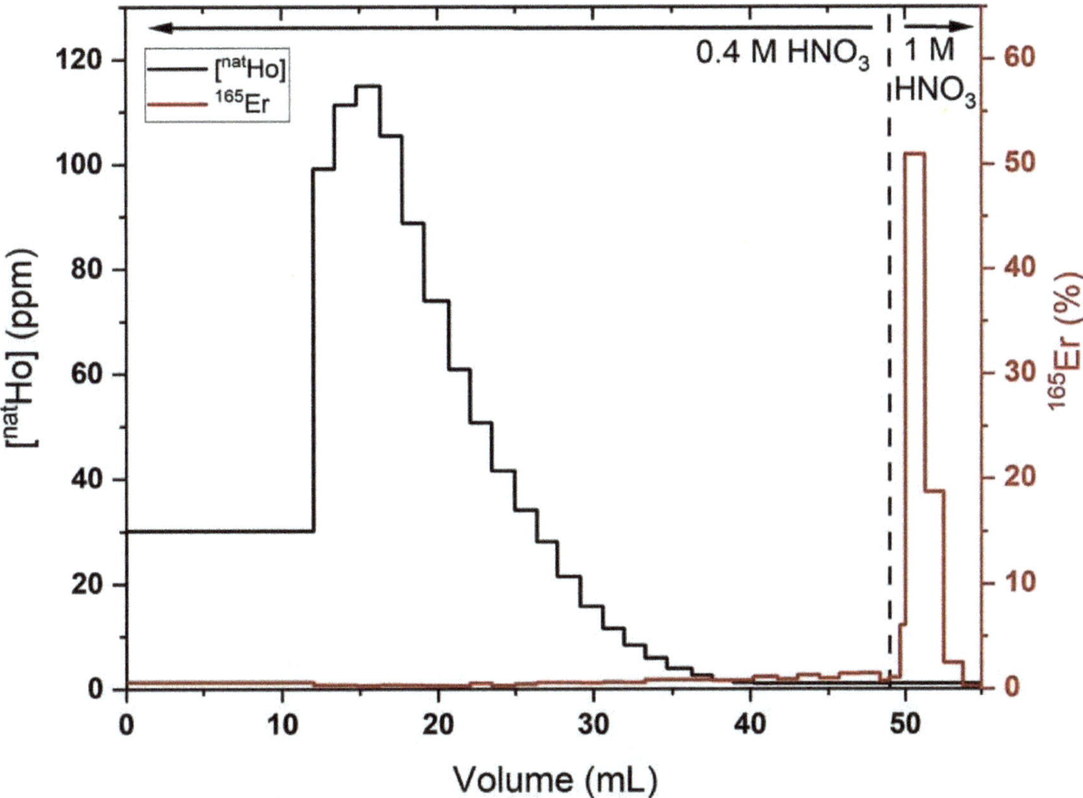

Figure 3. Holmium (black lines, quantified by MP-AES) and ^{165}Er (red lines, quantified by radioactivity dose calibrator) elution profiles from a representative 500 mg LN2 column loaded with ~1 mg holmium, ~2 MBq ^{165}Er in 0.1 M HNO$_3$ (200 mL), 70 mM αHIB and eluted with 0.4 M HNO$_3$ (50 mL), followed by 1 M HNO$_3$ (5 mL). Fractions with Ho/^{165}Er below limits of detection (Ho MP-AES: 1 ppm, ^{165}Er: 4 kBq) shown as upper limits.

In the optimized procedure, LN2 columns loaded with Ho (570 ± 370 µg) and rinsed with 0.4 M HNO$_3$ (52 ± 9 mL) resulted in 78 ± 6% ^{165}Er recovery and a Ho/Er SF of 1020 ± 320 (n = 4). The LN2 SF was estimated from the Ho mass in the final preparation of ^{165}Er, assuming no Ho/Er separation was achieved with the final bDGA column. Three additional replicates where holmium was below the MP-AES detection limit and four additional replicates with deviations from the above-described experimental procedure were excluded from this analysis. In two of these excluded replicates, a 0.4 M HNO$_3$ (42 mL) rinse resulted in 96% ^{165}Er recovery and a Ho/Er SF of 290 and a 0.4 M HNO$_3$ (52 mL) rinse resulted in a 63% ^{165}Er recovery and an SF of >2000. These results demonstrate the sensitivity of the procedure to HNO_3 concentration/volume and the interplay between Ho/Er SF and ^{165}Er recovery. The latter of the two results indicates that simply using a set volume of 0.4 M HNO_3 to remove Ho may lead to irreproducible ^{165}Er recovery and Ho/Er SF. To ensure a reproducible, optimal balance between recovery and SF, the radioactivity in the 0.4 M HNO_3 eluant is quantified by a dose calibrator as it is collected in 1–5 mL fractions. After ~20% of the loaded radioactivity had eluted, the mobile phase was changed to 1 M HNO_3 to elute the remaining high purity ^{165}Er with the optimized results reported above.

Step 3: bDGA EXC

The final step of the ^{165}Er isolation procedure reduced trace metal impurities while concentrating the product into a small volume, low acidity solution suitable for radiolabeling using a commercially available bDGA resin. The loading (5 M HNO_3, 5 mL) and rinsing (3 M HNO_3, 15 mL) solution concentrations were chosen due to literature studies showing high erbium and low transition metal (Fe, Co, Ni, Cu, Zn) affinities on a similar DGA resin [48]. A small volume rinse (0.5 M HNO_3, 2 mL) decreased column acidity and minimized subsequent ^{165}Er elution volume. Following this loading and rinsing routine, 97 ± 2 % of the loaded ^{165}Er was recovered in 0.01 M HCl (1.2 ± 0.2 mL) ($n = 18$). Smaller elution volumes were attainable by fractionation of the 0.01 M HCl elution solution, with the optimized ($n = 5$) elution profile shown in Table 3.

Table 3. Optimized ^{165}Er elution profile bDGA EXC column.

Fraction	Volume (μL)	^{165}Er Yield (%)
1	210 ± 20	0.1 ± 0.2
2	390 ± 30	88 ± 4
3	420 ± 40	9 ± 4
4	490 ± 30	1.0 ± 0.3
Column	dry	1.8 ± 0.5

Overall separation:

The overall chemical isolation procedure from start-of-dissolution to final ^{165}Er isolation in 0.01 M HCl (~400 μL) was 4.9 ± 0.7 h ($n = 13$). The optimized procedure had a decay-corrected ^{165}Er recovery of 64 ± 2% and a Ho/Er SF of $(2.8 \pm 1.1) \cdot 10^5$ and was validated for holmium target masses up to 180 mg, with the final ^{165}Er fraction containing 370 ± 180 ng of residual holmium and no detectable radionuclidic impurities (Supplementary Figure S6) after separation ($n = 4$).

3.3. DTPA/DOTA AMA Determination

Following successful ^{165}Er isolation from low-purity Ho foils (<100 ppm Er, Alfa Aesar, Haverhill, MA, USA) and high-purity Ho foils (0.5 ppm Er, DOE Ames Laboratory MPC, Ames, IA, USA), EoB-decay-corrected AMA values determined by DTPA and DOTA titration are shown in Table 4.

Table 4. Chelator-titration-based AMA results for ^{165}Er isolated from various qualities of Ho targets.

Ho Target Er Impurity (ppm)	DOTA AMA [†] (MBq/nmol)	DTPA AMA [†] (MBq/nmol)	n
<100	11 ± 4	10 ± 4	5
0.5	20 ± 24 *	92 ± 97	4

[†] AMA decay-corrected to end of bombardment. * $n = 3$ replicates.

3.4. Radiosynthesis and Characterization of [^{165}Er]PSMA-617

Following successful isolation, the [^{165}Er]PSMA-617 was synthesized with the labeling yields summarized in Table 5. Radioanalytical HPLC showed [^{165}Er]PSMA-617 (UV retention time (RT) = 6.45 min, radioactivity RT = 6.60), co-eluting with cold Er-PSMA-617 and Ho-PSMA-617 standards (230 nm RT = 6.45 min), shown in Supplementary Figure S7. Radiochemical purity of the [^{165}Er]PSMA-617 was > 95%. Visible also in the 230 nm absorbance chromatograph of the [^{165}Er]PSMA-617 radiopharmaceutical were multiple mass peaks with RT = 6.8–7.0 min, retention times corresponding to [^{nat}Zn]PSMA-617 (230 nm RT = 6.90 min), [^{nat}Fe]PSMA-617 (230 nm RT = 6.80 min), and [^{nat}Cu]PSMA-617 (230 nm RT = 6.95 min) (see Supplementary Figure S8). The [^{165}Er]PSMA-617 radiochemical purity remained > 95% after 12 h incubation in whole human serum, the longest time point assessed. The octanol-PBS distribution coefficient of [^{165}Er]PSMA-617 was -3.3 ± 0.3 ($n = 5$).

Table 5. [^{165}Er]PSMA-617 radiolabeling yields.

PSMA-617 (nmol)	^{165}Er Activity[†] (MBq)	Labeling Yield (%)	Labeled MA [†] (MBq/nmol)	n
1	170 ± 88	49 ± 7	82 ± 45	3
2	76	97	37	1
5	250	100	50	1

[†] Radioactivity and MA decay-corrected to end of bombardment.

4. Discussion

For application in receptor-targeted therapeutic radiopharmaceuticals, a high molar activity is necessary to achieve good radiolabeling yield of small pharmaceutical masses with large amounts of radioactivity. This high molar activity ensures that the biologically administered mass of radiopharmaceutical is sufficiently small to not saturate the targeted receptor on diseased cells. When prepared for human use, [^{177}Lu]PSMA-617 and [^{177}Lu]DOTATATE are radiolabeled with high yield at a molar activity of 60 MBq/nmol [26,27]. For cyclotron-produced ^{165}Er, achieving a high molar activity must begin with careful consideration of the holmium target material. Many commercial sources of holmium have significant erbium impurity, which, in turn, will limit the maximum attainable molar activity of ^{165}Er produced in the proton irradiation of an impure holmium target. Additionally, as shown in Table 2, the holmium target and irradiation parameters significantly affect the overall quantity of ^{165}Er that can be produced in a bombardment, with larger targets having greater overall yields compared to their smaller counterparts. However, larger holmium mass targets come with the added challenges of a more difficult ^{165}Er/Ho separation and a higher cold erbium and holmium burden in the purified ^{165}Er fraction.

In addition to sourcing Ho target material with low Er impurity content, an effective Ho/Er radiochemical isolation procedure is necessary. Because of the chemical similarity between the adjacent lanthanide elements, any residual holmium in the ^{165}Er final formulation will affect the molar activity of ^{165}Er. A high Ho/Er SF is accomplished in this work by a multi-step separation process utilizing cation exchange and extraction chromatography. Based on MP-AES analysis of the final ^{165}Er product, this ^{165}Er radiochemical isolation process gives a Ho/Er SF of $(2.8 \pm 1.1) \cdot 10^5$. The ^{165}Er radiochemical yield was $64 \pm 2\%$ calculated by the ratio of the decay-corrected ^{165}Er activity in the final product and the ^{165}Er activity produced in the irradiated target.

Two main sources of erbium/holmium—cold erbium target impurity and residual holmium due to incomplete separation—are the two dominating terms in the denominator of supplementary Equation (S1), which calculates a maximum achievable EoB MA for a given ^{165}Er preparation. With the experimental ^{165}Er production and radiochemical separation results presented above, supplementary Equation (S1) yields a calculated EoB ^{165}Er MA of 9.9 ± 0.5 MBq/nmol for a 1 h, 12.5 MeV, 130 mg, 9.5 mm Ø, low purity (100 ppm Er) holmium target PETtrace irradiation. This calculated MA$_{EoB}$ is in good agreement with the measured DOTA/DTPA ^{165}Er AMAs (Table 4). In this case, the calculated EoB MA is nearly entirely driven by the cold erbium impurity present in the holmium target material (second term in supplementary Equation (S1) denominator). This underscores the fact that under the investigated irradiation conditions, utilization of holmium targets with high erbium impurity content will limit the molar activity of the resulting ^{165}Er to values lower than typically acceptable for therapeutic radiopharmaceutical research applications. Producing high MA ^{165}Er requires holmium with extremely low erbium content, such as target material that has been prepurified from erbium, or the DOE Ames Laboratory MPC metal used in this work.

For an identical irradiation of a high-purity (0.5 ppm Er) holmium target, supplementary Equation (S1) yields an ^{165}Er EoB MA of 240 ± 60 MBq/nmol, with this molar activity being driven by cold holmium remaining in the ^{165}Er preparation after their separation by a factor of $(2.8 \pm 1.1) \cdot 10^5$ (third term in supplementary Equation (S1) denominator). However, the EoB ^{165}Er AMAs (Table 4) with DOTA (1.2–47 GBq/μmol$_{DOTA}$, n = 3) and DTPA (4.9–250 GBq/μmol$_{DTPA}$, n = 5) and radiolabeled [^{165}Er]PSMA-617 MAs (37–130 GBq/μmol$_{PSMA-617}$, n = 5, Table 5) do not reflect this high value of MA. This disagreement may be a result of nanomole, sub-ppm trace metal (Zn, Fe, Cu) impurities in the radiolabeling reactions, which are not considered in supplementary Equation (S1). This hypothesis is supported by the fact that, for ^{165}Er isolated from high-purity holmium targets, the titration-based AMA measured using DTPA were 5–22 times higher than for DOTA (n = 3). This difference in DTPA/DOTA AMA values, which has also been observed for cyclotron-produced ^{86}Y [49], is likely due to the non-selective nature of DOTA's metal binding properties. Compared with DTPA, DOTA binds 10,000 times stronger to Fe^{2+} (LogK$_{FeDOTA}$ = 20.22 ± 0.07 [50], LogK$_{FeDTPA}$ = 16.0 ± 0.1 [51]), 100 times stronger to Zn^{2+} (LogK$_{ZnDOTA}$ = 20.8 ± 0.2, LogK$_{ZnDTPA}$ = 18.6 ± 0.1 [51]), and 10 times stronger to Cu^{2+} (LogK$_{CuDOTA}$ = 22.3 ± 0.1, LogK$_{CuDTPA}$ = 21.5 ± 0.1 [51]), causing these three common trace metal impurities to be significantly more problematic in DOTA versus DTPA radiochemical labelings. Thus, a systematically higher DTPA-based AMA compared with DOTA-based AMA is indicative that Fe/Zn/Cu-based trace metal impurities in the radiolabeling solutions are significantly impacting the ^{165}Er AMA values. The impact of trace Zn, Fe, and Cu on the [^{165}Er]PSMA-617 radiolabeling experiments is supported by the presence of UV-absorbing impurities with analytical HPLC retention times equivalent to ^{nat}Zn-PSMA-617, ^{nat}Fe-PSMA-617, and ^{nat}Cu-PSMA-617 in the final radiopharmaceutical preparation (Supplementary Figure S7).

This work represents the first published radiosynthesis of [^{165}Er]PSMA-617, a radiopharmaceutical that could serve as a useful in vitro and in vivo tool that can be used to assess the role of AEs in the efficacy of PSMA-targeted radionuclide therapy of prostate cancer using [^{161}Tb]PSMA-617 [11]. The radiopharmaceutical is stable in serum for at least 12 h and has an octanol–water partition coefficient of LogD = -3.3 ± 0.3, less polar than [^{161}Tb]PSMA-617 (-3.9 ± 0.1) [11], [^{44}Sc]PSMA-617 (-4.21 ± 0.04), [^{177}Lu]PSMA-617 (-4.18 ± 0.06), and [^{68}Ga]PSMA-617 (-4.3 ± 0.1) [52].

5. Conclusions

A 2 h biomedical cyclotron irradiation and 5 h radiochemical separation can produce GBq-scale ^{165}Er suitable for high-yield radiolabelings of DOTA-based radiopharmaceuticals at molar activities befitting investigations of targeted radionuclide therapeutics. The separation utilizes column chromatography with commercially available resins and is well suited for automation. This significant step forward in the production and high holmium/erbium SF radiochemical isolation of ^{165}Er will enable fundamental radiation biology experiments of pure AE-emitting therapeutic radiopharmaceuticals. Proof-of-concept radiolabeling studies were successfully performed synthesizing [^{165}Er]PSMA-617, which will be utilized in vitro and in vivo to understand the role of AEs in PSMA-targeted radionuclide therapy of prostate cancer.

Supplementary Materials: Supplementary Figures S1–S8, Supplementary Material Sections S1–S3, and Supplementary Equation (S1),

Author Contributions: Conceptualization, I.D.S., J.C.M., J.W.E. and P.A.E.; methodology, I.D.S., T.R.J., J.C.M., E.A.-S., T.E.B., R.J.N., J.W.E. and P.A.E.; validation, I.D.S., T.R.J., J.C.M. and P.A.E.; formal analysis, I.D.S., T.R.J., J.C.M. and P.A.E.; investigation, I.D.S., T.R.J. and J.C.M.; resources, R.J.N., T.E.B., J.W.E. and P.A.E.; data curation, I.D.S., T.R.J., J.C.M. and P.A.E.; writing—original draft preparation, I.D.S., T.R.J., J.C.M. and P.A.E.; writing—review and editing, I.D.S., T.R.J., J.C.M., E.A.-S., T.E.B., R.J.N., J.W.E. and P.A.E.; supervision, J.W.E. and P.A.E.; project administration, P.A.E.; funding acquisition, J.W.E. and P.A.E.; All authors have read and agreed to the published version of the manuscript.

Funding: Research reported in this publication was supported by the National Cancer Institute of the National Institutes of Health under Award Number T32 CA009206 (J.C.M.). The content is solely the responsibility of the authors and does not necessarily represent the official views of the National Institutes of Health. This research is supported in part by the U.S. Department of Energy Isotope Program, managed by the Office of Science for Isotope R&D and Production, grant DE-SC0020955.

Acknowledgments: A portion of the LN2 resin used in this work was graciously provided by Triskem Int.

Conflicts of Interest: The authors declare no conflict of interest.

References

1. Strosberg, J.; El-Haddad, G.; Wolin, E.; Hendifar, A.; Yao, J.; Chasen, B.; Mittra, E.; Kunz, P.L.; Kulke, M.H.; Jacene, H.; et al. Phase 3 Trial of 177Lu-Dotatate for Midgut Neuroendocrine Tumors. *N. Engl. J. Med.* **2017**, *376*, 125–135. [CrossRef]
2. Hofman, M.S.; Emmett, L.; Sandhu, S.; Iravani, A.; Joshua, A.M.; Goh, J.C.; Pattison, D.A.; Tan, T.H.; Kirkwood, I.D.; Ng, S.; et al. [177Lu]Lu-PSMA-617 versus cabazitaxel in patients with metastatic castration-resistant prostate cancer (TheraP): A randomised, open-label, phase 2 trial. *Lancet* **2021**, *397*, 797–804. [CrossRef]
3. Sietmann, R. False Attribution. *Phys. Bull.* **1988**, *39*, 316–317. [CrossRef]
4. Matsakis, D.; Coster, A.; Laster, B.; Sime, R. A renaming proposal: "The Auger–Meitner effect". *Phys. Today* **2019**, *72*, 10–11. [CrossRef]
5. Cornelissen, B.; Vallis, K.A. Targeting the nucleus: An overview of Auger-electron radionuclide therapy. *Curr. Drug Discov. Technol.* **2010**, *7*, 263–279. [CrossRef]
6. Ku, A.; Facca, V.J.; Cai, Z.; Reilly, R.M. Auger electrons for cancer therapy–A review. *EJNMMI Radiopharm. Chem.* **2019**, *4*, 27. [CrossRef]
7. Pirovano, G.; Wilson, T.C.; Reiner, T. Auger: The future of precision medicine. *Nucl. Med. Biol.* **2021**, *96–97*, 50–53. [CrossRef] [PubMed]
8. Kassis, A.I. Molecular and cellular radiobiological effects of Auger emitting radionuclides. *Radiat. Prot. Dosimetry* **2011**, *143*, 241–247. [CrossRef]
9. Grünberg, J.; Lindenblatt, D.; Dorrer, H.; Cohrs, S.; Zhernosekov, K.; Köster, U.; Türler, A.; Fischer, E.; Schibli, R. Anti-L1CAM radioimmunotherapy is more effective with the radiolanthanide terbium-161 compared to lutetium-177 in an ovarian cancer model. *Eur. J. Nucl. Med. Mol. Imaging* **2014**, *41*, 1907–1915. [CrossRef] [PubMed]
10. Müller, C.; Reber, J.; Haller, S.; Dorrer, H.; Bernhardt, P.; Zhernosekov, K.; Türler, A.; Schibli, R. Direct in vitro and in vivo comparison of 161Tb and 177Lu using a tumour-targeting folate conjugate. *Eur. J. Nucl. Med. Mol. Imaging* **2014**, *41*, 476–485. [CrossRef]
11. Müller, C.; Umbricht, C.A.; Gracheva, N.; Tschan, V.J.; Pellegrini, G.; Bernhardt, P.; Zeevaart, J.R.; Köster, U.; Schibli, R.; van der Meulen, N.P. Terbium-161 for PSMA-targeted radionuclide therapy of prostate cancer. *Eur. J. Nucl. Med. Mol. Imaging* **2019**, *46*, 1919–1930. [CrossRef] [PubMed]
12. Eckerman, K.F.; Endo, A. Nuclear decay data for dosimetric calculations. A report of ICRP Committee 2. *Ann. ICRP* **2008**, *38*, 7–96.
13. Van de Voorde, M.; Van Hecke, K.; Cardinaels, T.; Binnemans, K. Radiochemical processing of nuclear-reactor-produced radiolanthanides for medical applications. *Coord. Chem. Rev.* **2019**, *382*, 103–125. [CrossRef]
14. Sadeghi, M.; Enferadi, M.; Tenreiro, C. Nuclear Model Calculations on the Production of Auger Emitter 165 Er for Targeted Radionuclide Therapy. *J. Mod. Phys.* **2010**, *1*, 217–225. [CrossRef]
15. Tárkányi, F.; Takács, S.; Hermanne, A.; Ditrói, F.; Király, B.; Baba, M.; Ohtsuki, T.; Kovalev, S.F.; Ignatyuk, A.V. Investigation of production of the therapeutic radioisotope ^{165}Er by proton induced reactions on erbium in comparison with other production routes. *Appl. Radiat. Isot.* **2009**, *67*, 243–247. [CrossRef] [PubMed]
16. Zandi, N.; Sadeghi, M.; Afarideh, H. Evaluation of the cyclotron production of 165Er by different reactions. *J. Radioanal. Nucl. Chem.* **2013**, *295*, 923–928. [CrossRef]
17. Tárkányi, F.; Hermanne, A.; Király, B.; Takács, S.; Ditrói, F.; Baba, M.; Ohtsuki, T.; Kovalev, S.F.; Ignatyuk, A.V. Study of activation cross-sections of deuteron induced reactions on erbium: Production of radioisotopes for practical applications. *Nucl. Instrum. Methods Phys. Res. Sect. B Beam Interact. Mater. Atoms* **2007**, *259*, 829–835. [CrossRef]

18. Tárkányi, F.; Hermanne, A.; Király, B.; Takács, S.; Ignatyuk, A.V. Study of excitation functions of alpha-particle induced nuclear reactions on holmium for 167Tm production. *Appl. Radiat. Isot.* **2010**, *68*, 404–411. [CrossRef]
19. Usman, A.R.; Khandaker, M.U.; Haba, H.; Otuka, N.; Murakami, M. Production cross sections of thulium radioisotopes for alpha-particle induced reactions on holmium. *Nucl. Instrum. Methods Phys. Res. Sect. B Beam Interact. Mater. Atoms* **2020**, *469*, 42–48. [CrossRef]
20. Tárkányi, F.; Hermanne, A.; Takács, S.; Ditrói, F.; Király, B.; Kovalev, S.F.; Ignatyuk, A.V. Experimental study of the ^{165}Ho(d,2n) and ^{165}Ho(d,p) nuclear reactions up to 20 MeV for production of the therapeutic radioisotopes 1^{65}Er and ^{166}gHo. *Nucl. Instrum. Methods Phys. Res. Sect. B Beam Interact. Mater. Atoms* **2008**, *266*, 3529–3534. [CrossRef]
21. Hermanne, A.; Adam-Rebeles, R.; Tarkanyi, F.; Takacs, S.; Csikai, J.; Takacs, M.P.; Ignatyuk, A. Deuteron induced reactions on Ho and La: Experimental excitation functions and comparison with code results. *Nucl. Instrum. Methods Phys. Res. Sect. B Beam Interact. Mater. Atoms* **2013**, *311*, 102–111. [CrossRef]
22. Beyer, G.J.; Zeisler, S.K.; Becker, D.W. The Auger-electron emitter ^{165}Er: Excitation function of the ^{165}Ho(p,n)165Er process. *Radiochim. Acta* **2004**, *92*, 219–222. [CrossRef]
23. Tárkányi, F.; Hermanne, A.; Takács, S.; Ditrói, F.; Király, B.; Kovalev, S.F.; Ignatyuk, A.V. Experimental study of the ^{165}Ho(p,n) nuclear reaction for production of the therapeutic radioisotope 165Er. *Nucl. Instrum. Methods Phys. Res. Sect. B Beam Interact. Mater. Atoms* **2008**, *266*, 3346–3352. [CrossRef]
24. Gracheva, N.; Carzaniga, T.S.; Schibli, R.; Braccini, S.; van der Meulen, N.P. ^{165}Er: A new candidate for Auger electron therapy and its possible cyclotron production from natural holmium targets. *Appl. Radiat. Isot.* **2020**, *159*, 109079. [CrossRef]
25. Schmor, P. Review of Cyclotrons for the Production of Radioactive Isotopes for Medical and Industrial Applications. *Rev. Accel. Sci. Technol.* **2011**, *04*, 103–116. [CrossRef]
26. Hennrich, U.; Kopka, K. Lutathera®: The First FDA- and EMA-Approved Radiopharmaceutical for Peptide Receptor Radionuclide Therapy. *Pharmaceuticals* **2019**, *12*, 114. [CrossRef]
27. Rahbar, K.; Bode, A.; Weckesser, M.; Avramovic, N.; Claesener, M.; Stegger, L.; Bögemann, M. Radioligand Therapy With 177Lu-PSMA-617 as A Novel Therapeutic Option in Patients With Metastatic Castration Resistant Prostate Cancer. *Clin. Nucl. Med.* **2016**, *41*, 522–528. [CrossRef]
28. Aslani, A.; Snowdon, G.M.; Bailey, D.L.; Schembri, G.P.; Bailey, E.A.; Pavlakis, N.; Roach, P.J. Lutetium-177 DOTATATE Production with an Automated Radiopharmaceutical Synthesis System. *Asia Ocean. J. Nucl. Med. Biol.* **2015**, *3*, 107–115.
29. Smith, H.L.; Hoffman, D.C. Ion-exchange separations of the lanthanides and actinides by elution with ammonium alpha-hydroxy-isobutyrate. *J. Inorg. Nucl. Chem.* **1956**, *3*, 243–247. [CrossRef]
30. Choppin, G.R.; Silva, R.J. Separation of the lanthanides by ion exchange with alpha-hydroxy isobutyric acid. *J. Inorg. Nucl. Chem.* **1956**, *3*, 153–154. [CrossRef]
31. Lehenberger, S.; Barkhausen, C.; Cohrs, S.; Fischer, E.; Grünberg, J.; Hohn, A.; Köster, U.; Schibli, R.; Türler, A.; Zhernosekov, K. The low-energy β-and electron emitter 161Tb as an alternative to 177Lu for targeted radionuclide therapy. *Nucl. Med. Biol.* **2011**, *38*, 917–924. [CrossRef] [PubMed]
32. Gracheva, N.; Müller, C.; Talip, Z.; Heinitz, S.; Köster, U.; Zeevaart, J.R.; Vögele, A.; Schibli, R.; van der Meulen, N.P. Production and characterization of no-carrier-added 161Tb as an alternative to the clinically-applied 177Lu for radionuclide therapy. *EJNMMI Radiopharm. Chem.* **2019**, *4*, 12. [CrossRef] [PubMed]
33. Mocko, V.; Taylor, W.A.; Nortier, F.M.; Engle, J.W.; Barnhart, T.E.; Nickles, R.J.; Pollington, A.D.; Kunde, G.J.; Rabin, M.W.; Birnbaum, E.R. Isolation of 163Ho from dysprosium target material by HPLC for neutrino mass measurements. *Radiochim. Acta* **2015**, *103*, 577–585. [CrossRef]
34. Schwantes, J.M.; Taylor, W.A.; Rundberg, R.S.; Vieira, D.J. Preparation of a one-curie 171Tm target for the detector for advanced neutron capture experiments (DANCE). *J. Radioanal. Nucl. Chem.* **2008**, *276*, 533–542. [CrossRef]
35. Gharibyan, N.; Bene, B.J.; Sudowe, R. Chromatographic separation of thulium from erbium for neutron capture cross section measurements—Part I: Trace scale optimization of ion chromatography method with various complexing agents. *J. Radioanal. Nucl. Chem.* **2017**, *311*, 179–187. [CrossRef]
36. Horwitz, E.P.; Bloomquist, C.A.A. Chemical separations for super-heavy element searches in irradiated uranium targets. *J. Inorg. Nucl. Chem.* **1975**, *37*, 425–434. [CrossRef]
37. McAlister, D.R.; Horwitz, E.P. Characterization of Extraction of Chromatographic Materials Containing Bis(2-ethyl-1-hexyl)Phosphoric Acid, 2-Ethyl-1-Hexyl (2-Ethyl-1-Hexyl) Phosphonic Acid, and Bis(2,4,4-Trimethyl-1-Pentyl)Phosphinic Acid. *Solvent Extr. Ion Exch.* **2007**, *25*, 757–769. [CrossRef]
38. Aziz, A.; Artha, W.T. Radiochemical Separation of 161Tb from Gd/Tb Matrix Using Ln Resin Column. *Indones. J. Chem.* **2018**, *16*, 283–288. [CrossRef]
39. Jiang, J.; Davies, A.V.; Britton, R.E. Measurement of 160Tb and 161Tb in nuclear forensics samples. *J. Radioanal. Nucl. Chem.* **2017**, *314*, 727–736. [CrossRef]
40. Horwitz, E.P.; McAlister, D.R.; Bond, A.H.; Barrans, R.E.; Williamson, J.M. A process for the separation of 177Lu from neutron irradiated 176Yb targets. *Appl. Radiat. Isot.* **2005**, *63*, 23–36. [CrossRef]
41. Vaudon, J.; Frealle, L.; Audiger, G.; Dutillly, E.; Gervais, M.; Sursin, E.; Ruggeri, C.; Duval, F.; Bouchetou, M.-L.; Bombard, A.; et al. First Steps at the Cyclotron of Orléans in the Radiochemistry of Radiometals: 52Mn and 165Er. *Instruments* **2018**, *2*, 15. [CrossRef]

42. Malikidogo, K.P.; Da Silva, I.; Morfin, J.-F.; Lacerda, S.; Barantin, L.; Sauvage, T.; Sobilo, J.; Lerondel, S.; Tóth, É.; Bonnet, C.S. A cocktail of 165Er(iii) and Gd(iii) complexes for quantitative detection of zinc using SPECT and MRI. *Chem. Commun.* **2018**, *54*, 7597–7600. [CrossRef]

43. Tapio, S. Studies towards Purification of Auger Electron Emitter 165Er-a Possible Radiolanthanide for Cancer Treatment. Masters's Thesis, University of Helsinki, Helsinki, Finland, 2018.

44. Ziegler, J.F.; Ziegler, M.D.; Biersack, J.P. SRIM–The stopping and range of ions in matter (2010). *Nucl. Instrum. Methods Phys. Res. Sect. B Beam Interact. Mater. Atoms* **2010**, *268*, 1818–1823. [CrossRef]

45. Ellison, P.A.; Valdovinos, H.F.; Graves, S.A.; Barnhart, T.E.; Nickles, R.J. Spot-welding solid targets for high current cyclotron irradiation. *Appl. Radiat. Isot.* **2016**, *118*, 350–353. [CrossRef]

46. Aluicio-Sarduy, E.; Hernandez, R.; Olson, A.P.; Barnhart, T.E.; Cai, W.; Ellison, P.A.; Engle, J.W. Production and in vivo PET/CT imaging of the theranostic pair 132/135La. *Sci. Rep.* **2019**, *9*, 10658. [CrossRef]

47. Jain, A.K.; Ghosh, A.; Singh, B. Nuclear Data Sheets for A = 165. *Nucl. Data Sheets* **2006**, *107*, 1075–1346. [CrossRef]

48. Pourmand, A.; Dauphas, N. Distribution coefficients of 60 elements on TODGA resin: Application to Ca, Lu, Hf, U and Th isotope geochemistry. *Talanta* **2010**, *81*, 741–753. [CrossRef]

49. Aluicio-Sarduy, E.; Hernandez, R.; Valdovinos, H.F.; Kutyreff, C.J.; Ellison, P.A.; Barnhart, T.E.; Nickles, R.J.; Engle, J.W. Simplified and automatable radiochemical separation strategy for the production of radiopharmaceutical quality 86Y using single column extraction chromatography. *Appl. Radiat. Isot.* **2018**, *142*, 28–31. [CrossRef] [PubMed]

50. Chaves, S.; Delgado, R.; Da Silva, J.J.R.F. The stability of the metal complexes of cyclic tetra-aza tetra-acetic acids. *Talanta* **1992**, *39*, 249–254. [CrossRef]

51. Anderegg, G.; Arnaud-Neu, F.; Delgado, R.; Felcman, J.; Popov, K. Critical evaluation of stability constants of metal complexes of complexones for biomedical and environmental applications* (IUPAC Technical Report). *Pure Appl. Chem.* **2005**, *77*, 1445–1495. [CrossRef]

52. Umbricht, C.A.; Benešová, M.; Schmid, R.M.; Türler, A.; Schibli, R.; van der Meulen, N.P.; Müller, C. (44)Sc-PSMA-617 for radiotheragnostics in tandem with (177)Lu-PSMA-617-preclinical investigations in comparison with (68)Ga-PSMA-11 and (68)Ga-PSMA-617. *EJNMMI Res.* **2017**, *7*, 9. [CrossRef] [PubMed]

 *molecules* 

Article

Safety Evaluation of an Alpha-Emitter Bismuth-213 Labeled Antibody to (1→3)-β-Glucan in Healthy Dogs as a Prelude for a Trial in Companion Dogs with Invasive Fungal Infections

Muath Helal [1,†], Kevin J. H. Allen [1,†], Hilary Burgess [2], Rubin Jiao [1], Mackenzie E. Malo [1], Matthew Hutcheson [3], Ekaterina Dadachova [1,*,‡] and Elisabeth Snead [2,‡]

[1] College of Pharmacy and Nutrition, University of Saskatchewan, Saskatoon, SK S7N 5E5, Canada; muh166@mail.usask.ca (M.H.); kja782@mail.usask.ca (K.J.H.A.); ruj501@mail.usask.ca (R.J.); mem510@mail.usask.ca (M.E.M.)

[2] Western College of Veterinary Medicine, Saskatoon, SK S7N 5B4, Canada; hilary.burgess@usask.ca (H.B.); ecs212@mail.usask.ca (E.S.)

[3] Safety Resources, University of Saskatchewan, Saskatoon, SK S7N 5C5, Canada; matt.hutcheson@usask.ca

* Correspondence: ekaterina.dadachova@usask.ca; Tel.: +1-306-966-5163

† These authors contributed equally to this work.

‡ These authors shared senior authorship.

Academic Editors: Alessandra Boschi and Petra Martini

Received: 3 July 2020; Accepted: 6 August 2020; Published: 8 August 2020

Abstract: Background: With the limited options available for therapy to treat invasive fungal infections (IFI), radioimmunotherapy (RIT) can potentially offer an effective alternative treatment. Microorganism-specific monoclonal antibodies have shown promising results in the experimental treatment of fungal, bacterial, and viral infections, including our recent and encouraging results from treating mice infected with *Blastomyces dermatitidis* with ^{213}Bi-labeled antibody 400-2 to (1→3)-β-glucan. In this work, we performed a safety study of ^{213}Bi-400-2 antibody in healthy dogs as a prelude for a clinical trial in companion dogs with acquired invasive fungal infections and later on in human patients with IFI. **Methods**: Three female beagle dogs (≈6.1 kg body weight) were treated intravenously with 155.3, 142.5, or 133.2 MBq of ^{213}Bi-400-2 given as three subfractions over an 8 h period. RBC, WBC, platelet, and blood serum biochemistry parameters were measured periodically for 6 months post injection. **Results**: No significant acute or long-term side effects were observed after RIT injections; only a few parameters were mildly and transiently outside reference change value limits, and a transient atypical morphology was observed in the circulating lymphocyte population of two dogs. **Conclusions**: These results demonstrate the safety of systemic ^{213}Bi-400-2 administration in dogs and provide encouragement to pursue evaluation of RIT of IFI in companion dogs.

Keywords: radioimmunotherapy; bismuth-213; invasive fungal infections; 1-3-beta-glucan; dogs

1. Introduction

For immunosuppressed patients, such as those suffering from cancer or are post organ transplant, invasive fungal infections (IFI) can be devastating. As causes of both morbidity and mortality, cases of IFI have more than tripled since 1979, highlighting the need for treatment [1]. IFI can also have devastating effects on dogs and cats [2]. New effective treatments for IFI are needed for human and veterinary medicine to overcome resistance issues, issues associated with poor bioavailability, and the high cost of oral antifungals in dogs. Radioimmunotherapy (RIT) uses antigen–antibody interaction to deliver sufficient activities of ionizing radiation to cells to induce DNA double-strand breaks and alter cell membrane and intracellular components for cell apoptosis while preserving healthy tissues. This treatment has demonstrated efficacy in primarily nonsolid tumors such as non-Hodgkin lymphomas [3]. RIT offers several advantages over other therapeutics: (1) Due to the nature of ionizing radiation, RIT physically destroys the desired cells, and does not just abolish a single pathway; (2) the total cell destruction makes it very difficult for a drug resistance mechanism to develop; (3) RIT does not rely on a healthy immune system to be effective, making the immune status of the patient less important; (4) the specific nature of RIT means there are less off-target effects, which results in lower toxicity compared to that of conventional chemotherapies. However, RIT is still subject to radioresistance phenomena, especially for big solid tumors, and induces hematoxicity in some patients. Additionally, we were the first to show that microorganism-specific monoclonal antibodies can be a viable therapy for viral, bacterial, and fungal infections in an experimental environment ([4,5]). RIT for infections potentially offers higher specificity and lower toxicity than cancer RIT as it targets microbial antigens which have no or very little homology with human proteins. In contrast to cancer, where every cancerous cell must be destroyed either directly by crossfire and bystander effects or indirectly via abscopal effect to prevent the resurgence of the disease, low numbers of microbial cells remaining post RIT can be eliminated by the immune system. Previously, we theorized that beta-glucan, a surface-expressed antigen that is shared by many major IFI-causing pathogens, would make an ideal target for RIT [6]. To show this, we utilized *Cryptococcus neoformans* and *Candida albicans*, as they are highly present. The in vitro results demonstrated that when labeled with the alpha-particle-emitting radionuclide Bismuth-213 (^{213}Bi) the antibodies to the pan-antigens killed a very high percentage (80–100%) of the fungal cells [7]. ^{213}Bi, an alpha-emitter with a short physical half-life of 46 min, delivers its radiation impact within a short period of time, which can match the doubling time of the majority of the fungal pathogens. It gives ^{213}Bi an advantage over other longer-lived alpha emitters such as Actinium-225. Next, we were able to show that this approach can be translated in vivo by infecting mice with *Blastomyces dermatitidis* and targeting the pan-antigen beta-glucan [8]. *B. dermatitidis* was chosen as an IFI model as it is endemic to parts of Canada and causes devastating infections in immunocompetent people, companion dogs, and immunocompromised patients [2]. In this work, we performed a safety study of ^{213}Bi-400-2 antibody in healthy dogs as a prelude for a clinical trial in companion dogs with invasive fungal infections.

2. Results

On separate days, three dogs were intravenously administered 155.3, 142.5, or 133.2 MBq (1→3)-β-glucan-targeting antibody 400-2 labeled with ^{213}Bi, respectively. Figure 1 shows the values for red blood cell (RBC), white blood cell (WBC), and platelet counts throughout the study. These parameters are of particular importance for radiolabeled antibody-based treatment as they reflect possible reversible or nonreversible toxicity to the bone marrow. In all three dogs, these parameters were within laboratory reference intervals at all time points and generally within the reference change value (RCV) limits throughout the duration of the study.

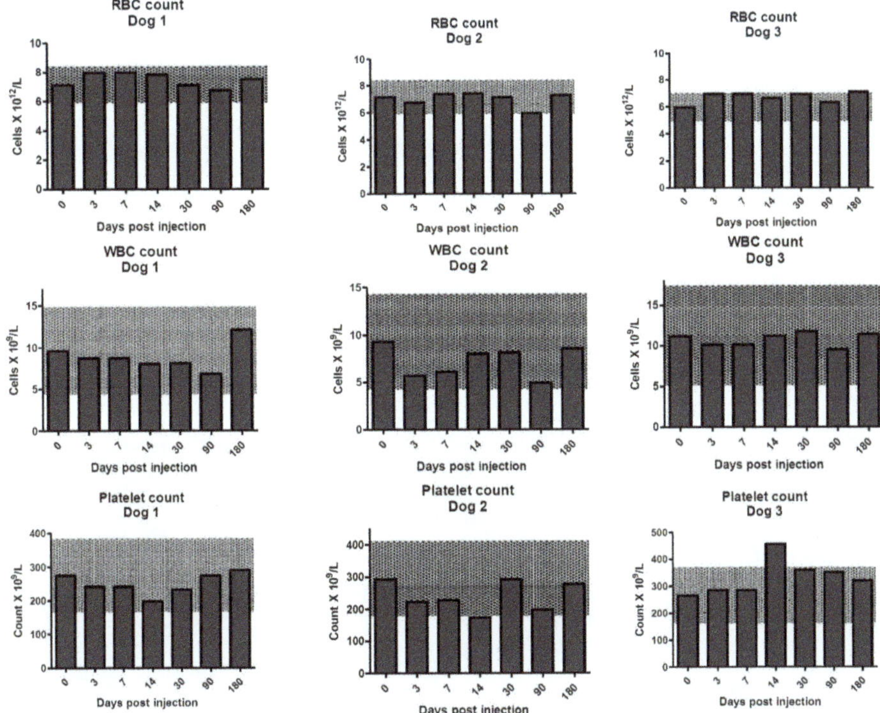

Figure 1. Red blood cell (RBC) counts (**upper row**), white blood cell (WBC) counts (**middle row**), and platelet counts (**lower row**) for the dogs on days 0–180 post treatment with ^{213}Bi-400-2 antibody. Shaded areas show reference change values (RCVs).

In two of the patients (dogs 1 and 2), atypical cells, most consistent with cells of lymphoid origin, were noted on days 7 and 14, respectively (Figure 2). These cells were intermediate to large in size and round to mildly irregular in shape. Cytoplasm was of low to moderate volume, basophilic, and occasionally had few pinpoint clear vacuoles or pale areas. Nuclei were round, but in rare cells could appear binucleate or possibly lobular, with ropey, occasionally clumped, chromatin (Figure 2).

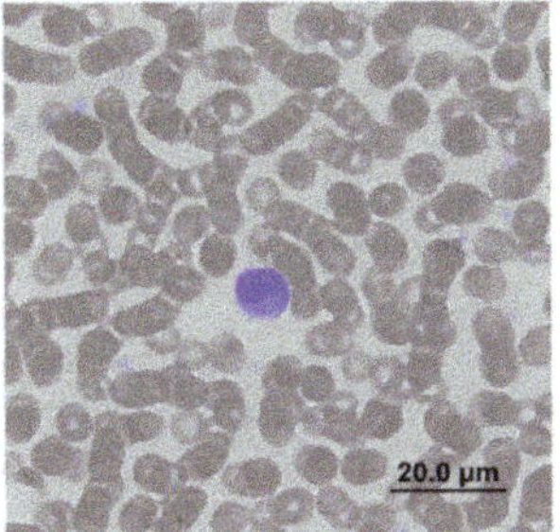
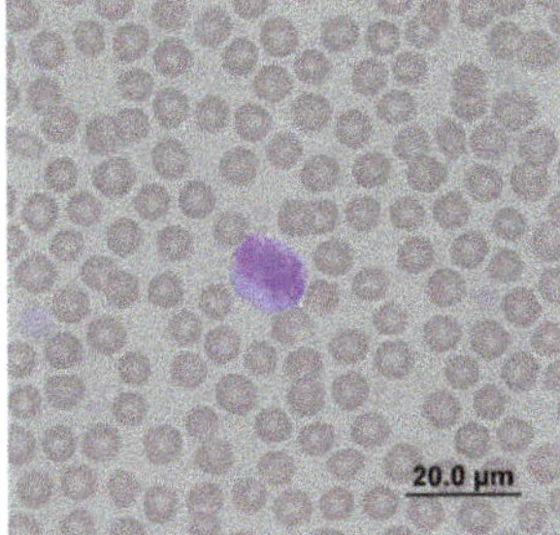

Figure 2. Direct smears of peripheral blood from dogs 1 (**left micrograph**) and 2 (**right micrograph**) showing examples of atypical cells, most consistent with lymphoid origin. These cells are intermediate to large in size with low volumes of basophilic cytoplasm, occasionally containing few small clear vacuoles. Nuclei in these two cells are cleaved to binucleate or lobular with clumped chromatin. Modified Wright stain. Bar = 20.0 μm.

These atypical cells were not a persistent finding, and they were not discernible in any of the follow-up complete blood counts (CBCs) in the two affected dogs. The mild decrease in lymphocyte numbers could reflect a transient elevation in cortisol (stress); however, given the brief appearance of atypical cells, a very transient influence of the treatment on lymphoid populations in the bone marrow and other organs leading to lympholysis cannot be entirely ruled out. The changes in monocyte numbers did not follow a consistent pattern between subjects, and while decreased numbers are typically considered clinically insignificant, the mild elevations could reflect an increased demand for tissue macrophages. Although within the RCV limits, mean corpuscular volume (MCV) showed a consistent downward trend from day 3 to day 14 in all three dogs, and continued to 30 days in dogs 1 and 3 (Figure 3). While there was an upward trend in all dogs by day 90, the values again decreased at 180 days. This finding can be associated with iron-limited erythropoiesis, although its connection to the test therapy is unclear. The dogs were all on a low-protein, plant-based diet which may have influenced iron stores independent of the test therapy.

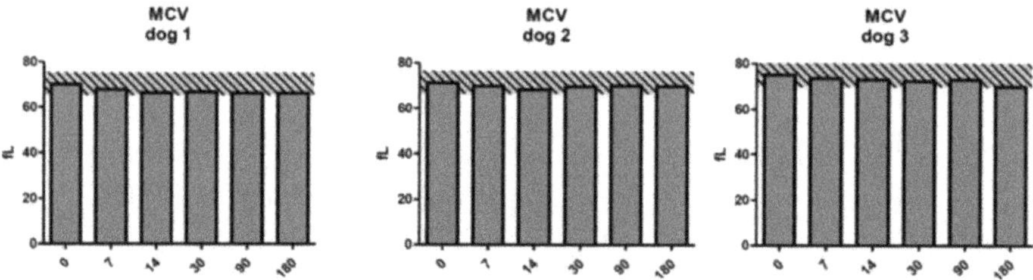

Figure 3. Mean corpuscular volume (MCV) for the dogs on days 0–180 post treatment with ^{213}Bi-400-2 antibody. Shaded areas show reference change values (RCVs). X axis shows days post injection of radiolabeled antibody.

A selection of serum biochemical analytes are shown in Figures 4–7. With the exception of a temporary decrease in creatinine at day 90 for dog 3, and mild elevations in urea relative to the RCV at day 180 in dogs 1 and 2, urea and creatinine were within RCV throughout the study (Figure 4).

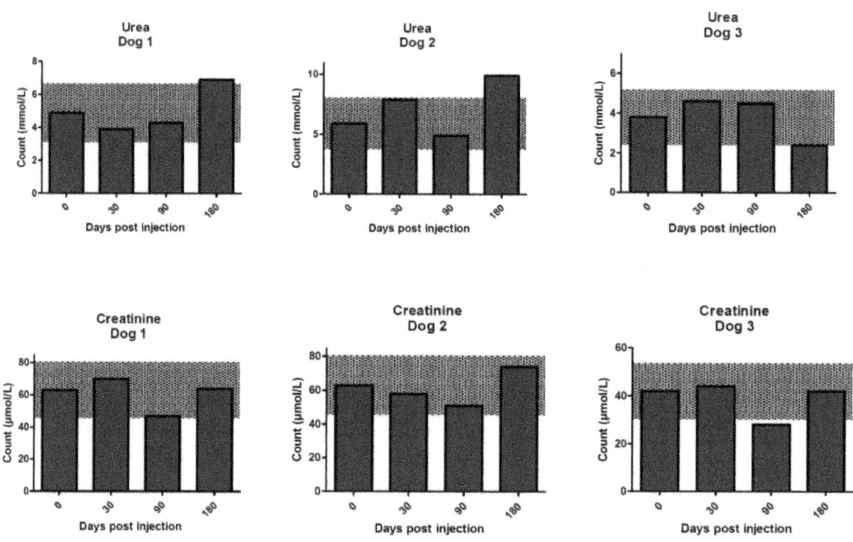

Figure 4. Renal analytes for the dogs on days 0–180 post treatment with ^{213}Bi-400-2 antibody. **Upper row**—urea; **lower row**—creatinine. Shaded areas show reference change values (RCVs).

Chloride, phosphorus, and potassium were concurrently elevated relative to the RCV in dog 2, and potassium was elevated relative to the RCV in dog 1, suggesting altered hydration status was the likely cause of the urea elevation in both dogs (Figure 5). Hydration status was also the likely cause of elevations in potassium at day 30 in dog 2 and day 90 in dogs 1 and 3, as again sodium showed a similar increase or was above the RCV at these same time points and was accompanied by a chloride level above the RCV in dog 3 at day 90. There was a slight decrease in potassium relative to the RCV in dogs 1 and 3 at day 30 and dog 2 on day 90, which may have been related to decreased intake. Despite these changes, relative to the RCV, potassium was within laboratory reference intervals throughout the study.

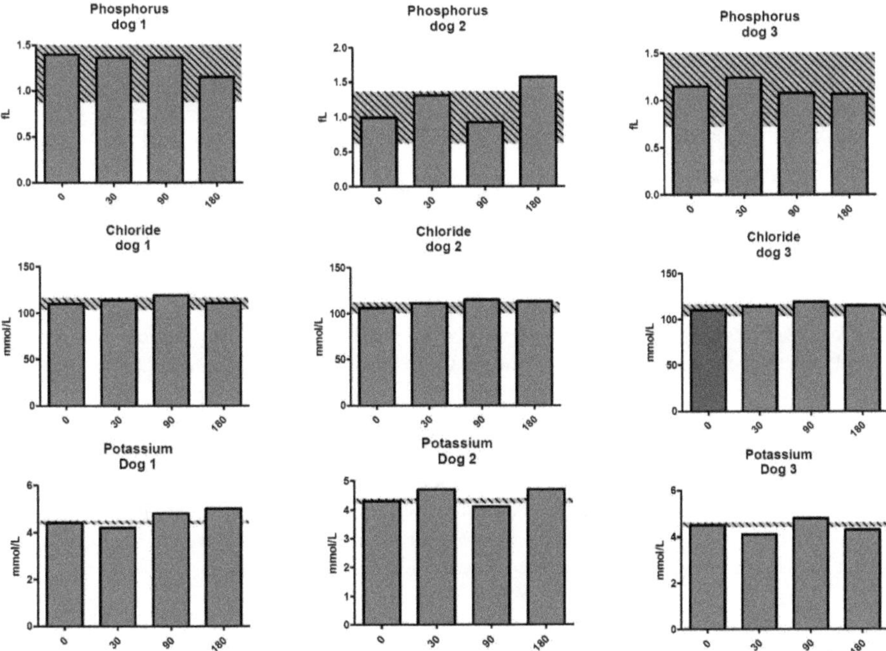

Figure 5. Phosphorus, chloride, and potassium for the dogs on days 0–180 post treatment with ^{213}Bi-400-2 antibody. **Upper row**—phosphorous; **middle row**—chloride; **lower row**—potassium. Shaded areas show reference change values (RCVs). X axis shows days post injection of radiolabeled antibody.

With the exception of alkaline phosphatase (ALP) in dog 2, which significantly increased above RCV at day 30 but corrected by day 90, and slight increases in alanine aminotransferase ALT at day 180 in dogs 2 and 3, all hepatobiliary enzymes remained within the RCV limits throughout the study (Figure 6).

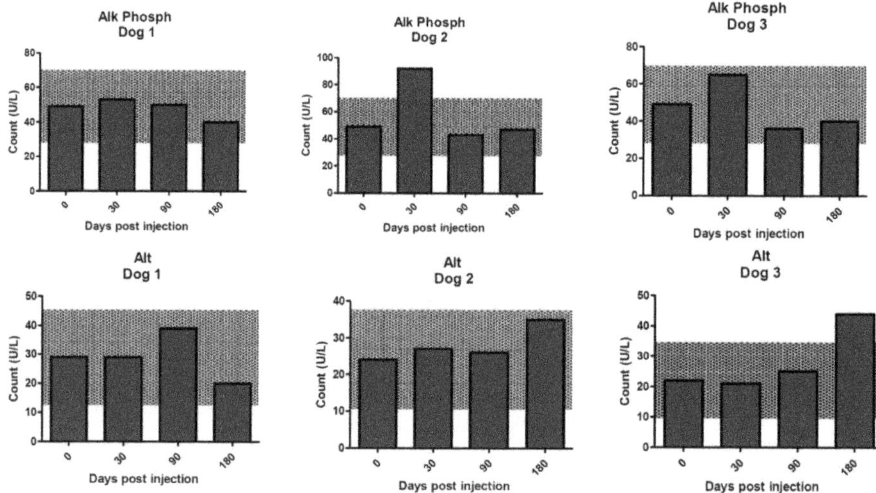

Figure 6. Liver enzyme measurements for the dogs on days 0–180 post treatment with [213]Bi-400-2 antibody. **Upper row**—alkaline phosphatase; **lower row**—alanine aminotransferase. Shaded areas show reference change values (RCVs).

The bilirubin and albumin were somewhat low, relative to laboratory reference intervals, in all three dogs at baseline and throughout the 6-month study (Figure 7). Total bilirubin was within RCV limits for all dogs throughout the study. Albumin was decreased relative to the RCV in dog 1 (at 90 days), dog 2 (at 30 days), and dog 3 (at 90 and 180 days). In dogs 1 and 2, these were transient changes. However, in dog 3, who had the lowest albumin at the start of the study, there was suspicion of a pre-existing protein-losing enteropathy (PLE). The albumin results were consistently and significantly lower than laboratory reference intervals at baseline and throughout the study in dog 3. Suspicion of PLE was further supported by the development of ascites, characterized as a pure transudate, at day 90 which was responsive to a hypoallergenic diet and prednisolone therapy.

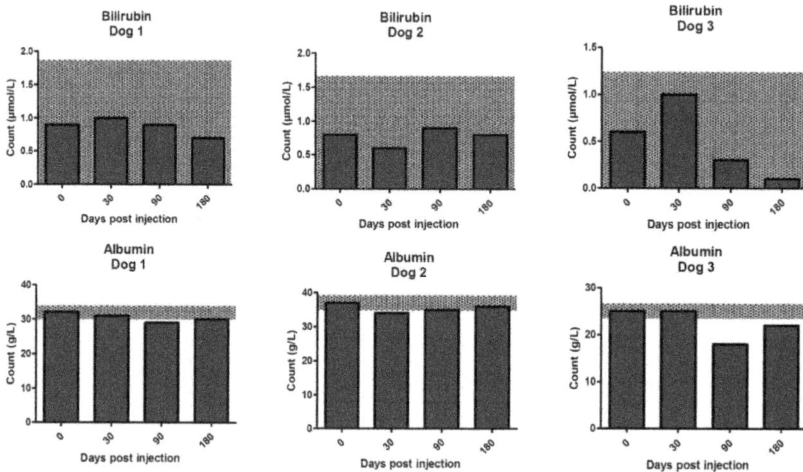

Figure 7. Bilirubin and albumin concentrations for the dogs on days 0–180 post treatment with [213]Bi-400-2 antibody. **Upper row**—bilirubin; **lower row**—albumin. Shaded areas show reference change values (RCVs).

Cholesterol was decreased concurrently, relative to RCV, in dog 1 at 90 days and in dog 3 at 180 days (Figure 8). There was a mild increase in cholesterol in dog 2 at 30 days, with a concurrent mild elevation of ALP relative to the RCV. This likely reflected a transient cholestasis, although insufficient to influence bilirubin concentrations. In the same dog, cholesterol had decreased relative to the RCV at 90 days and showed a slight elevation above the RCV at 180 days.

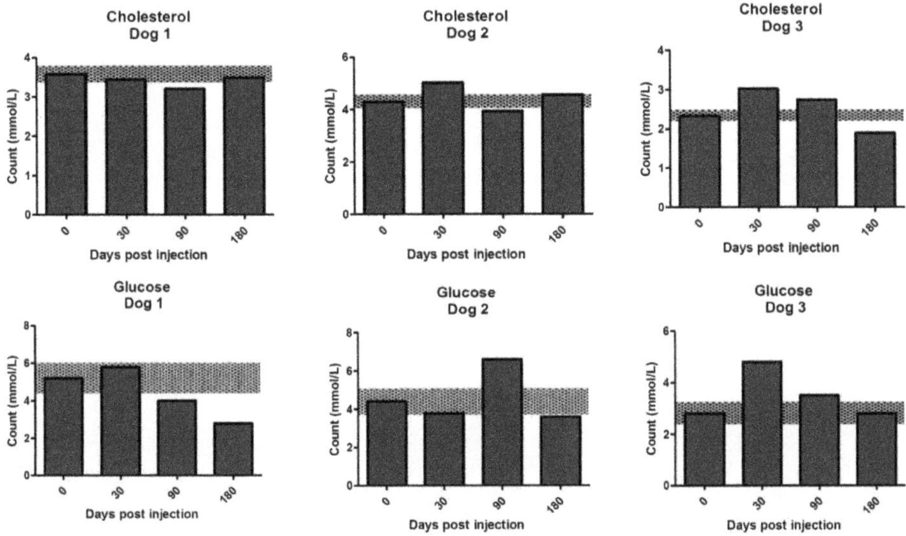

Figure 8. Cholesterol and glucose concentrations for the dogs on days 0–180 post treatment with ^{213}Bi-400-2 antibody. **Upper row**—cholesterol; **lower row**—glucose. Shaded areas show reference change values (RCVs).

ALP corrected by 90 days. Elevations in cholesterol, relative to the RCV, were also noted in dog 3 at 30 and 90 days. These elevations were transient and nonspecific. Glucose was found to decrease relative to the RCV in dog 1 on days 90 and 180 (Figure 8). Delayed serum separation was thought to be the most likely explanation for this change. Mild elevations in glucose relative to the RCV were noted in dog 2 at day 90 and dog 3 at days 30 and 90, which could be explained by a cortisol response, further supported in dog 3 at 30 days by a concurrent decrease in lymphocyte numbers. However, in dog 3, the baseline glucose was low relative to the laboratory reference intervals (likely storage artifact). As the RCV was based on this artifactually low value, any future sample not influenced by storage artifact would be expected to exceed the upper limit of the RCV. CK was elevated above the RCV, but remained within laboratory reference intervals, at day 30 in dog 1, likely reflecting mild myocyte leakage related to restraint during sample collection.

3. Discussion

Overall, throughout the study, changes outside the RCV were few, mild, and transient, attesting to the absence of systemic toxicity of ^{213}Bi-400-2 antibody. In regard to the radiation safety, the research and veterinary staff administering the ^{213}Bi-labeled antibody and providing animal care received a combined dose of no more than 6 μSv personal dose equivalent at the depth of 10 mm (Hp(10)) per handled dog as measured by electronic dosimeters due to the 440 keV gamma emissions from the ^{213}Bi. The staff exposures indicate that, with proper precautions, personnel exposure will not be limiting in a clinical trial setting. The short half-life of ^{213}Bi makes the required isolation period less burdensome than using longer-lived radionuclides.

B. dermatitidis has relatively high prevalence in different areas in Canada and the United States [2,9], and the endemic regions may be increasing, as evidenced by reports in New York, Vermont, Texas, Nebraska, and Kansas [10], which makes it a relevant fungal pathogen to be treated with radiolabeled antibodies to a pan-antigen beta-glucan. Treatment of mice infected intratracheally with *B. dermatitidis* by ^{213}Bi-400-2 antibody given as a single 5.6 MBq (277.5 MBq/kg body weight) intraperitoneal injection reduced the fungal burden in the lungs by two logs [8]. For treatment of the dogs in this study, approximately 18.5 MBq ^{213}Bi-400-2 antibody per kg body weight was administered, reflecting the difference in body weight to body surface area ratios between mice and dogs. In this regard, the studies of T-cell ablation in dogs with ^{213}Bi-labeled anti-TCRαβ antibody showed that the injected activities in the range of 137 to 207 MBq/kg did not cause normal organ damage [11,12]. In addition, because (1→3)-β-glucan is a fungal antigen with no homology to human proteins, the antibodies to (1→3)-β-glucan are expected to have very low cross-reactivity with human and canine cells, thus decreasing the potential side effects from off-target antibody mediated injury associated with radiolabeled antibody administration. ^{213}Bi-labeled whole antibodies have been successfully used in treatment of patients with acute myeloid leukemia and melanoma [3]. Currently, there is a worldwide effort to increase the availability of ^{213}Bi, which will enable the expansion of ^{213}Bi-labeled whole antibodies' use for a variety of oncological and infectious disease applications.

4. Materials and Methods

4.1. Radionuclides, Radiolabeling, and Quality Control of the Radiolabeled Antibody

The (1→3)-β-glucan-targeting 400-2 antibody (Biosupplies Australia Pty. Ltd., Parkville, Australia) was first conjugated to the chelating agent *C*-functionalized *trans*-cyclohexyldiethylene-triamine penta-acetic acid derivative (CHXA″) (Macrocyclics, San Antonio, TX, USA). Conjugation and labeling were carried out according to a previously published method [13]. To summarize, conjugation was carried out at 37 °C for 1.5 h in a sodium carbonate buffer at pH 8.5 that had been previously run through a chelex column to remove any trace metal impurities. After the conjugation, the reaction was exchanged into a chelexed 0.15 M ammonium acetate buffer at pH 6.5. The ^{213}Bi/^{225}Ac radionuclide generator was purchased from Oak Ridge National Laboratory (Oak Ridge, TN, USA). ^{213}Bi was eluted from the ^{213}Bi/^{225}Ac radionuclide generator using 300 μL 0.1 M hydroiodic acid (HI) solution followed by 300 μL milliQ H$_2$O [14]. The elution pH was adjusted to 6.5 with 80 μL of 5 M ammonium acetate buffer (chelexed) prior to the addition of the desired amount of CHXA″ conjugated antibody (370:1 kBq/μg specific activity, 55.5 kBq/pmol) and the reaction mixture was heated for 5 min at 37 °C with shaking, followed by an addition of 3 μL of 0.05 M EDTA solution. The ^{213}Bi-labeled antibody was then purified on a 0.5 mL Amicon disposable size exclusion filter (30K MW cut-off, Fisher, Ottawa, ON, Canada. The percentage of radiolabeling was measured by silica gel instant thin-layer chromatography (SG-iTLC, Agilent Technologies, Santa Clara, CA, USA). using 0.15 M ammonium acetate buffer as the eluent. SG-iTLCs were cut in half, and each half was read on a Wizard2 2470 Automatic Gamma Counter (R_f = 0 containing radiolabeled antibody, R_f = 1 containing free ^{213}Bi (Perkin Elmer, Waltham, MA, USA). Only the batches with >98% radiochemical purity were used for administration to the dogs.

4.2. Administration of Radiolabeled Antibody to Dogs and Follow up Analyses

The study design was approved by the University of Saskatchewan's Animal Research Ethics Board and adhered to the Canadian Council on Animal Care guidelines for humane animal use (Animal Use Protocol 20190091, approved on August 7, 2019). Three 1.5 years old, spayed female beagle dogs (5.91–6.1 kg body weight) who were deemed healthy and suitable for radiolabeled antibody administration were utilized. Dogs were allowed to acclimate for one week prior to use, and their appetite and activity levels were monitored before the initiation of the administration of radiolabeled antibody. All dogs were eating a homemade, balanced, low-protein diet that met all minimum Association of American Feed Control Officials (AFFCO) standards. On the injection day, vital parameters such as normal body temperature, body weight, and heart rate were recorded and monitored. Ethylenediamine tetra-acetic acid (EDTA)-anticoagulated blood, serum, and urine were submitted to Prairie Diagnostic Services (PDS; Saskatoon, SK, Canada) for baseline complete blood count (CBC) (Advia Hematology Analyzer, Siemens Healthcare, Germany), serum biochemistry (Cobas 311, Hitachi High-Technologies Corporation, Tokyo, Japan), and urinalysis, respectively. On separate days, three dogs were intravenously administered 155.3, 142.5, or 133.2 MBq $(1\rightarrow3)$-β-glucan-targeting antibody 400-2 labeled with ^{213}Bi (specific activity 370 kBq/μg), respectively. The injected activity was divided into three subfractions separated by 2–3 h for each dog. The dogs were closely monitored between the injections and after the last subfraction for 12–14 h. EDTA-anticoagulated blood from each dog was submitted to PDS 3, 7, 14, 30, 90, and 180 days post injection for CBC monitoring. Serum collected on 30, 90, and 180 days post injection was used to measure serum biochemistry levels. Urine samples were collected by ultrasound-guided cystocentesis on days 0, 90, and 180. CBC evaluation included total white blood cell count (WBC), with manual differential leukocyte count, red blood cell count (RBC), hemoglobin concentration (Hgb), hematocrit (Hct), mean corpuscular volume (MCV), mean corpuscular hemoglobin concentration (MCHC), mean corpuscular hemoglobin (MCH), platelet count, and evaluation of red blood cell morphology. Serum biochemical analytes evaluated were sodium, potassium, chloride, calcium, phosphorus, magnesium, urea, creatinine, lipase, glucose, cholesterol, total bilirubin, alkaline phosphatase (ALP), gamma glutamyltransferase (GGT), alanine aminotransferase (ALT), creatinine kinase (CK), total protein, albumin, and calculated globulin. Upper and lower limits for individual variation were calculated from the baseline value for each analyte using canine CV_I data from VetBiologicalVariation.org. For those analytes where one or more values were outside of these limits, a reference change value was calculated using RCV (95%) data from VetBiologicalVariation.org.

4.3. Radiation Safety for the Research and Veterinary Staff

Research staff administering the ^{213}Bi-labeled antibody and providing animal care wore electronic dosimeters (DMC 3000, Mirion Technologies, Smyrna, GA, USA). Animals were housed in an isolation area suitable for work with radioisotopes for approximately 8 h after the final administration and released from isolation when less than 0.1% of the administered ^{213}Bi remained.

5. Conclusions

In conclusion, our results have demonstrated an acceptable safety profile of $(1\rightarrow3)$-β-glucan-targeting ^{213}Bi-400-2 antibody in dogs. RIT may be a useful adjunct therapy for invasive fungal infections and have a role to play in speeding resolution of an infection by helping the body to reduce the infectious load to a lower level where either the immune system or other therapies may be more effective. The clinical trial will recruit canine patients with invasive fungal infections and will also provide the information needed for clinical translation into human patients with IFI.

Author Contributions: Conceptualization, E.D. and E.S.; methodology, E.D., E.S., and H.B.; software, M.H. (Matthew Hutcheson); validation, E.D. and E.S.; formal analysis, M.H. (Muath Helal) and H.B.; investigation, M.H. (Muath Helal), K.J.H.A., M.E.M., R.J., and M.H. (Matthew Hutcheson); resources, E.D. and E.S.; data curation, E.D. and E.S.; writing—original draft preparation, M.H. (Muath Helal); writing—review and editing, E.D., E.S., and H.B.; visualization, M.H. (Muath Helal); supervision, E.D. and E.S.; project administration, E.D.; funding acquisition, E.D. and E.S. All authors have read and agreed to the published version of the manuscript.

Funding: This research was funded by the Saskatchewan Health Research Foundation, grant number 4172.

References

1. Enoch, D.A.; Yang, H.; Aliyu, S.H.; Micallef, C. The Changing Epidemiology of Invasive Fungal Infections. *Methods Mol. Biol.* **2017**, *1508*, 17–65. [PubMed]
2. Davies, J.L.; Epp, T.; Burgess, H.J. Prevalence and geographic distribution of canine and feline blastomycosis in the Canadian prairies. *Can. Vet. J.* **2013**, *54*, 753–760. [PubMed]
3. Larson, S.M.; Carrasquillo, J.A.; Cheung, N.K.; Press, O.W. Radioimmunotherapy of human tumours. *Nat. Rev. Cancer* **2015**, *15*, 347–360. [CrossRef] [PubMed]
4. Dadachova, E.; Nakouzi, A.; Bryan, R.A.; Casadevall, A. Ionizing radiation delivered by specific antibody is therapeutic against a fungal infection. *Proc. Natl. Acad. Sci. USA* **2003**, *100*, 10942–109427. [CrossRef] [PubMed]
5. Helal, M.; Dadachova, E. Radioimmunotherapy as a Novel Approach in HIV, Bacterial, and Fungal Infectious Diseases. *Cancer Biother. Radiopharm.* **2018**, *33*, 330–335. [CrossRef] [PubMed]
6. Nosanchuk, J.D.; Dadachova, E. Radioimmunotherapy of fungal diseases: The therapeutic potential of cytocidal radiation delivered by antibody targeting fungal cell surface antigens. *Front. Microbiol.* **2012**, *12*, 283. [CrossRef] [PubMed]
7. Bryan, R.A.; Guimaraes, A.J.; Hopcraft, S.; Jiang, Z.; Bonilla, K.; Morgenstern, A.; Bruchertseifer, F.; Del Poeta, M.; Torosantucci, A.; Cassone, A.; et al. Toward developing a universal treatment for fungal disease using radioimmunotherapy targeting common fungal antigens. *Mycopathologia* **2011**, *173*, 463–471. [CrossRef] [PubMed]
8. Helal, M.; Allen, K.J.H.; van Dijk, B.; Nosanchuk, J.D.; Snead, E.; Dadachova, E. Radioimmunotherapy of blastomycosis in a mouse model with a $(1{\rightarrow}3)$-β-glucans targeting antibody. *Front. Microbiol* **2020**, *11*, 147. [CrossRef] [PubMed]
9. Gray, N.A.; Baddour, L.M. Cutaneous inoculation blastomycosis. *Clin. Infect. Dis.* **2002**, *34*, 44–49. [CrossRef] [PubMed]
10. McDonald, R.; Dufort, E.; Jackson, B.R.; Tobin, E.H.; Newman, A.; Benedict, K.; Blog, D. Notes from the field: Blastomycosis cases occurring outside of regions with known endemicity-New York, 2007–2017. *MMWR Morb. Mortal Wkly. Rep.* **2018**, *67*, 1077–1078. [CrossRef] [PubMed]
11. Bethge, W.A.; Wilbur, D.S.; Storb, R.; Hamlin, D.K.; Santos, E.B.; Brechbiel, M.W.; Fisher, D.R.; Sandmaier, B.M. Selective T-cell ablation with bismuth-213-labeled anti-TCRalphabeta as nonmyeloablative conditioning for allogeneic canine marrow transplantation. *Blood* **2003**, *101*, 5068–5075. [CrossRef] [PubMed]
12. Bethge, W.A.; Wilbur, D.S.; Sandmaier, B.M. Radioimmunotherapy as non-myeloablative conditioning for allogeneic marrow transplantation. *Leuk Lymphoma* **2006**, *47*, 1205–1214. [CrossRef] [PubMed]
13. Allen, K.J.H.; Jiao, R.; Malo, M.E.; Frank, C.; Dadachova, E. Biodistribution of a radiolabeled antibody in mice as an approach to evaluating antibody pharmacokinetics. *Pharmaceutics* **2018**, *10*, 262. [CrossRef] [PubMed]
14. Allen, K.J.H.; Jiao, R.; Malo, M.E.; Frank, C.; Fisher, D.R.; Rickles, D.; Dadachova, E. Comparative radioimmunotherapy of experimental melanoma with novel humanized antibody to melanin labeled with 213Bismuth and 177Lutetium. *Pharmaceutics* **2019**, *11*, 348. [CrossRef]

Sample Availability: Samples of the compounds are not available from the authors.

 *molecules*

Article

Hybrid Imaging Agents for Pretargeting Applications Based on Fusarinine C—Proof of Concept

Dominik Summer [1], Milos Petrik [2], Sonja Mayr [1], Martin Hermann [3], Piriya Kaeopookum [1], Joachim Pfister [1], Maximilian Klingler [1], Christine Rangger [1], Hubertus Haas [4] and Clemens Decristoforo [1,*]

1. Department of Nuclear Medicine, Medical University Innsbruck, A-6020 Innsbruck, Austria; summer.dominik@gmail.com (D.S.); sonja-mayr@hotmail.com (S.M.); gamsuk@hotmail.com (P.K.); joachim.pfister@i-med.ac.at (J.P.); Maximilian.klingler@i-med.ac.at (M.K.); Christine.Rangger@i-med.ac.at (C.R.)
2. Institute of Molecular and Translational Medicine, Faculty of Medicine and Dentistry, Palacky University Olomouc, 772-00 Olomouc, Czech Republic; MilosPetrik@upol.cz
3. Department of Anaesthesia and Intensive Care, Medical University Innsbruck, A-6020 Innsbruck, Austria; Martin.Hermann@i-med.ac.at
4. Institute of Molecular Biology, Medical University Innsbruck, A-6020 Innsbruck, Austria; Hubertus.Haas@i-med.ac.at
* Correspondence: Clemens.decristoforo@i-med.ac.at; Tel.: +43-512-504-80951

Academic Editors: Alessandra Boschi and Petra Martini
Received: 17 April 2020; Accepted: 29 April 2020; Published: 1 May 2020

Abstract: Hybrid imaging combining the beneficial properties of radioactivity and optical imaging within one imaging probe has gained increasing interest in radiopharmaceutical research. In this study, we modified the macrocyclic gallium-68 chelator fusarinine C (FSC) by conjugating a fluorescent moiety and tetrazine (Tz) moieties. The resulting hybrid imaging agents were used for pretargeting applications utilizing click reactions with a *trans*-cyclooctene (TCO) tagged targeting vector for a proof of principle both in vitro and in vivo. Starting from FSC, the fluorophores Sulfocyanine-5, Sulfocyanine-7, or IRDye800CW were conjugated, followed by introduction of one or two Tz motifs, resulting in mono and dimeric Tz conjugates. Evaluation included fluorescence microscopy, binding studies, logD, protein binding, in vivo biodistribution, μPET (micro-positron emission tomography), and optical imaging (OI) studies. ^{68}Ga-labeled conjugates showed suitable hydrophilicity, high stability, and specific targeting properties towards Rituximab-TCO pre-treated CD20 expressing Raji cells. Biodistribution studies showed fast clearance and low accumulation in non-targeted organs for both SulfoCy5- and IRDye800CW-conjugates. In an alendronate-TCO based bone targeting model the dimeric IRDye800CW-conjugate resulted in specific targeting using PET and OI, superior to the monomer. This proof of concept study showed that the preparation of FSC-Tz hybrid imaging agents for pretargeting applications is feasible, making such compounds suitable for hybrid imaging applications.

Keywords: fusarinine C; click chemistry; fluorescence; optical imaging; PET; gallium-68

1. Introduction

Various imaging modalities have evolved as valuable tools for molecular imaging of human diseases. When used as a stand-alone technique, every method shows benefits and limitations [1] so combining different modalities, namely dual-modality- (DMI) or hybrid imaging (HI), is a field with increasing attraction. Optical imaging (OI), for example, is characterized by excellent sensitivity to acquire morphological information but poor tissue penetration limits its applicability for non-invasive imaging purposes. In contrast, single photon emission- (SPECT) and positron emission tomography (PET) show excellent tissue penetration and allow to obtain functional information with high sensitivity. However, due to the lack of morphological information unsurprisingly computed tomography (CT) or magnetic resonance (MR) are used to add morphological details. Although hybrid imaging systems like PET–CT/MRI and SPECT–CT are well established in clinical routine for high resolution imaging their value for real-time intra-operative guided surgery is limited. Despite recent advancements regarding Cerenkov luminescence imaging (CLI) [2] and the development of directional gamma probes (DGP) [3] for radio-guided surgery, optical imaging, particularly near-infrared fluorescence (NIRF) enables the delineation of tumor margins with high accuracy, thus bearing an enormous potential for image-guided surgery [4–7]. Developing targeted imaging probes to combine the complementary nature of PET and NIRF has gained increasing interest and various studies have shown promising results [8–13]. However, most of these studies rely on the use of small peptides as targeting structures being characterized by fast blood clearance and short biological half-life, thus exhibiting fast pharmacokinetics. In contrast, monoclonal antibodies (mAbs) are characterized by their prolonged circulation time exhibiting slow pharmacokinetics but their high affinity and excellent selectivity towards molecular targets bear great potential as imaging agents. The inverse electron-demand Diels-Alder (IEDDA) [14] reaction between 1,2,4,5-tetrazine (Tz) and trans-cyclooct-2-en (TCO) has been established successfully as a highly promising pretargeting strategy, where the TCO-modified mAb is administered prior to the injection of the Tz-bearing radioactive payload to form the radioimmunoconjugate in vivo with exquisite selectivity and tremendously rapid reaction kinetics [15]. With this the applicability of mAbs as targeting vectors has reached a new dimension. Zeglis and co-workers went one step further and have recently shown the suitability of mAbs as targeting vectors for PET/OI hybrid-imaging either using direct labelling [16] or following the IEDDA approach [17]. However, in both studies the fluorescent dye was directly attached to the mAb. In contrast, we herein report on the design of dual-modality imaging agents for pretargeting applications by chelator scaffolding i.e., the fluorescent residue was conjugated to the chelator. Therefore, we utilized the macrocyclic chelator fusarinine C (FSC), which has been already proven to be a suitable scaffold for the development of targeted hybrid imaging agents [18] and allows straight forward radiolabelling with Gallium-68 for PET applications without further modifications, providing advantages over, e.g., F-18 labelling. FSC provides three amine functionalities at the chelating backbone for site specific modification and despite the optical signalling residue we conjugated one or two Tz-motifs for potentially improved IEDDA-based pretargeting as presented in Scheme 1. We used Cyanine based fluorescent dyes as they provide a choice of wavelengths both for microscopy (Cy5) and near-infrared for OI (Cy7, IRDye800CW). A Polyethleneglycol-5 (PEG$_5$)-Tetrazine was chosen for conjugation to FSC, which has provided suitable properties in a previous approach [19].

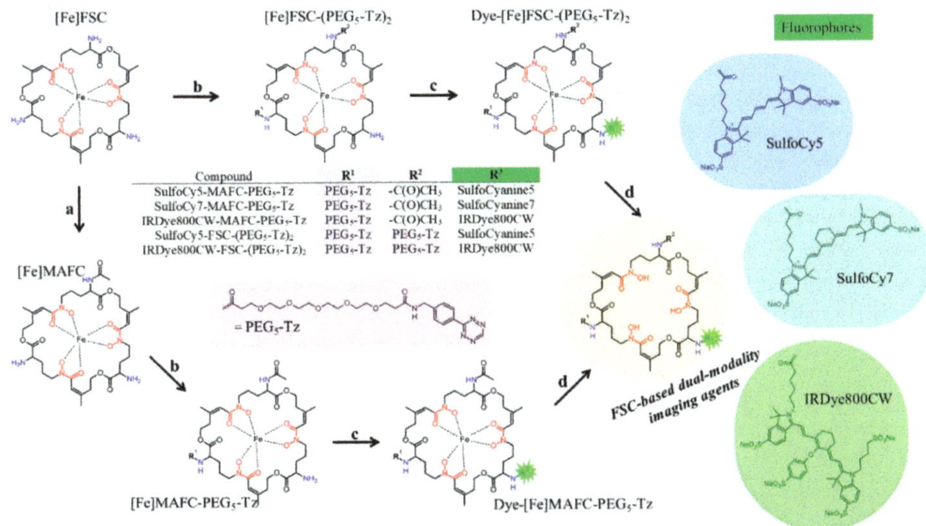

Scheme 1. Route of synthesis for fusarinine C (FSC)-based clickable dual-modality imaging agents (**a**: MetOH/Ac$_2$O; **b**: PEG$_5$-Tz-NHS, DMF/DIPEA; **c**: fluorophore, DMF/DIPEA and *O*-(7-Azabenzotriazol-1-yl)-*N*,*N*,*N*′,*N*′-tetramethyluronium-hexafluorphosphate (HATU); **d**: EDTA).

2. Results

2.1. Synthesis

The FSC-based Tz-bearing hybrid imaging agents were accessible by a straightforward three–four-step synthesis starting from the macrocyclic chelator [Fe]FSC in a similar approach as described in [18,19]. The monomeric conjugates were prepared by initial coupling of the Tetrazine-PEG$_5$-NHS ester to the mono acetyl protected form of FSC, [Fe]MAFC, the dimeric conjugates by starting from [Fe]FSC. Using equal molar amounts the predominant products were [Fe]FSC-(Tetrazine-PEG$_5$)$_2$ or [Fe]MAFC-Tetrazine-PEG$_5$, which were isolated by high-performance liquid chromatography (HPLC), whereby for [Fe]FSC yields were lower due to formation of [Fe]FSC-(Tetrazine-PEG$_5$). The resulting conjugates were straight forward coupled with the respective Dyes by *O*-(7-Azabenzotriazol-1-yl)-*N*,*N*,*N*′,*N*′-tetramethyluronium-hexafluorphosphate (HATU) activation and finally iron was removed. The conjugates could be obtained in acceptable yield i.e., 30–60% for monomeric and 40–60% for dimeric FSC-based Tz hybrid imaging agents in sufficient purity (>90%) determined by analytical reversed phase (RP)-HPLC using UV absorption at λ = 220 nm.

2.2. Radiolabeling

Radiolabeling with the ^{68}Ge/^{68}Ga-generator derived radiometal gallium-68 was conducted at ambient temperature within five minutes and radiochemical yields greater than 98% determined by radio-RP-HPLC and radio-ITLC (instant thin layer chromatography) could be achieved—as exemplarily shown in Figure 1.

2.3. In Vitro Characterization

The results of the distribution coefficient logD and protein binding studies are presented in Table 1. All ^{68}Ga-labeled conjugates showed reasonable hydrophilicity with values ranging from −1.92 for the most lipophilic (SulfoCy7-MAFC-PEG$_5$-Tz) to −2.85 for the most hydrophilic conjugate (SulfoCy5-MAFC-PEG$_5$-Tz). Interestingly, by replacing the acetyl residue by a second PEG$_5$-Tz moiety the hydrophilicity was altered for the SulfoCy5-conjugate while for the corresponding

IRDye800CW-conjugate it remained the same. The protein binding of the ^{68}Ga-labeled FSC-based hybrid imaging agents was consistent over time and ranged from intermediate (30–50%) to high (50–70%). Introducing a SulfoCy5- or -Cy7 residue to the MAFC-Tz scaffold showed a comparable protein binding while the conjugation of the IRDye800CW led to significantly increased values. Interestingly, the replacement of the acetyl moiety by a PEG$_5$-Tz residue did not increase the protein binding compared to the monomeric Tz conjugates. However, this is consistent with the results of a previous study where increasing the number of PEG$_5$-Tz by chelator scaffolding did not increase the protein binding [19].

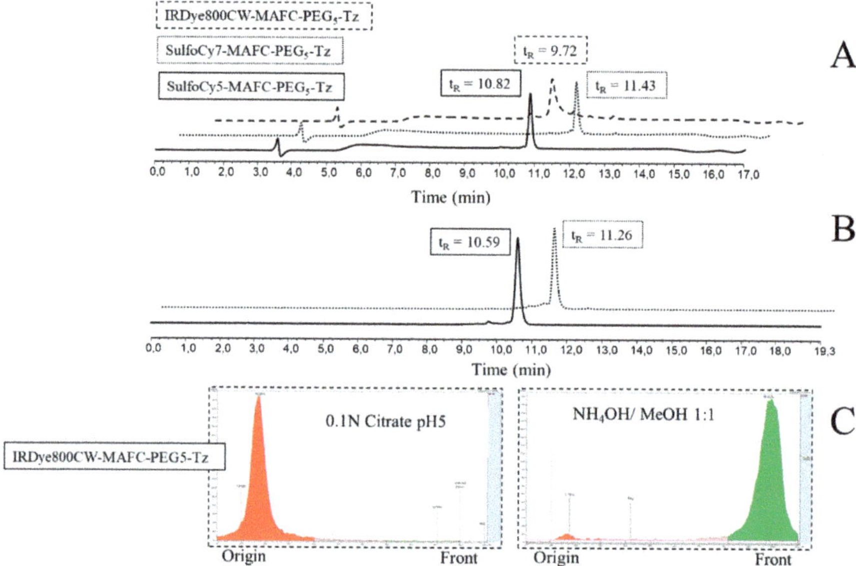

Figure 1. Representative chromatograms of monomeric FSC-based dual-modality imaging (DMI) agents. (**A**) shows UV-detection at λ = 220 nm, (**B**) represents radio detection of [^{68}Ga]Ga-SulfoCy5-MAFC-PEG$_5$-Tz and [^{68}Ga]Ga-SulfoCy7-MAFC-PEG$_5$-Tz, whereas (**C**) shows radio-ITLC (instant thin layer chromatography) analysis of [^{68}Ga]Ga-IRDye800CW-MAFC-PEG$_5$-Tz using citrate (left) and ammonium acetate/ethanol (right) as mobile phase.

Table 1. Distribution coefficient (logD) and protein binding of ^{68}Ga-labeled FSC-based Tz-bearing hybrid imaging agents.

[68]Ga-Labeled Conjugate	Distribution Coefficient	Protein Binding (%)		
	logD (pH 7.4)	1 h	2 h	4 h
SulfoCy5-MAFC-PEG$_5$-Tz	−2.85 ± 0.08	37.0 ± 0.3	37.4 ± 2.3	37.6 ± 1.2
SulfoCy7-MAFC-PEG$_5$-Tz	−1.92 ± 0.05	36.9 ± 0.8	40.4 ± 1.5	41.6 ± 2.7
IRDye800CW-MAFC-PEG$_5$-Tz	−2.40 ± 0.05	65.7 ± 1.3	67.3 ± 1.4	67.7 ± 0.2
SulfoCy5-FSC-(PEG$_5$-Tz)$_2$	−2.29 ± 0.10	35.7 ± 0.7	39.1 ± 0.4	40.8 ± 0.6
IRDye800CW-FSC-(PEG$_5$-Tz)$_2$	−2.46 ± 0.09	47.3 ± 0.2	50.3 ± 0.3	54.9 ± 1.1

Data are presented as mean ± SD (n = 3).

The results of cell binding studies of the ^{68}Ga-labeled FSC-based hybrid imaging agents on CD20-expressing Raji cells pre-treated with Rituximab(RTX)–TCO-modified or non-modified RTX are summarized in Figure 2. All conjugates showed highly specific targeting properties with ratios of specifically to non-specifically bound radioligand ranging from 3 to 5. The binding of the ^{68}Ga-labeled monomeric DMI agents was 3.09 ± 0.58% for [^{68}Ga]Ga-SulfoCy5-MAFC-PEG$_5$-Tz, 4.12 ± 0.88% for [^{68}Ga]Ga-SulfoCy7-MAFC-PEG$_5$-Tz and 2.88 ± 0.53% for [^{68}Ga]Ga-IRDye800CW-MAFC-PEG$_5$-Tz which was comparable to the non-fluorescent conjugate [^{68}Ga]Ga-DAFC-PEG$_5$-Tz (4.01 ± 0.36%) [19]. The binding of the dimeric DMI agents radiolabeled with gallium-68 was significantly higher (p < 0.005) in comparison to their monomeric counterparts and resulted to be 5.91 ± 1.62% for [^{68}Ga]Ga-SulfoCy5-FSC-(PEG$_5$-Tz)$_2$ and 4.59 ± 1.45% for [^{68}Ga]Ga-IRDye800CW-FSC-(PEG$_5$-Tz)$_2$ but was significantly lower (p < 0.0005) compared to the ^{68}Ga-labeled non-fluorescent conjugate [^{68}Ga]Ga-MAFC-(PEG$_5$-Tz)$_2$ (7.35 ± 0.50%) [19]. Furthermore, the unspecific binding was significantly increased (p < 0.005) also when comparing ^{68}Ga-labeled mono- and dimeric FSC-based DMI agents.

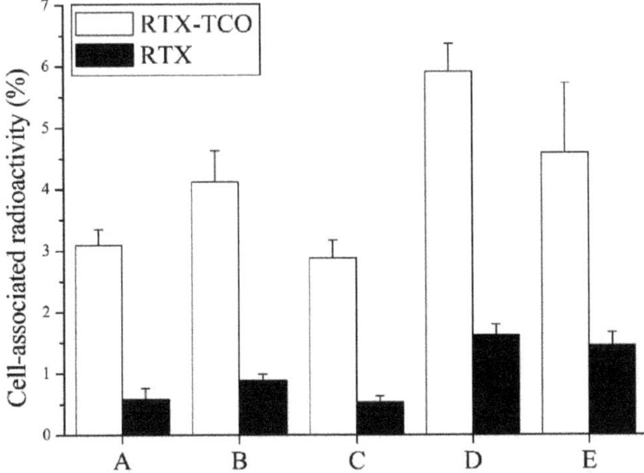

Figure 2. Cell binding studies of [^{68}Ga]Ga-SulfoCy5-MAFC-PEG$_5$-Tz (A), [^{68}Ga]Ga-SulfoCy7-MAFC-PEG$_5$-Tz (B), [^{68}Ga]Ga-IRDye800CW-MAFC-PEG$_5$-Tz (C), [^{68}Ga]Ga-SulfoCy5-FSC-(PEG$_5$-Tz)$_2$ (D), and [^{68}Ga]Ga-IRDye800CW-FSC-(PEG$_5$-Tz)$_2$ (E) on CD20-expressing Raji cells pre-treated with *trans*-cyclooctene (TCO) modified rituximab (white bars) and non-modified antibody (black bars).

Fluorescence microscopy of SulfoCy5-conjugates is presented in Figure 3 and clearly shows the high target specificity as the RTX–TCO pre-treated Raji cells revealed SulfoCy5-specific signaling whereas the negative control, lacking the TCO functionality, did not.

2.4. Biodistribution Studies

The ex vivo biodistribution profile of the ^{68}Ga-labeled FSC-based hybrid imaging agents in healthy BALB/C mice 1 h post injection (p.i.) is summarized in Figure 4. Except from the SulfoCy7-conjugate all imaging probes showed relatively fast blood clearance and low retention in muscle and bone indicating suitable properties for imaging applications matching the short half-life of Gallium-68. The SulfoCy7-conjugate also revealed higher activities in all organs compared to their SulfoCy5 and IRDye800CW counterparts, matching the highest lipophilicity of this conjugate, which was therefore not further explored in imaging studies. Comparable tissue activity was found comparing Sulfo-Cy5 and IRdye800CW conjugates for most organs, except for kidneys where values for IRDye800CW were considerably higher (>20%IA/g). Monomeric MAFC-PEG$_5$-Tz conjugates showed comparable biodistribution to non-dye conjugated FSC-based tetrazines [19] and considerably lower activities especially in liver and spleen (~2%IA/g) as compared to FSC-(PEG$_5$-Tz)$_2$ conjugates with around 10%IA/g. Lung retention was relatively high for all compounds (~5–10%IA/g), no specific reason for this phenomenon could be found.

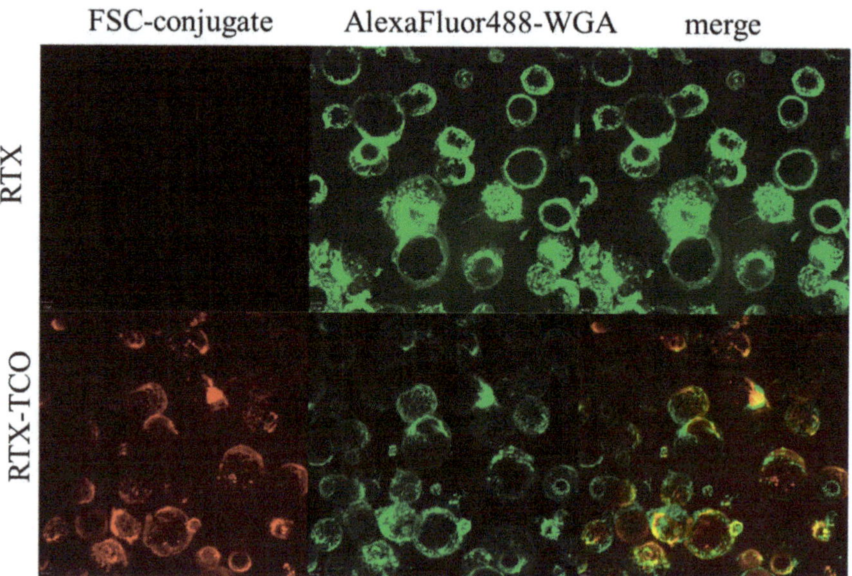

Figure 3. Fluorescence microscopy of CD20-expressing Raji cells pre-treated with RTX (upper lane) and with RTX–TCO (lower lane, negative control) prior to incubation with SulfoCy5-MAFC-PEG$_5$-Tz. Each lane shows the fluorescence signal of the FSC-based conjugate (red), optical signal from AlexaFlour488-WGA labeled cells (green) and both images merged (from left to right).

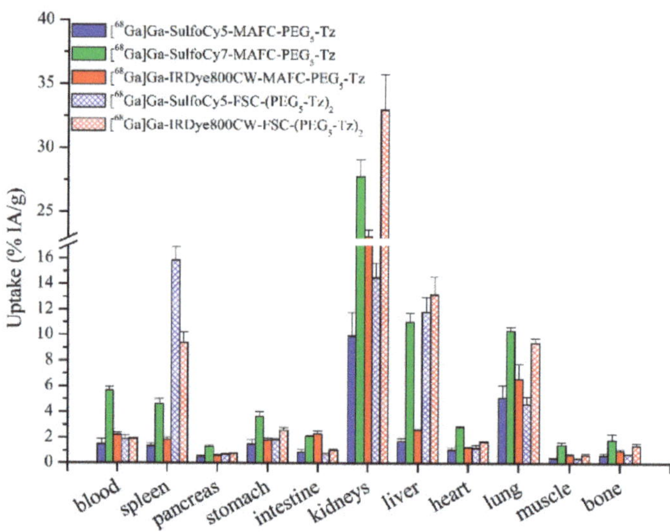

Figure 4. Ex vivo biodistribution profile studies of ^{68}Ga-labeled FSC-based Tz-bearing dual-modality imaging agents in healthy BALB/c mice 1 h after administration.

2.5. Imaging Studies

A simple pretargeting model using bone targeting TCO-alendronate was chosen for a straight forward proof of concept in vivo. PET/CT images revealed specific uptake of both IRDye800CW conjugates in bone structures of TCO-alendronate pre-treated mice, whereas no uptake was seen in mice treated with alendronate alone, proving the in vivo specificity of both constructs. Both conjugates revealed predominant renal excretion and good contrast 1 h p.i. Figure 5 shows images of [68Ga]Ga-IRDdye800CW-FSC-(PEG5-Tz)2 revealing high accumulation in joints and spine with excellent contrast and no uptake in the alendronate control images. This specific accumulation could be confirmed by optical imaging of excised bones with a high accumulation in the joint area, which is not seen in the alendronate control animal. Comparable images of [68Ga]Ga- IRDye800CW-MAFC-PEG5-Tz are shown in supplementary data (Figure S1) indicating less pronounced uptake in comparison with its monomeric counterpart both in PET/CT and optical images, however without being statistically significant. This is in line with the higher in vitro binding (Figure 2) and our findings when comparing non derivatized mono- and dimeric Tetrazine-FSC conjugates [19], showing that dimerization has a positive effect on in vivo targeting properties. Overall, this proves that dye-conjugated tetrazines based on the FSC scaffold can be used for pretargeting applications combining PET and optical imaging. Further studies, ideally in appropriate pretargeting tumor models, are required to judge the full potential of these compounds for tumor imaging and image-guided procedures.

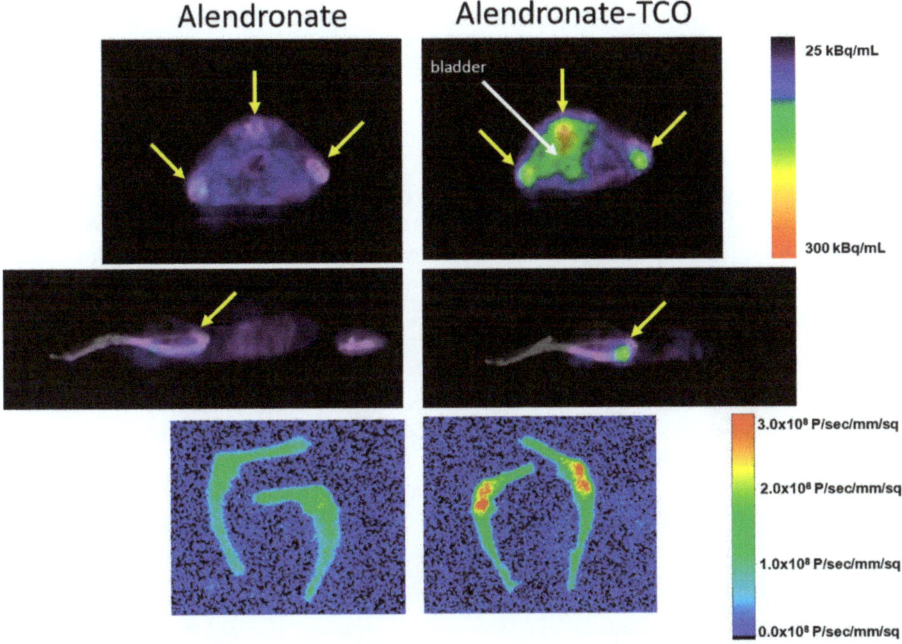

Figure 5. Imaging of 68Ga-labeled IRDdye800CW-FSC-(PEG5-Tz)2 in mice receiving alendronate alone (left row) or alendronate-TCO (right), PET/CT images: transverse slices (top) and sagittal slices (middle), yellow arrows indicate uptake in joints and spine, white arrow indicates the bladder. Bottom image: Optical image of excised bones of the lower limbs. Both PET and OI indicate higher accumulation in bone of alendronate-TCO pre-treated mice as compared to controls.

This is to our knowledge the first report on combining PET and NIRF imaging based on a single tetrazine based pretargeting vector. The FSC scaffold, which is a versatile chelator for Gallium-68, was well suited to develop this approach. In contrast to other attempts, where the dye was conjugated to an antibody [16,17], our approach allows a DMI application without modifying the actual targeting vector, thereby avoiding separate development and characterization of the two versions with and without the dye conjugate.

3. Materials and Methods

3.1. Analytics

Analytical [radio]-RP-HPLC. Reversed-phase high-performance liquid chromatography analysis was performed with the following instrumentation: UltiMate 3000 RS UHPLC pump, UltiMate 3000 autosampler, Ultimate 3000 column compartment (25 °C oven temperature), UltiMate 3000 Variable Wavelength Detector (Dionex, Germering, Germany; UV detection at λ = 220 nm) a radio detector (GabiStar, Raytest; Straubenhardt, Germany), Jupiter 5 µm C_{18} 300 Å 150 × 4.6 mm (Phenomenex Ltd., Aschaffenburg, Germany) column with acetonitrile (ACN)/H_2O/0.1% trifluoroacetic acid (TFA) as mobile phase; flow rate of 1 mL/min; gradient: 0.0–1.0 min 10% ACN, 1.0–12.0 min 10–60% ACN, 13.0–15.0 min 60–80% ACN, 15.0–16.0 min 80–10% ACN, and 16.0–20.0 min 10% ACN.

Preparative RP-HPLC. Sample purification via RP-HPLC was carried out as follows: Gilson 322 Pump with a Gilson UV/VIS-155 detector (UV detection at λ = 220 nm) using a PrepFC™ automatic fraction collector (Gilson, Middleton, WI, USA), Eurosil Bioselect Vertex Plus 30 × 8mm 5 µm C_{18A} 300 Å pre-column and Eurosil Bioselect Vertex Plus 300 × 8mm 5 µm C_{18A} 300 Å column (Knauer, Berlin, Germany) and following ACN/H_2O/0.1% TFA gradients with a flow rate of 2 mL/min: gradient A: 0.0–5.0min 0% ACN, 5.0–35 min 0–50% ACN, 35.0–38.0 min 50% ACN, 38.0–40.0 min 50%–0 ACN. gradient B: 0.0–5.0 min 10% ACN, 5.0–40.0 min 10–60% ACN, 41.0–45.0 min 60% ACN, 46.0–50.0 min 60–80% ACN, and 51.0–55.0 min 80–10% ACN.

Mass Spectrometry

Mass analysis was conducted on a Bruker microflex™ bench-top MALDI-TOF MS (Bruker Daltonics, Bremen, Germany) using dried-droplet method on a micro scout target (MSP96 target ground steel BC, Bruker Daltonics) with α-cyano-4-hydroxycinnamic acid (HCCA, Sigma-Aldrich, Handels GmbH, Vienna, Austria) as matrix. Flex Analysis 2.4 software (Bruker Daltonics, Bremen, Germany) was used for data processing.

3.2. Synthesis

General. All chemicals and solvents were obtained as reagent grade from commercial sources unless otherwise stated. *Trans*-Cyclooctene-NHS ester and Tetrazine-PEG$_5$-NHS ester were purchased from Click Chemistry Tools (Scottsdale, AZ, USA). Rituximab (MabThera®, Roche Pharma AG, Grenzach-Wyhlen, Germany) was a kind gift from the University Hospital of Innsbruck. The fluorescent dyes, SulfoCyanine5-NHS ester and SulfoCyanine7 carboxylic acid were obtained from Lumiprobe GmbH (Hannover, Germany) while IRDye800CW carboxylic acid was purchased from LI-COR Biosciences GmbH (Bad Homburg, Germany).

3.2.1. [Fe]fusarinine C ([Fe]FSC) and [Fe]*N*-Monoacetylfusarinine C ([Fe]MAFC)

The macrocyclic iron protected precursors were obtained according to previously published procedures [19]. Briefly, [Fe]FSC was directly isolated from fungal culture in high yield and sufficient chemical purity (>90%, UV–Vis, λ = 220 nm). The monoacetylated derivative [Fe]MAFC could be isolated with reasonable yield and in high purity (>99% UV-detection, λ = 220 nm) via preparative RP-HPLC from a mixture of mono- and multiple acetylated [Fe]FSC-derivatives when reacting [Fe]FSC with acetic anhydride. Both derivatives gave a red–brown colored solid after lyophilization.

- [Fe]FSC: analytical RP-HPLC t_R = 6.95 min; MALDI TOF-MS: *m/z* [M + H]$^+$ = 779.93 [$C_{33}H_{51}FeN_6O_{12}$; M_r = 779.63 (calculated)]
- [Fe]MAFC analytical RP-HPLC t_R = 7.67 min; MALDI TOF-MS: *m/z* [M + H]$^+$ = 822.04 [$C_{35}H_{53}FeN_6O_{13}$; M_r = 821.67 (calculated)]

3.2.2. Conjugation of PEGylated tetrazine (PEG$_5$-Tz)

Either [Fe]FSC (8.0 mg, 10.3 µmol) or [Fe]MAFC (11.0 mg, 13.4 µmol) was dissolved in 500 µL dry DMF and after pH adjustment with DIPEA (pH 9) the mixture was stirred for 30 min at RT. Tetrazine-PEG$_5$-NHS (1 equivalent) was dissolved in 200 µL anhydrous DMF and was slowly added dropwise to the solution over a period of 15 min. After 2 h reaction time at ambient temperature the organic solvent was concentrated in vacuo and purified by preparative RP-HPLC using gradient B to give [Fe]FSC-(PEG$_5$-Tz)$_2$ (t$_R$ = 32.9 min) and [Fe]MAFC-PEG$_5$-Tz (t$_R$ = 26.9 min) as red-brown colored solid after lyophilization. Analytical data:

- [Fe]MAFC-PEG$_5$-Tz: 12.5 mg [9.5 µmol, 71%], RP-HPLC t$_R$ = 10.2 min; MALDI TOF-MS: *m/z* [M + H]$^+$ = 1312.21 [C$_{58}$H$_{84}$FeN$_{11}$O$_{20}$; M$_r$ = 1311.19 (calculated)].
- [Fe]FSC-(PEG$_5$-Tz)$_2$: 4.76 mg [2.71 µmol, 33%], RP-HPLC t$_R$ = 11.4 min; MALDI TOF-MS: *m/z* [M + H]$^+$ = 1759.03 [C$_{79}$H$_{113}$FeN$_{16}$O$_{26}$; M$_r$ = 1758.68 (calculated)].

3.2.3. Synthesis of Monomeric FSC-based Tz Hybrid Imaging Agents

For conjugation of the fluorescent dyes to the monomeric FSC-based Tz-ligand, 2.3 mg of [Fe]MAFC-PEG$_5$-Tz (1.75 µmol) were each dissolved in 500 µL dry DMF, pH was adjusted with DIPEA (pH 9) and 1.1 equivalent of the corresponding dye dissolved in 500 µL DMF was added. SulfoCyanine5-NHS ester (1.50 mg, 1.93 µmol) was added directly while the carboxylic acid of SulfoCyanine7 (1.41 mg, 1.93 µmol), as well as IRDye800CW (2.1 mg, 1.93 µmol) were pre-activated with 1.5 equivalent of *O*-(7-Azabenzotriazol-1-yl)-*N,N,N',N'*-tetramethyluronium-hexafluorphosphate (HATU, 1.1 mg, 2.89 µmol) for 10 min at ambient temperature. The reaction mixtures were maintained for complete conjugation at RT for 6 h followed by evaporation of DMF in vacuo. The crude bioconjugates were re-dissolved in 1 mL 50% ACN/H$_2$O (*v/v*). Half of the solution (500 µL) was purified by preparative RP-HPLC (gradient B) to obtain SulfoCyanine5-[Fe]MAFC-Tz (t$_R$ = 32.0 min) as dark blue and SulfoCyanine7-[Fe]MAFC-Tz (t$_R$ = 34.9 min) as well as IRDye800CW-[Fe]MAFC-Tz (t$_R$ = 28.8 min) as dark green colored solids after lyophilization. Analytical data:

- SulfoCyanine5-[Fe]MAFC-PEG$_5$-Tz: 0.61 mg [0.32 µmol, 36%], RP-HPLC t$_R$ = 10.6 min; MALDI TOF-MS: *m/z* [M + H]$^+$ = 1936.99 [C$_{90}$H$_{120}$FeN$_{13}$O$_{27}$S$_2$; M$_r$ = 1935.96 (calculated)]
- SulfoCyanine7-[Fe]MAFC-PEG$_5$-Tz: 0.85 mg [0.42 µmol, 49%], RP-HPLC t$_R$ = 11.2 min; MALDI TOF-MS: *m/z* [M + H]$^+$ = 2002.85 [C$_{95}$H$_{126}$FeN$_{13}$O$_{27}$S$_2$; M$_r$ = 2002.06 (calculated)]
- IRDye800CW-[Fe]MAFC-PEG$_5$-Tz: 1.11 mg [0.48 µmol, 55%], RP-HPLC t$_R$ = 9.5 min; MALDI TOF-MS: *m/z* [M + H]$^+$ = 2297.02 [C$_{104}$H$_{136}$FeN$_{13}$O$_{34}$S$_4$; M$_r$ = 2296.36 (calculated)]

The remaining half of the reaction mixture (500 µL) was used for demetallation. Therefore, 1 mL of disodium EDTA (Na$_2$EDTA, 200 mM) was added and the mixture was stirred for 4 h at RT to completely remove the iron from the conjugates followed by preparative RP-HPLC purification to give intensively green to blue colored solids after freeze drying. Analytical data:

- SulfoCyanine5-MAFC-PEG$_5$-Tz: 0.55 mg [0.29 µmol, 34%], gradient B (t$_R$ = 32.5 min); Analytical data: RP-HPLC t$_R$ = 10.8 min; MALDI TOF-MS: *m/z* [M + H]$^+$ = 1883.75 [C$_{90}$H$_{123}$N$_{13}$O$_{27}$S$_2$; M$_r$ = 1883.14 (calculated)]
- SulfoCyanine7-MAFC-PEG$_5$-Tz: 0.70 mg [0.36 µmol, 41%], gradient B (t$_R$ = 35.5 min); Analytical data: RP-HPLC t$_R$ = 11.4 min; MALDI TOF-MS: *m/z* [M + H]$^+$ = 1949.70 [C$_{95}$H$_{129}$N$_{13}$O$_{27}$S$_2$; M$_r$ = 1949.24 (calculated)]
- IRDye800CW-MAFC-PEG$_5$-Tz: 1.23 mg [0.55 µmol, 63%], gradient B (t$_R$ = 29.2 min); Analytical data: RP-HPLC t$_R$ = 9.7 min; MALDI TOF-MS: *m/z* [M + H]$^+$ = 2244.26 [C$_{104}$H$_{139}$N$_{13}$O$_{34}$S$_4$; M$_r$ = 2243.54 (calculated)]

3.2.4. Synthesis of Dimeric FSC-based Tz Hybrid Imaging Agents

Conjugation of the fluorescent dye to the dimeric FSC-based Tz-ligand was performed as described above using 1.0 mg of [Fe]FSC-(PEG$_5$-Tz)$_2$ (0.57 µmol) as starting material. After successful conjugation, demetallation was performed as described above followed by purification via preparative RP-HPLC. Analytical data:

- SulfoCyanine5-FSC-(PEG$_5$-Tz)$_2$: 0.80 mg [0.34 µmol, 60%], gradient B (t_R = 36.7 min); Analytical data: RP-HPLC t_R = 11.61 min; MALDI TOF-MS: *m/z* [M + H]$^+$ = 2332.75 [C$_{111}$H$_{153}$N$_{18}$O$_{33}$S$_2$; M$_r$ = 2331.63 (calculated)]
- IRDye800CW-FSC-(PEG$_5$-Tz)$_2$: 0.61 mg [0.22 µmol, 40%], gradient B (t_R = 31.5 min); Analytical data: RP-HPLC t_R = 11.10 min; MALDI TOF-MS: m/z [M + H]$^+$ = 2693.40 [C$_{125}$H$_{169}$N$_{18}$O$_{40}$S$_4$; M$_r$ = 2692.03 (calculated)]

3.2.5. Modification of Rituximab (RTX)

Anti-CD20 monoclonal antibody Rituximab was modified as previously reported [19]. Briefly, the antibody solution (Mabthera®, 10 mg/mL, Roche Pharma AG, Grenzach-Wyhlen, Germany) was allowed to react at 4 °C overnight in 0.1 M NaHCO$_3$ solution with 20 molar equivalent of TCO–NHS ester dissolved in DMSO followed by size exclusion column purification via PD-10 (GE Healthcare Vienna, Austria) to give RTX–TCO in PBS (2.5 mg/mL).

3.3. Radiochemistry

Radiolabelling with Gallium-68 was performed by mixing 500 µL [^{68}Ga]gallium chloride ([^{68}Ga]GaCl$_3$) eluate, obtained by fractioned elution of a ^{68}Ge/^{68}Ga-generator (IGG100, nominal activity 1850 MBq, Eckert and Ziegler, Berlin, Germany) with 0.1 M hydrochloric acid (HCl, Rotem Industries Ltd., Beer-Sheva, Israel), with 100 µL sodium acetate solution (1.14 M) to give a final pH of 4.5 and adding 10 µg (3.71-5.31 nmol) of the corresponding fluorescent FSC-Tz derivative. After incubation for 5 min at ambient temperature radio-RP-HPLC- as well as radio-ITLC-analysis was performed.

Radio-ITLC

Instant thin layer chromatography (ITLC) analysis was performed using TLC-SG strips (Varian, Lake Forest, CA, USA) as stationary phase and 0.1 M sodium citrate solution (pH 5) or 1 M ammonium acetate/methanol (1:1, *v/v*) with a pH of 6.8 as mobile phase. The strips were analyzed using a TLC scanner (Scan-RAM™, LabLogic, Sheffield, UK).

3.4. In Vitro Characterization

3.4.1. Distribution Coefficient (logD)

To determine the hydrophilicity the distribution of the ^{68}Ga-labeled conjugates between an organic (octanol) and aqueous (PBS) layer was assessed. Aliquots (50 µL) of the ^{68}Ga-labeled tracers (~5 µM) were diluted in 1 mL of octanol/PBS (1:1, *v/v*) and the mixture was vortexed at 1400 rpm (MS 3 basic vortexer, IKA, Staufen, Germany) for 15 min at RT. Subsequently the mixture was centrifuged for 2 min at 4500 rpm followed by measuring aliquots (50 µL) of both layers in the gamma counter (Wizard^2 3″, Perkin Elmer, Waltham, MA, USA). LogD was calculated using Microsoft Excel (*n* = 3, six replicates).

3.4.2. Protein Binding

The ability of the ^{68}Ga-labeled conjugates to bind to serum proteins was determined by size exclusion chromatography. Aliquots (50 µL, *n* = 3) of the radioligand solution (~10 µM) were incubated in 450 µL freshly prepared human serum as well as 450 µL PBS as control and the mixtures were maintained at 37 °C. After 1, 2, and 4 h, aliquots (25 µL) were transferred to illustra MicroSpin G-50 columns (GE Healthcare Vienna, Austria) and the percentage of protein-bound (eluate) and non-bound (column) conjugate was calculated.

3.4.3. Cell-Binding Studies

CD20-expressing Raji cells (humanoid lymphoblast-like B-lymphocyte cells) were obtained from the American Type Culture Collection (ATCC, Manassas, Virginia, USA). The cells were cultured in tissue culture flasks (Cellstar; Greiner Bio-One, Kremsmuenster, Austria) using RPMI-1640 medium with 10% (*v/v*) fetal bovine serum (FBS) as supplement (Invitrogen Corporation, Lofer, Austria). For cell binding studies 10×10^6 cells were washed twice with fresh medium, diluted with PBS to a final concentration of 1×10^6 cells per mL and 500 μL of cell suspension was transferred to Eppendorf tubes. Hereafter, 50 μL of RTX–TCO or non-modified RTX as negative control (both 0.5 μM) was added and the cell suspension was maintained at 37 °C under gentle shaking. After 1 h the suspension was centrifuged (2 min, 11×10^3 rcf), the supernatant was discarded and the cells were washed twice with 600 μL PBS and finally resuspended with 450 μL PBS. Subsequently 50 μL of the radioligand solution (22 nM) was added and the suspension was incubated for 30 min at 37 °C. After centrifugation and two washing steps with 600 μL PBS, the cells were resuspended in 500 μL PBS and transferred to polypropylene vials for gamma counter measurement followed by calculation of cell-associated activity in comparison to the total activity applied ($n = 3$, six replicates).

3.4.4. Fluorescence Microscopy

CD20-expressing Raji cells were prepared and pre-treated with modified or non-modified RTX as described above (cell-binding studies). Instead of the radioligand, 50 μL (22 nM) of iron protected SulfoCyanine5-[Fe]MAFC-PEG$_5$-Tz was added and the cells were treated according to the cell-binding studies with the ^{68}Ga-labeled counterparts. After resuspension fluorescence microscopy was performed using a laser ($\lambda = 561$ nm) for excitation of the fluorescent dye. The imaging was carried out with a spinning-disc confocal microscopic system (Ultra VIEW VoX, PerkinElmer, Waltham, MA, USA) linked to a Zeiss AxioObserver Z1 inverted microscope (Zeiss, Oberkochen, Germany). Images were acquired with Volocity software (PerkinElmer) utilizing a 63× oil immersion objective (numerical aperture 1.42). The images show z-stacks ($n = 4$; 1 μm spacing). Cell morphology was visualized by adding fluorescently labeled wheat germ agglutinin (Alexa Fluor® 488 WGA, ThermoFischer Scientific, Vienna, Austria).

3.5. In Vivo Characterization

Ethics statement: All animal experiments were performed in accordance with regulations and guidelines of the Austrian animal protection laws and the Czech Animal Protection Act (No. 246/1992), with approval of the Austrian Ministry of Science (BMWF-66.011/0161-WF/V/3b/2016) and the Czech Ministry of Education, Youth, and Sports (MSMT-18724/2016-2), and the institutional Animal Welfare Committee of the Faculty of Medicine and Dentistry of Palacky University in Olomouc, Czech Republic.

3.5.1. Biodistribution Studies

Biodistribution studies of the FSC-based hybrid imaging agents radiolabeled with Gallium-68 were performed using healthy five-week-old female BALB/c mice (Charles River Laboratories, Sulzfeld, Germany). Animals ($n = 3$) were injected via lateral tail vein with 1 nmol of conjugate and a total activity of approximately 6 MBq. Mice were sacrificed by cervical dislocation 1 h p.i. followed by collection of the main organs and tissue, subsequent gamma counter measurement and calculation of the percentage of injected activity per gram tissue (% IA/g).

3.5.2. Imaging Studies

Pretargeting Model

For the proof of concept of pretargeting in vivo a model described by Yazdani et al. was applied [20], based on TCO-modified alendronate for bone targeting. For preparation of TCO–alendronate, briefly, 20 mg sodium alendronate trihydrate (Sigma Aldrich) was dissolved in 400 μL 0.1N sodium-bicarbonate solution pH 8.5. Then 0.6 mL TCO–NHS ester (Click chemistry tools, Scottsdale, AZ, USA), 20 mg/mL in DMSO, were slowly added to the solution and stirred overnight in the dark, followed by freeze drying. As a control the same preparation was prepared, replacing TCO–NHS solution by the same volume of DMSO. For injection in mice the residue was dissolved in water and adjusted to pH 7 (30 mM).

PET/CT Imaging

MicroPET/CT images were acquired with an Albira PET/SPECT/CT small animal imaging system (Bruker Biospin Corporation, Woodbridge, CT, USA). Healthy 10-week-old female BALB/c mice (Envigo, Horst, The Netherlands) (n = 4) were pre-treated by intraperitoneal injection of 250 μL corresponding to 4 mg of alendronate-TCO or 2.5 mg alendronate alone, respectively. Pre-treated mice were retro-orbitally injected with radiolabeled tracer in a dose of 5–10 MBq corresponding to 1–2 μg of conjugate per animal 5 h after the pre-treatment. Anaesthetized (2% isoflurane (FORANE, Abbott Laboratories, Abbott Park, IL, USA)) animals were placed in a prone position in the Albira system before the start of imaging. Static PET/CT images were acquired over 30 min starting 1 h p.i.. A 10-min PET scan (axial FOV 148 mm) was performed, followed by a double CT scan (axial FOV 2 × 65 mm, 45 kVp, 400 μA, at 400 projections). Scans were reconstructed with the Albira software (Bruker Biospin Corporation, Woodbridge, CT, USA) using the maximum likelihood expectation maximization (MLEM) and filtered back projection (FBP) algorithms. After reconstruction, acquired data was viewed and analyzed with PMOD software (PMOD Technologies Ltd., Zurich, Switzerland).

Optical Imaging

Near-infrared in vivo fluorescence imaging was performed with an In-Vivo MS FX PRO small animal imaging system (Bruker Biospin Corporation, Woodbridge, CT, USA) after the PET/CT investigation. Mice injected with the different conjugates were sacrificed 24 h p.i. and excised bones of the lower limbs were imaged. An appropriate filter set (λ_{ex} = 710 nm and λ_{em} = 790 nm) was used for acquiring the fluorescence of the IRDye800CW-conjugates ex vivo. Identical illumination settings (acquisition time = 30 s, filters = 710/790 nm, f-stop = 2.8, field of view = 100 mm, and binning = 2 × 2) were used for image acquisition and fluorescence emission was normalized to photons/s/mm^2. Acquired images were analyzed using Bruker MI SE software (Bruker Biospin Corporation, Woodbridge, CT, USA).

3.6. Statistical Analysis

Statistical analysis was performed using independent two-tailed Student's T-Test with P-value < 0.05 indicating significance.

4. Conclusions

We synthesized a series of tetrazines combining radiolabeling for PET with optical NIR imaging on a single scaffold. A Sulfo-Cy7 conjugate was shown to have suboptimal properties but versions based on Sulfo-Cy5 and IRDye800CW revealed specific binding to TCO-modified targets in vitro and specific accumulation in vivo exemplified by imaging. A dimeric version showed enhanced and more specific signal in vivo, making this a promising candidate to further investigate multimodality imaging techniques based on pretargeting strategies.

Supplementary Materials: The following are available online, Figure S1: OI and PET Imaging of ^{68}Ga-labeled IRDdye800CW-MAFC-PEG$_5$-Tz in mice.

Author Contributions: Conceptualization, C.D. and D.S.; methodology, D.S., M.P., M.H., and C.R.; validation, S.M., P.K., and M.K.; formal analysis, D.S., J.P., S.M., and P.K.; investigation, J.P., S.M., D.S., M.P., M.K., and P.K.; resources, H.H., M.P., C.D., C.R., and M.H.; data curation, D.S., C.D., M.P., and M.H.; writing—original draft preparation, D.S.; writing—review and editing, C.D. and C.R.; visualization, D.S., M.P., and M.H.; supervision, C.D. and H.H.; project administration, C.D. and M.P.; funding acquisition, C.D. and M.P. All authors have read and agreed to the published version of the manuscript.

Funding: This research was funded by the Austrian Science Foundation (FWF) grant P 25899-B23 to D.S. and C.D. and the European Regional Development Fund - Project ENOCH (No. CZ.02.1.01/0.0/0.0/16_019/0000868) to M.P. Open Access Funding by the Austrian Science Fund (FWF).

Acknowledgments: Claudia Manzl is acknowledged for providing Raji cells.

Conflicts of Interest: The authors declare no conflict of interest.

References

1. Youn, H.; Chung, J.K. Reporter gene imaging. *Am. J. Roentgenol.* **2013**, *201*, 206–214. [CrossRef] [PubMed]

2. Grootendorst, M.R.; Cariati, M.; Kothari, A.; Tuch, D.S.; Purushotham, A. Cerenkov luminescence imaging (CLI) for image-guided cancer surgery. *Clin. Transl. Imaging* **2016**, *4*, 353–366. [CrossRef] [PubMed]

3. Massari, R.; Ucci, A.; D'Elia, A.; Campisi, C.; Bertani, E.; Soluri, A. Directional probe for radio-guided surgery: A pilot study: A. *Med. Phys.* **2018**, *45*, 622–628. [CrossRef] [PubMed]

4. Gibbs, S.L. Near infrared fluorescence for image-guided surgery. *Quant. Imaging Med. Surg.* **2012**, *2*, 177–187. [CrossRef] [PubMed]

5. Te Velde, E.A.; Veerman, T.; Subramaniam, V.; Ruers, T. The use of fluorescent dyes and probes in surgical oncology. *Eur. J. Surg. Oncol.* **2010**, *36*, 6–15. [CrossRef] [PubMed]

6. Nagaya, T.; Nakamura, Y.A.; Choyke, P.L.; Kobayashi, H. Fluorescence-Guided Surgery. *Front. Oncol.* **2017**, *7*. [CrossRef] [PubMed]

7. Wang, C.; Wang, Z.; Zhao, T.; Li, Y.; Huang, G.; Sumer, B.D.; Gao, J. Optical molecular imaging for tumor detection and image-guided surgery. *Biomaterials* **2018**, *157*, 62–75. [CrossRef] [PubMed]

8. Nahrendorf, M.; Keliher, E.; Marinelli, B.; Waterman, P.; Feruglio, P.F.; Fexon, L.; Pivovarov, M.; Swirski, F.K.; Pittet, M.J.; Vinegoni, C.; et al. Hybrid PET-optical imaging using targeted probes. *Proc. Natl. Acad. Sci. USA* **2010**, *107*, 7910–7915. [CrossRef] [PubMed]

9. Azhdarinia, A.; Ghosh, P.; Ghosh, S.; Wilganowski, N.; Sevick-Muraca, E.M. Dual-labeling strategies for nuclear and fluorescence molecular imaging: A review and analysis. *Mol. Imaging Biol.* **2012**, *14*, 261–276. [CrossRef] [PubMed]

10. Lütje, S.; Rijpkema, M.; Helfrich, W.; Oyen, W.J.G.; Boerman, O.C. Targeted Radionuclide and Fluorescence Dual-modality Imaging of Cancer: Preclinical Advances and Clinical Translation. *Mol. Imaging Biol.* **2014**, *16*, 747–755. [CrossRef] [PubMed]

11. Welling, M.M.; Bunschoten, A.; Kuil, J.; Nelissen, R.G.H.H.; Beekman, F.J.; Buckle, T.; Van Leeuwen, F.W.B. Development of a Hybrid Tracer for SPECT and Optical Imaging of Bacterial Infections. *Bioconjug. Chem.* **2015**, *26*, 839–849. [CrossRef] [PubMed]

12. Kang, C.M.; Koo, H.-J.; An, G.I.; Choe, Y.S.; Choi, J.Y.; Lee, K.-H.; Kim, B.-T. Hybrid PET/optical imaging of integrin αVβ3 receptor expression using a 64Cu-labeled streptavidin/biotin-based dimeric RGD peptide. *EJNMMI Res.* **2015**, *5*, 60. [CrossRef] [PubMed]

13. Baranski, A.; Schäfer, M.; Bauder-Wüst, U.; Roscher, M.; Schmidt, J.; Stenau, E.; Simpfendörfer, T.; Teber, D.; Maier-Hein, L.; Hadaschik, B.; et al. PSMA-11 Derived Dual-labeled PSMA-Inhibitors for Preoperative PET Imaging and Precise Fluorescence-Guided Surgery of Prostate Cancer. *J. Nucl. Med.* **2018**, *59*, 639–645. [CrossRef] [PubMed]

14. Knall, A.-C.; Slugovc, C. Inverse electron demand Diels-Alder (iEDDA)-initiated conjugation: A (high) potential click chemistry scheme. *Chem. Soc. Rev.* **2013**, *42*, 5131–5142. [CrossRef] [PubMed]

15. Karver, M.R.; Weissleder, R.; Hilderbrand, S.A. Synthesis and evaluation of a series of 1,2,4,5-tetrazines for bioorthogonal conjugation. *Bioconjug. Chem.* **2011**, *22*, 2263–2270. [CrossRef] [PubMed]

16. Zeglis, B.M.; Davis, C.B.; Abdel-Atti, D.; Carlin, S.D.; Chen, A.; Aggeler, R.; Agnew, B.J.; Lewis, J.S. Chemoenzymatic strategy for the synthesis of site-specifically labeled immunoconjugates for multimodal PET and optical imaging. *Bioconjug. Chem.* **2014**, *25*, 2123–2128. [CrossRef] [PubMed]

17. Adumeau, P.; Carnazza, K.E.; Brand, C.; Carlin, S.D.; Reiner, T.; Agnew, B.J.; Lewis, J.S.; Zeglis, B.M. A pretargeted approach for the multimodal PET/NIRF imaging of colorectal cancer. *Theranostics* **2016**, *6*, 2267–2277. [CrossRef] [PubMed]

18. Summer, D.; Grossrubatscher, L.; Petrik, M.; Michalcikova, T.; Novy, Z.; Rangger, C.; Klingler, M.; Haas, H.; Kaeopookum, P.; Von Guggenberg, E.; et al. Developing Targeted Hybrid Imaging Probes by Chelator Scaffolding. *Bioconjug. Chem.* **2017**, *28*, 1722–1733. [CrossRef] [PubMed]

19. Summer, D.; Mayr, S.; Petrik, M.; Rangger, C.; Schoeler, K.; Vieider, L.; Matuszczak, B.; Decristoforo, C. Pretargeted Imaging with Gallium-68 — Improving the Binding Capability by Increasing the Number of Tetrazine Motifs. *Pharmaceuticals* **2018**, *11*, 102. [CrossRef] [PubMed]

20. Yazdani, A.; Bilton, H.; Vito, A.; Genady, A.R.; Rathmann, S.M.; Ahmad, Z.; Janzen, N.; Czorny, S.; Zeglis, B.M.; Francesconi, L.C.; et al. A Bone-Seeking trans-Cyclooctene for Pretargeting and Bioorthogonal Chemistry: A Proof of Concept Study Using ^{99m}Tc- and ^{177}Lu-Labeled Tetrazines. *J. Med. Chem.* **2016**, *59*, 9381–9389. [CrossRef] [PubMed]

Sample Availability: Not available.

Article

Yields of Photo-Proton Reactions on Nuclei of Nickel and Separation of Cobalt Isotopes from Irradiated Targets

Andrey G. Kazakov [1,*], Julia S. Babenya [1], Taisya Y. Ekatova [1], Sergey S. Belyshev [2,3], Vadim V. Khankin [3], Alexander A. Kuznetsov [2,3], Sergey E. Vinokurov [1] and Boris F. Myasoedov [1]

[1] Radiochemistry Laboratory, Vernadsky Institute of Geochemistry and Analytical Chemistry, The Russian Academy of Sciences, Kosygin St., 19, 119991 Moscow, Russia; fanta2000-10@mail.ru (J.S.B.); ekatova.t@gmail.com (T.Y.E.); vinokurov.geokhi@gmail.com (S.E.V.); bfmyas@mail.ru (B.F.M.)

[2] Department of Physics, Lomonosov Moscow State University, Leninskie Gory, 1, Bld. 2, 119991 Moscow, Russia; belyshev@depni.sinp.msu.ru (S.S.B.); kuznets@depni.sinp.msu.ru (A.A.K.)

[3] Skobeltsyn Institute of Nuclear Physics, Lomonosov Moscow State University, Leninskie Gory, 1, Bld. 2, 119991 Moscow, Russia; v-k32@yandex.ru

* Correspondence: adeptak92@mail.ru

Abstract: Nowadays, cobalt isotopes ^{55}Co, ^{57}Co, and ^{58m}Co are considered to be promising radionuclides in nuclear medicine, with ^{55}Co receiving the most attention as an isotope for diagnostics by positron emission tomography. One of the current research directions is dedicated to its production using electron accelerators (via photonuclear method). In our work, the yields of nuclear reactions occurring during the irradiation of ^{nat}Ni and ^{60}Ni by bremsstrahlung photons with energy up to 55 MeV were determined. A method of fast and simple cobalt isotopes separation from irradiated targets using extraction chromatography was developed.

Keywords: photonuclear method; cobalt isotopes; cobalt-55; extraction chromatography; nuclear medicine

Citation: Kazakov, A.G.; Babenya, J.S.; Ekatova, T.Y.; Belyshev, S.S.; Khankin, V.V.; Kuznetsov, A.A.; Vinokurov, S.E.; Myasoedov, B.F. Yields of Photo-Proton Reactions on Nuclei of Nickel and Separation of Cobalt Isotopes from Irradiated Targets. *Molecules* 2022, 27, 1524. https://doi.org/10.3390/molecules27051524

Academic Editors: Alessandra Boschi and Petra Martini

Received: 25 January 2022
Accepted: 21 February 2022
Published: 24 February 2022

Publisher's Note: MDPI stays neutral with regard to jurisdictional claims in published maps and institutional affiliations.

1. Introduction

Two main methods for producing radioisotopes or their generators for nuclear medicine are widely used today: in nuclear reactors and cyclotrons. Another possible way of their production is photonuclear method. Production of radioactive isotopes for different purposes by this method was widely investigated in the 1970–1980s, and today the growing number of studies on medical isotopes production by photonuclear method can be observed. Due to the development of this method, nowadays ^{47}Sc, ^{67}Cu, and ^{99}Mo/^{99m}Tc generator as well as light isotopes ^{11}C, ^{13}N, ^{15}O, ^{18}F for positron emission tomography (PET) are already obtained in electron accelerators on a regular basis, and the production of ^{225}Ac, ^{177}Lu, ^{111}In, ^{105}Rh, and ^{44}Ti/^{44}Sc generator is currently being investigated [1].

Radioactive isotopes ^{55}Co, ^{57}Co, and ^{58m}Co are considered to be used in nuclear medicine, as they are not too well known and well-studied but are promising. The most attention is paid to ^{55}Co ($T_{1/2}$ = 17.5 h, 77% β^+, $E_{max\beta}{}^+$ = 1498 keV), which is auspicious for studying slow processes in an organism by PET. At the dawn of nuclear medicine, it was reported that ^{55}Co complexes could be used for the diagnostics of lung cancer and for the visualization of tumors [2–4]. It was shown in contemporary studies that ^{55}Co-EDTA is suitable for use in nephrological research [5] and for the visualization of prostate cancer [6] as well as for the detection of distant metastasis in close vicinity of the bladder and kidneys [7]. It is important to mention that the chemical properties of ^{55}Co(II) and its behavior in an organism are similar to that of PET-isotope ^{64}Cu(II) ($T_{1/2}$ = 12.7 h, 17.4% β^+, $E_{max\beta}{}^+$ = 653 keV) and also of Ca(II), the latter being present in body but lacking suitable radioactive isotopes for its visualization [8–10]. In comparison to ^{64}Cu(II), ^{55}Co has the following advantages: first, compounds labeled with ^{55}Co tend to be absorbed less by the liver than compounds labeled with ^{64}Cu. Second, the higher yield of positrons produced by ^{55}Co results in less activity of the drug and/or less time required for PET diagnostics. Using ^{55}Co in PET as an indicator of calcium allows one to visualize the affected tissue in patients with traumatic brain injury and to estimate neuronal damage with strokes and brain tumors [11–13]. As for other medical isotopes of Co, ^{58m}Co ($T_{1/2}$ = 9.04 h) is Auger emitter and thus is suitable for Auger therapy, and ^{57}Co ($T_{1/2}$ = 271.8 d, E_γ = 122 keV) is suitable for preclinical research and the study of pharmacokinetics of drugs based on cobalt due to the long half-life of this isotope and the high yield of produced gamma-quanta [14].

In spite of the advantages listed above, the use of cobalt isotopes in medicine is limited by the difficulties of their production (Figure 1). ^{55}Co is mainly produced in cyclotrons by nuclear reactions ^{54}Fe(d,n)^{55}Co, ^{56}Fe(p,2n)^{55}Co, and ^{58}Ni(p,α)^{55}Co [15]. However, all listed ways of production require enriched targets, which would also prevent long-lived radioactive impurities ^{56}Co ($T_{1/2}$ = 77.27 d) and ^{57}Co from forming [16]. According to calculations, when a target made of 100% ^{54}Fe is irradiated with deuterons, a yield of up to 30 MBq/μA·h can be achieved, while the content of long-lived cobalt isotopes is minimal [15]. In the case of irradiation of 100% ^{56}Fe with protons, a significantly higher yield can be achieved up to 180 MBq/μA·h; however, the content of ^{56}Co will also be higher. Finally, for the ^{58}Ni(p,α)^{55}Co reaction using an enriched target, the maximum yield is 13 MBq/μA·h, and the impurity content is minimal. Thus, ^{54}Fe(d,n)^{55}Co is the most promising reaction for use in nuclear medicine among cyclotron ones.

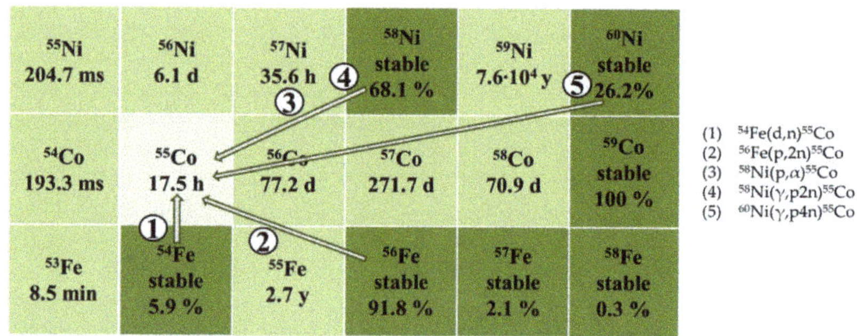

Figure 1. Studied methods of ^{55}Co production.

Cobalt isotopes (including ^{55}Co) can also be obtained using an electron accelerator—by irradiation of nickel. Currently, data on yields of photo-proton reactions on nickel nuclei, leading to the formation of medical isotopes of cobalt, is limited. In works [17,18], flux-weighted average cross-sections of reactions ^{nat}Ni(γ,pxn) in energy range of 55 to 75 MeV were determined. It was established that cross-sections of ^{nat}Ni(γ,pxn)^{55}Co in this range varied insignificantly. However, there are no data on the yields of nuclear reactions in these works, which makes it impossible to evaluate the possibility of obtaining cobalt isotopes in sufficient quantities for nuclear medicine using electron accelerators.

No methods of separation of cobalt isotopes produced in an electron accelerator can be found in the literature. At the same time, there are works dedicated to the separation of cobalt isotopes from cyclotron-irradiated nickel targets. In these works, Ni(II) and Co(II) were separated using anion exchange resin Dowex AG-1X8 [11,14,19–27] and using extraction chromatography sorbent based on diglycolamide, where Co(II) was obtained in 3 M HCl [28]. The best results were achieved in the last case, the yield of cobalt was 92%, separation factors of cobalt from different impurities varied from $8 \cdot 10^2$ to $2 \cdot 10^4$, and the process lasted for 2 h. It is worth mentioning that the main task of these works was more difficult than just separation of cobalt and nickel: first, irradiation of nickel also results in the formation of copper isotopes; second, nickel was usually applied via electrodeposition as a coating to a metal plate, irradiation of which also led to impurities. To produce cobalt isotopes for nuclear medicine purposes, it is necessary to develop a technique with higher yield, higher Ni/Co separation factors, and less time of separation, allowing one to obtain cobalt in diluted HCl medium.

Thus, the purpose of this work was to determine yields of photonuclear reactions on nickel nuclei and also to develop a fast, simple, and effective method of carrier-free cobalt isotopes separation from nickel targets.

2. Results and Discussion

2.1. Radionuclide Composition of Irradiated Targets and Yields of Nuclear Reactions

Gamma-spectra of irradiated targets made of ^{nat}Ni and ^{60}Ni are presented in Figure 2A,B; yields of photonuclear reactions leading to the formation of nickel and cobalt isotopes during the irradiation of ^{nat}Ni, ^{60}Ni, and ^{58}Ni are presented in Table 1. It was established that in each case, ^{55}Co was produced along with significant quantities of long-lived impurities ^{56,57,58}Co. Irradiation of ^{nat}Ni and ^{58}Ni resulted in the yield of ^{56,57,58}Co being 1.2–1.5 times higher than the yield of ^{55}Co, and the irradiation of ^{60}Ni led to the yield of ^{55}Co being no more than 3% of the yield of all cobalt isotopes. Obviously, ^{55}Co with such low radionuclide purity is not suitable for PET. On the other hand, isotopes ^{56,57,58}Co are gamma-emitters and can be used in preclinical research of radiopharmaceuticals based on cobalt, including in vivo experiments. For these purposes, we recommend irradiating ^{nat}Ni; in this case, the yield of ^{57}Co is 66 kBq/(μA·h·g/cm^2) on a thin plate, and this value can be increased by using massive target. ^{58}Ni can also be used as a target material increasing the yield of ^{57}Co up to 82 kBq/(μA·h·g/cm^2) on thin foil; however, targets made of ^{58}Ni are more expensive than ones made of ^{nat}Ni. Therefore, photonuclear method allows one to produce cobalt isotopes with sufficient activity for preclinical research.

Table 1. Yields of photonuclear reactions on ^{nat}Ni, ^{60}Ni (obtained experimentally), and ^{58}Ni (calculated using yields on ^{nat}Ni, ^{60}Ni nuclei) with maximum energy of bremsstrahlung photons being 55 MeV. Values obtained by TALYS [29] are presented in brackets.

Isotope	$T_{1/2}$	Y_{EOB}, kBq/(μA·h·g/cm^2)		
		^{nat}Ni	^{60}Ni	^{58}Ni
^{56}Ni	6.08 d	20.7 ± 0.3 (32.5)	0.16 ± 0.01 (0.06)	29.6 ± 0.3 (47.7)
^{57}Ni	35.6 h	3883 ± 76 (4404)	48.3 ± 1.0 (21)	5440 ± 93 (6460)
^{55}Co	17.53 h	60.2 ± 4.2 (72.9)	1.1 ± 0.1 (0.02)	82.4 ± 5.1 (107)
^{56}Co	77.27 d	16.5 ± 0.1 (17.4)	0.48 ± 0.03 (0.16)	21.6 ± 0.2 (25.4)
^{57}Co	271.8 d	66.1 ± 0.4 (46.5)	2.75 ± 0.04 (1.35)	81.8 ± 0.6 (67.8)
^{58}Co	70.86 d	10.3 ± 0.1 (3.6)	39.6 ± 0.4 (13.6)	0

Table 1 also compares the experimentally measured yields with theoretical calculations using the TALYS program, taking into account the bremsstrahlung spectrum. On the whole, we can see a satisfactory agreement between the experimental yields and the theoretical calculations. The difference in values can be due to two main factors: TALYS uses default photoabsorption cross-sections, and also does not take into account the isospin splitting of the giant dipole resonance, which has a significant effect on the yields of photo-proton reactions.

2.2. Separation of Co(II) Isotopes without a Carrier Using Extraction Chromatography

Obtained chromatograms of Ni(II) and Co(II) separation using DGA resin are presented in Figure 3. It was established that elution of Co(II) by HCl solutions with concentration varying from 0.01 to 3 M resulted in similar elution profiles; yield of Co(II) was close to quantitative in each case, and the process lasted for no longer than 0.5 h. To determine the separation factor of Ni(II) and Co(II), fractions marked in Figure 3 were selected for gamma-spectra registration for 24 h. Peaks of isotopes ^{56,57}Ni were absent in registered gamma-spectra, and the separation factor of Ni/Co was calculated using detection limit of these radionuclides and was $2.8 \cdot 10^5$, which is one order of magnitude higher than the factor during the separation using sorbent with similar composition in work [28]. Thus, the method of carrier-free Co(II) separation was developed; it allows one to obtain necessary isotopes in different solutions of HCl, including the solutions with low concentration, which is preferable for nuclear medicine.

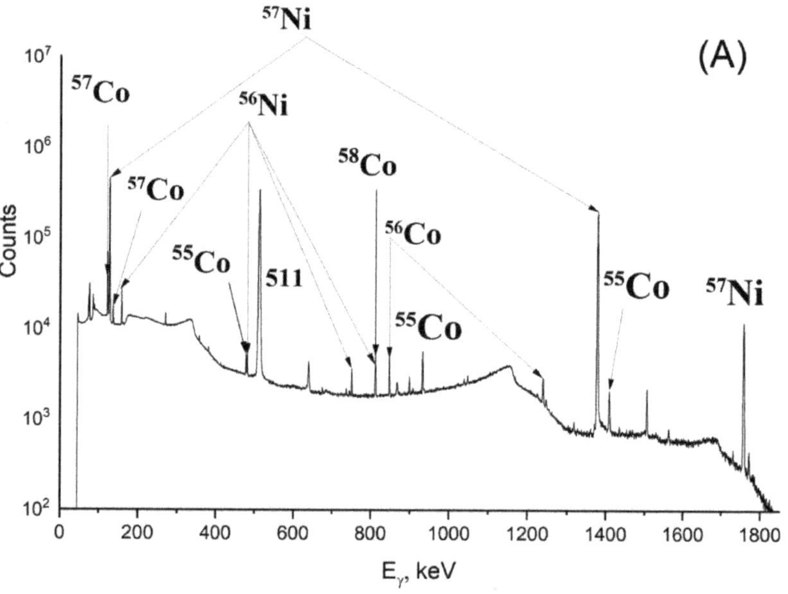

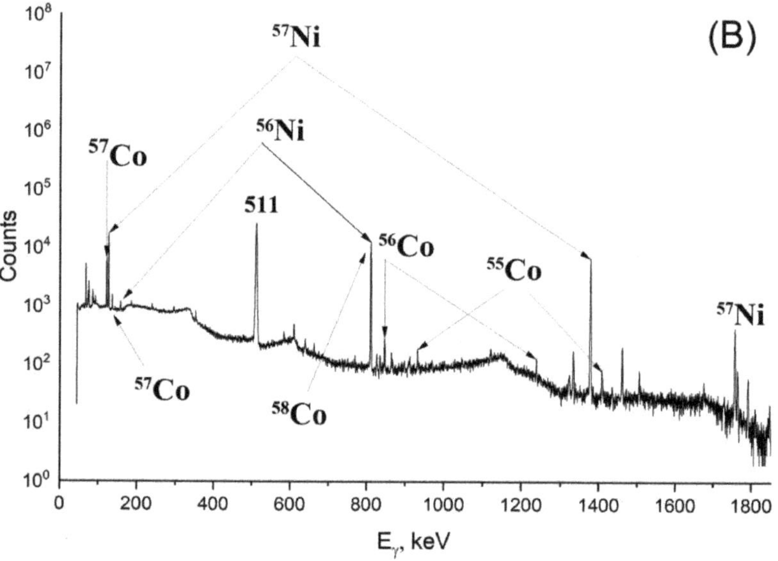

Figure 2. Gamma-spectra of ^{nat}Ni irradiated by bremsstrahlung photons with energy up to 55 MeV during 1 h after EOB (**A**), and of irradiated ^{60}Ni for 19 h after EOB (**B**). The most intense peaks of each isotope are labeled in the figures.

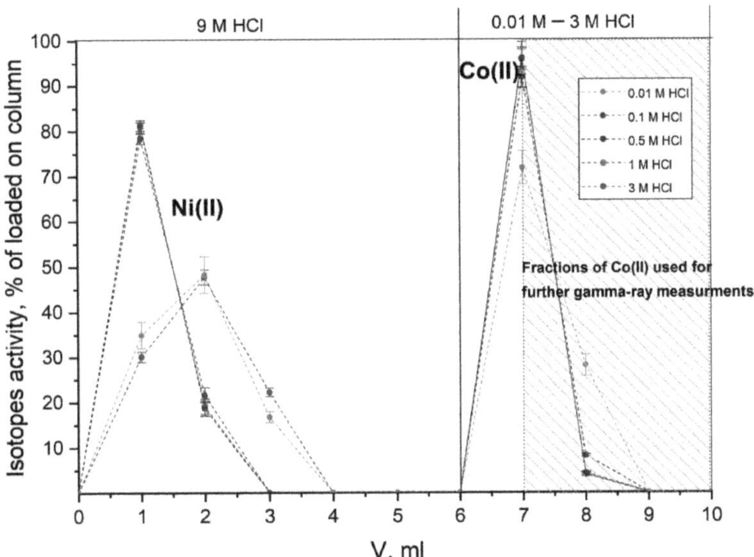

Figure 3. Elution curves of Ni(II) and Co(II) during separation on DGA resin column in HCl solutions.

As stated above in Section 2.1, the highest yields of cobalt isotopes are achieved during the irradiation of enriched ^{58}Ni targets. Obviously, the expensive target material should be regenerated after separation. For this purpose, it is possible to evaporate eluate-containing Ni(II) to dryness to irradiate NiCl$_2$ to produce cobalt isotopes one more time. In this case, there will be no need to dissolve the target during prolonged heating, since NiCl$_2$ is water-soluble. Another possible way is the irradiation of Ni(II) solution separated from the column, which can be eluted through the column again without any preparative treatment. In any case, the regeneration of enriched nickel is not complicated. It is also worth noting that formed isotopes of cobalt have $T_{1/2}$ no longer than 272 d, allowing one to utilize samples with general waste after prolonged storage, i.e., no radioactive waste requiring special treatment and disposal is produced after the irradiation. Thus, this studied method of production and separation of cobalt isotopes is environmentally friendly due to the regeneration of the target and the absence of radioactive waste.

3. Materials and Methods

3.1. Theoretical Calculations of Cross-Sections and Yields of Photonuclear Reactions

The cross-sections were calculated using the TALYS program, while the total photoabsorption cross-section was calculated based on the parameters from the RIPL-2 experimental database [30]. To calculate the cross-sections for photonuclear reactions, TALYS uses a combination of the evaporative and exciton preequilibrium decay mechanism of a compound nucleus with the emission of nucleons and gamma-quanta. The obtained cross-sections of the main photonuclear reactions leading to the formation of ^{55,56,57,58}Co and ^{55,56,57}Ni isotopes are shown in Figure 4.

The theoretical yield of isotope formation, taking into account all possible reactions leading to the formation of the selected isotope, was calculated by Equation (1):

$$Y = \frac{\lambda \alpha}{\rho q_e} \sum_i \eta_i \int_{E_i}^{E_m} \phi(E_\gamma, E_m) \sigma_i(E_\gamma) dE_\gamma \tag{1}$$

where λ is the decay constant, α is the number of studied nuclei per 1 cm^2 of target, ρ is the surface density of the target, q_e is the electron charge in μA·h, index i corresponds to the number of the reaction contributing to the formation of studied isotope, η_i is the percentage of the nickel isotope on which the reaction occurs in a natural mixture of isotopes, E_i is the threshold of the corresponding reaction, E_m is the maximum energy of the bremsstrahlung spectrum, $\sigma_i(E_\gamma)$ is the cross-section of the corresponding photonuclear reaction, and $\phi(E_\gamma, E_m)$ is the bremsstrahlung spectrum on the target.

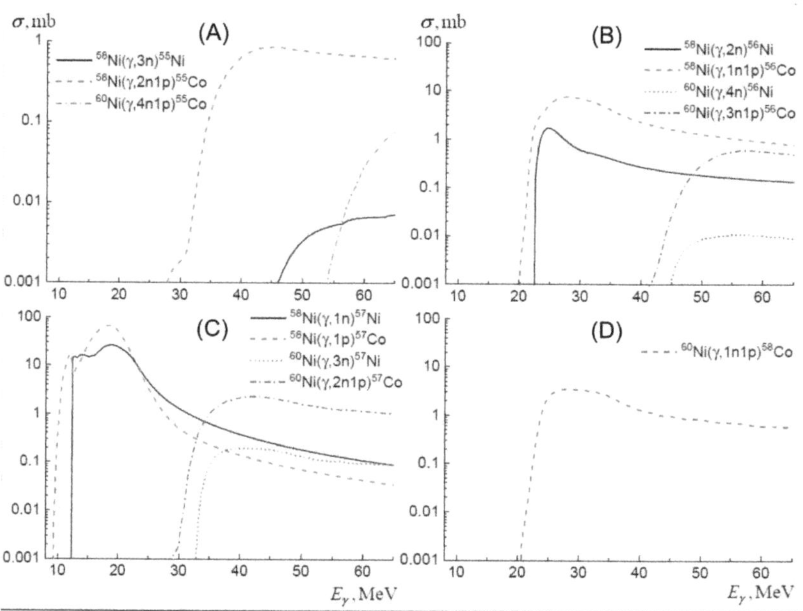

Figure 4. Calculated cross-sections of the main photonuclear reactions leading to the formation of ^{55}Co and ^{55}Ni (**A**), ^{56}Co and ^{56}Ni (**B**), ^{57}Co and ^{57}Ni (**C**), and ^{58}Co (**D**).

The bremsstrahlung spectrum calculated using a full 3D simulation of the irradiation process using the Geant4 program, taking into account the formation of gamma-quanta both in the converter and in the target, is presented on Figure 5.

3.2. Irradiation of Targets and Determination of Yields of Photonuclear Reactions

To study the yields of photonuclear reactions on nickel nuclei, two targets were irradiated in RTM-55 microtron with maximum energy of electron beam being 55 MeV [31]. The first target was a plate made of ^{nat}Ni with the size of 1 cm × 1 cm, thickness of 500 μm, and weight of 440 mg. Purity of ^{nat}Ni was determined in our work by atomic emission spectroscopy using Thermo Scientific ICAP-6500 Duo (Horsham and Loughborough, England) and was 99.78%. The second target made of ^{60}Ni purchased from Federal State Unitary Enterprise Combine "Elektrokhimpribor" (Lesnoy, Russia) was thin trapezoidal foil 67 μm thick and with a weight of 86.5 mg. Isotope composition of ^{60}Ni-target is presented in Table 2, purity of ^{60}Ni—99.9224% according to manufacturer's data. Tungsten plates 1 mm thick were used as convertors. Monitor targets were a 0.11 mm thick cobalt plates and were located directly after the targets for irradiation. Bremsstrahlung targets, nickel targets, and monitor targets were fully overlapping the beam. Current fluctuations during the irradiations were measured using Faraday cup. Normalization of current was carried out by the processing of bremsstrahlung spectrum and by comparing experimentally measured yield of ^{59}Co(γ,n)^{58}Co reaction to the yield calculated using known cross-sections. The duration of irradiation of each target was 1 h, average currents were 73 and 48 nA for ^{nat}Ni and ^{60}Ni accordingly.

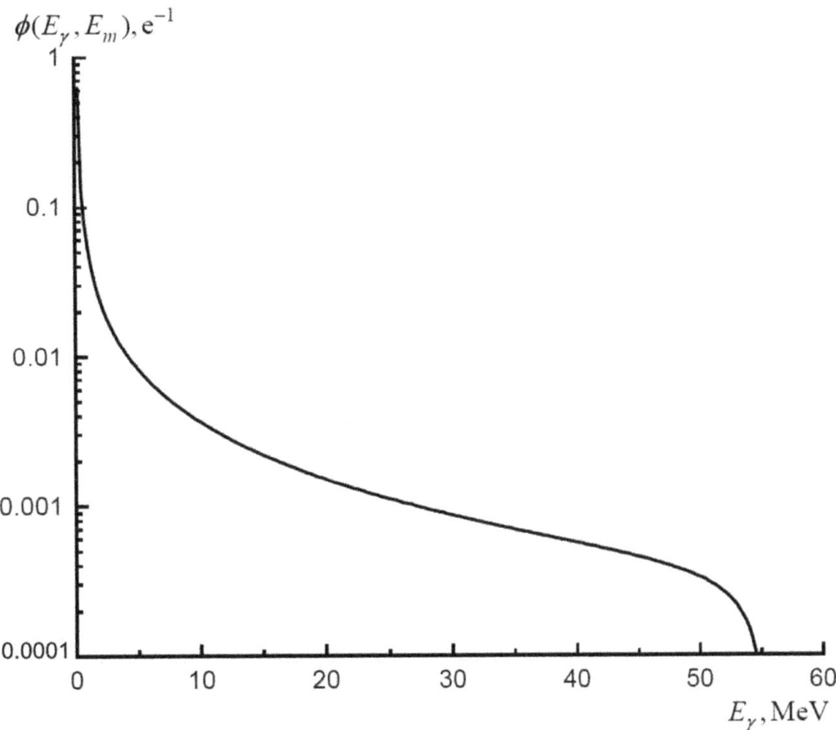

Figure 5. Bremsstrahlung spectrum per one beam electron for used convertors.

Table 2. Isotope composition of ^{60}Ni foil according to manufacturer's data.

Isotope	Content, %
^{58}Ni	0.31
^{60}Ni	99.6 ± 0.1
^{61}Ni	0.05
^{62}Ni	0.04
^{64}Ni	<0.05

The residual activity of nickel targets after irradiation was registered using gamma-ray spectrometer with high-purity germanium detector GC3019 (Canberra Ind, Meridan, CT, USA). Relative efficiency of detector was 30%, energy resolution of detector was 0.9 keV for 122 keV and 1.9 keV for 1.33 MeV. Efficiency calibration of spectrometer was conducted using measurements of activity of certified point sources (^{152}Eu, ^{137}Cs, ^{60}Co, ^{241}Am) in different location geometries of source and detector and was also modeled in GEANT4. The selection of the peak maximum in the spectra was carried out using an automatic system for registration and analysis of spectra specially designed for this purpose. Spectra with the duration of 3.5 s each were saved into the database, and the analysis system allowed us to summarize them and display the total spectrum with assigned duration [32]. Activity and yields of the produced isotopes were determined using areas of the most intensive peaks corresponding to the decay of the resulting isotope in spectra of residual activity, taking into account duration of irradiation, duration of transportation and registration of spectrum, and also efficiency of gamma-quanta registration and quantum yield of gamma-transition. Gamma-spectra of each irradiated target were registered three times for 2 days during the 2 months to exactly determine both short-lived and long-lived isotopes.

Yields of photonuclear reactions on ^{nat}Ni and ^{60}Ni in kBq/(μA·h·g/cm^2), normalized by electron beam charge and surface density of the target, were calculated using Equation (2):

$$Y = \frac{\lambda S}{C k \rho \left(e^{-\lambda(t_3 - t_1)} - e^{-\lambda(t_2 - t_1)}\right)} \tag{2}$$

where S is the area of photopeak in spectra of residual activity, corresponding to the gamma-transition during the decay of the resulting nucleus, occurring during the registration, t_1 is the irradiation time, t_2 is the starting time of the registration, t_3 is the ending time of the registration, λ is the exponential decay constant, k is the coefficient equal to the multiplication of detector efficiency, coefficient of cascade summation, and quantum yield of gamma-quant during gamma-transition, ρ is the target surface area, and C is the coefficient taking into account the change in accelerator current during the irradiation (Equation (3), Figure 6).

$$C = \int_0^{t_1} I(t) e^{-\lambda t} dt \tag{3}$$

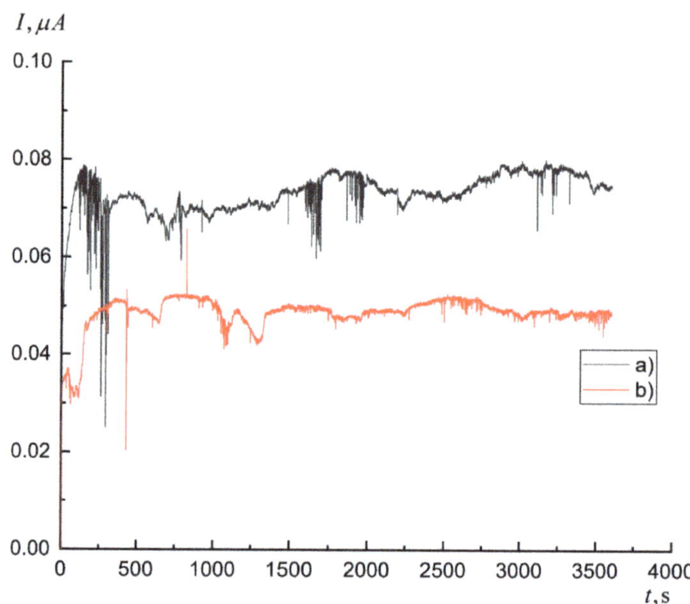

Figure 6. Accelerator current during irradiation of ^{nat}Ni (**a**) and ^{60}Ni (**b**).

Yields of reactions on ^{58}Ni were calculated as a difference between the yields of isotope production on natural mix and on ^{60}Ni, taking into account the percentage of ^{58}Ni and ^{60}Ni in natural mix. The accuracy of the selected calculation method was confirmed by the ratio of formation yields after the irradiation of ^{60}Ni and ^{nat}Ni (3.85 $\pm$ 0.05) coincided within the margin of error with the ratio of ^{60}Ni content in targets—3.81.

3.3. Separation of Co(II) Isotopes from Irradiated Nickel Target

DGA resin Normal (base–N,N,N′,N′-tetraoctyl-1,5-diglycolamide, particle size 100–150 μm, TrisKem Int., Bruz, France) was used for separation. This resin sorbs Co(II) in solutions of HCl with concentration more than 6 M [33]. In these solutions, the distribution coefficient (K_d) for Co(II) reaches 20, while K_d(Ni) does not exceed 1 in solutions of HCl with concentration less than 10 M, which allows one to separate Co(II) and Ni(II) using this sorbent. K_d(Co) decreases with quantity of HCl decreasing: in a more diluted solution of HCl, i.e., less than 4 M, Co(II) is not retained on the column.

To optimize the separation method of carrier-free Co(II) from irradiated nickel, experiments with the plate made of ^{nat}Ni (1 cm $\times$ 1 cm), similar to the plate from Section 3.2 were carried out. The irradiated plate was dissolved in 12 M HCl by prolonged heating, then the solution was evaporated to dryness in 9 M HCl. The sorbent was preliminarily held in 0.001 M HCl for 1 h, then the column (height 4 cm, diameter 0.6 cm, volume 2 mL) was filled with it. After the dissolution of the irradiated target, 0.5–1 mL of obtained solution was placed in the column with DGA resin; Co(II), unlike Ni(II), was sorbed onto the column. The remaining Ni(II) was eluted with 5 mL of 9 M HCl, then Co(II) was eluted with 5 mL of HCl solution with the concentration varying from 0.01 M to 3 M. Fractions of 1 mL were collected during the separation, the content of Co(II) and Ni(II) was determined using gamma-spectroscopy with high-purity germanium detector GC1020 (Canberra Ind.). Co(II) was identified by peaks of ^{57}Co (122 keV, 85.6%), and Ni(II) was identified by peaks of ^{57}Ni (127 keV, 16.7%) and ^{56}Ni (158 keV, 98.8%).

4. Conclusions

Targets made of ^{nat}Ni and ^{60}Ni were irradiated by bremsstrahlung photons with energy up to 55 MeV; the radionuclide composition and yields of nuclear reactions on ^{nat}Ni, ^{60}Ni, and ^{58}Ni were determined. It was established that in every case, the activities of produced ^{56,57,58}Co were higher than the activity of ^{55}Co, and therefore enough for preclinical research of radiopharmaceuticals based on cobalt. It was also demonstrated that the radionuclide purity of ^{55}Co produced by the photonuclear method at 55 MeV is not sufficient for PET. A fast, simple, and effective method of cobalt isotopes separation without a carrier from irradiated targets by extraction chromatography was developed; it was demonstrated that the separation of Co(II) is possible in wide range of HCl concentrations (from 0.01 to 3 M). The separation factor of Ni/Co was $2.8 \cdot 10^5$, the yield of Co(II) was close to quantitative, and separation lasted for no longer than 0.5 h.

Author Contributions: Conceptualization, A.G.K., A.A.K., S.E.V. and B.F.M.; methodology, A.G.K., S.S.B., V.V.K. and A.A.K.; software, S.S.B. and A.A.K.; validation, A.G.K., A.A.K., S.E.V. and B.F.M.; formal analysis, A.G.K., S.S.B. and A.A.K.; investigation, A.G.K., J.S.B., T.Y.E., S.S.B., A.A.K. and V.V.K.; resources, V.V.K., S.E.V. and B.F.M.; data curation, S.S.B.; writing—original draft preparation, A.G.K., J.S.B. and T.Y.E.; writing—review and editing, S.S.B., A.A.K., S.E.V. and B.F.M.; visualization, A.G.K., J.S.B., T.Y.E., S.S.B. and A.A.K.; supervision, B.F.M.; project administration, B.F.M.; funding acquisition, A.G.K., S.E.V. and B.F.M. All authors have read and agreed to the published version of the manuscript.

Funding: The study was supported by the Russian Science Foundation (project no. 21-13-00449).

Institutional Review Board Statement: Not applicable.

Informed Consent Statement: Not applicable.

Data Availability Statement: Not applicable.

Conflicts of Interest: The authors declare no conflict of interest.

References

1. Kazakov, A.G.; Ekatova, T.Y.; Babenya, J.S. Photonuclear production of medical radiometals: A review of experimental studies. *J. Radioanal. Nucl. Chem.* **2021**, *328*, 493–505. [CrossRef]
2. Lagunas-Solar, M.C.; Jungerman, J.A. Cyclotron production of carrier-free cobalt-55, a new positron-emitting label for bleomycin. *Int. J. Appl. Radiat. Isot.* **1979**, *30*, 25–32. [CrossRef]
3. Nieweg, O.E.; Beekhuis, H.; Paans, A.M.J.; Piers, D.A.; Vaalburg, W.; Welleweerd, J.; Wiegman, I.; Woldring, M.G. Detection of lung cancer with ^{55}Co-bleomycin using a positron camera. A comparison with ^{57}Co-bleomycin and ^{55}Co-bleomycin single photon scintigraphy. *Eur. J. Nucl. Med.* **1982**, *7*, 104–107. [CrossRef] [PubMed]
4. Sharma, H.; Zweit, J.; Smith, A.M.; Downey, S. Production of cobalt-55, a short-lived, positron emitting radiolabel for bleomycin. *Int. J. Radiat. Appl. Instrumentation. Part A* **1986**, *37*, 105–109. [CrossRef]
5. Goethals, P.; Volkaert, A.; Vandewielle, C.; Dierckx, R.; Lameire, N. ^{55}Co-EDTA for renal imaging using positron emission tomography (PET): A feasibility study. *Nucl. Med. Biol.* **2000**, *27*, 77–81. [CrossRef]

6. Andersen, T.L.; Baun, C.; Olsen, B.B.; Dam, J.H.; Thisgaard, H. Improving contrast and detectability: Imaging with [^{55}Co]Co-DOTATATE in comparison with [^{64}Cu]Cu-DOTATATE and [^{68}Ga]Ga-DOTATATE. *J. Nucl. Med.* **2020**, *61*, 228–235. [CrossRef] [PubMed]

7. Dam, J.H.; Olsen, B.B.; Baun, C.; Høilund-Carlsen, P.F.; Thisgaard, H. A PSMA ligand labeled with cobalt-55 for PET imaging of prostate cancer. *Mol. Imaging Biol.* **2017**, *19*, 915–922. [CrossRef] [PubMed]

8. Zhou, Y.; Baidoo, K.E.; Brechbiel, M.W. Mapping biological behaviors by application of longer-lived positron emitting radionuclides. *Adv. Drug Deliv. Rev.* **2013**, *65*, 1098–1111. [CrossRef] [PubMed]

9. Blower, P.J. A nuclear chocolate box: The periodic table of nuclear medicine. *Dalt. Trans.* **2015**, *44*, 4819–4844. [CrossRef]

10. Radford, L.L.; Fernandez, S.; Beacham, R.; El Sayed, R.; Farkas, R.; Benešová, M.; Müller, C.; Lapi, S.E. New ^{55}Co-labeled albumin-binding folate derivatives as potential PET agents for folate receptor imaging. *Pharmaceuticals* **2019**, *12*, 166. [CrossRef]

11. Jansen, H.M.L.; Knollema, S.; Van Der Duin, L.V.; Willemsen, A.T.M.; Wiersma, A.; Franssen, E.J.F.; Russel, F.G.M.; Korf, J.; Paans, A.M.J. Pharmacokinetics and dosimetry of cobalt-55 and cobalt-57. *J. Nucl. Med.* **1996**, *37*, 2082–2086.

12. Stevens, H.; Jansen, H.M.L.; De Reuck, J.; Lemmerling, M.; Strijckmans, K.; Goethals, P.; Lemahieu, I.; De Jong, B.M.; Willemsen, A.T.M.; Korf, J. ^{55}Cobalt (Co) as a PET-tracer in stroke, compared with blood flow, oxygen metabolism, blood volume and gadolinium-MRI. *J. Neurol. Sci.* **1999**, *171*, 11–18. [CrossRef]

13. Jansen, H.M.L.; Van Der Naalt, J.; Van Zomeren, A.H.; Paans, A.M.J.; Veenma-van Der Duin, L.; Hew, J.M.; Pruim, J.; Minderhoud, J.M.; Korf, J. Cobalt-55 positron emission tomography in traumatic brain injury: A pilot study. *J. Neurol. Neurosurg. Psychiatry* **1996**, *60*, 221–224. [CrossRef] [PubMed]

14. Heppeler, A.; André, J.P.; Buschmann, I.; Wang, X.; Reubi, J.C.; Hennig, M.; Kaden, T.A.; Maecke, H.R. Metal-ion-dependent biological properties of a chelator-derived somatostatin analogue for tumour targeting. *Chem. A Eur. J.* **2008**, *14*, 3026–3034. [CrossRef]

15. Amjed, N.; Hussain, M.; Aslam, M.N.; Tárkányi, F.; Qaim, S.M. Evaluation of nuclear reaction cross sections for optimization of production of the emerging diagnostic radionuclide ^{55}Co. *Appl. Radiat. Isot.* **2016**, *108*, 38–48. [CrossRef]

16. Sitarz, M.; Cussonneau, J.P.; Matulewicz, T.; Haddad, F. Radionuclide candidates for β+γ coincidence PET: An overview. *Appl. Radiat. Isot.* **2020**, *155*, 108898. [CrossRef]

17. Zaman, M.; Kim, G.; Naik, H.; Kim, K.; Shahid, M.; Nadeem, M.; Shin, S.G.; Cho, M.H. Flux weighted average cross-sections of ^{nat}Ni(γ,x) reactions with the bremsstrahlung end-point energies of 55, 59, 61 and 65 MeV. *Nucl. Phys. A* **2018**, *978*, 173–186. [CrossRef]

18. Naik, H.; Kim, G.; Nguyen, T.H.; Kim, K.; Shin, S.G.; Kye, Y.C.M. Measurement of ^{nat}Ni(γ,xn)^{57,56}Ni and ^{nat}Ni(γ,pxn)$^{58-55}$Co reaction cross sections in bremsstrahlung with end-point energies of 65 and 75 MeV. *J. Radioanal. Nucl. Chem.* **2020**, *324*, 837–846. [CrossRef]

19. Maziére, B.; Stulzaft, O.; Verret, J.; Comar, D.; Syrota, A. [^{55}Co]- and [^{64}Cu]-DTPA: New radiopharmaceuticals for quantitative tornocisternography. *Int. J. Appl. Radiat. Isot.* **1983**, *34*, 595–601. [CrossRef]

20. Spellerberg, S.; Reimer, P.; Blessing, G.; Coenen, H.H.; Qaim, S.M. Production of ^{55}Co and ^{57}Co via proton induced reactions on highly enriched ^{58}Ni. *Appl. Radiat. Isot.* **1998**, *49*, 1519–1522. [CrossRef]

21. Jalilian, A.R.; Rowshanfarzad, P.; Akhlaghi, M.; Sabet, M.; Kamali-Dehghan, M.; Pouladi, M. Preparation and biological evaluation of a [^{55}Co]-2-acetylpyridine thiosemicarbazone. *Sci. Pharm.* **2009**, *77*, 567–578. [CrossRef]

22. Jalilian, A.R.; Rowshanfarzad, P.; Yari-Kamrani, Y.; Sabet, M.; Majdabadi, A. Preparation and evaluation of [^{55}Co](II)-DTPA for blood cell labeling. *Open Inorg. Chem. J.* **2009**, *3*, 21–25. [CrossRef]

23. Valdovinos, H.F.; Graves, S.; Barnhart, T.; Nickles, R.J. ^{55}Co separation from proton irradiated metallic nickel. *AIP Conf. Proc.* **2014**, *1626*, 217–220. [CrossRef]

24. Reimer, P.; Qaim, Q.S. Excitation functions of proton induced reactions on highly enriched ^{58}Ni with special relevance to the production of ^{55}Co and ^{57}Co. *Radiochim. Acta* **1998**, *80*, 113–120. [CrossRef]

25. Mastren, T.; Marquez, B.V.; Sultan, D.E.; Bollinger, E.; Eisenbeis, P.; Voller, T.; Lapi, S.E. Cyclotron production of high-specific activity ^{55}Co and in vivo evaluation of the stability of ^{55}Co metal-chelate-peptide complexes. *Mol. Imaging* **2015**, *14*, 526–532. [CrossRef]

26. Mastren, T.; Sultan, D.; Lapi, S.E. Production and separation of ^{55}Co via the ^{58}Ni(p,α)^{55}Co reaction. *AIP Conf. Proc.* **2012**, *1509*, 96–100. [CrossRef]

27. Valdovinos, H.F.; Hernandez, R.; Graves, S.; Ellison, P.A.; Barnhart, T.E.; Theuer, C.P.; Engle, J.W.; Cai, W.; Nickles, R.J. Cyclotron production and radiochemical separation of ^{55}Co and ^{58m}Co from ^{54}Fe, ^{58}Ni and ^{57}Fe targets. *Appl. Radiat. Isot.* **2017**, *130*, 90–101. [CrossRef]

28. Valdovinos, H.F.; Graves, S.; Barnhart, T.; Nickles, R.J. Simplified and reproducible radiochemical separations for the production of high specific activity ^{61}Cu, ^{64}Cu, ^{86}Y and ^{55}Co. *AIP Conf. Proc.* **2017**, *1845*, 020021. [CrossRef]

29. Koning, A.J.; Duijvestijn, M.C.; Hilarie, S. Talys 1.0. In Proceedings of the International Conference on Nuclear Data for Science and Technology, Nice, France, 17 June 2007; pp. 211–214.

30. Belgya, T.; Bersillon, O.; Capote, R.; Fukahori, T.; Zhigang, G.; Goriely, S.; Herman, M.; Ignatyuk, A.V.; Kailas, S.; Koning, A.; et al. *Handbook for Calculations of Nuclear Reaction Data, RIPL-2*; IAEA-TECDOC-1506; IAEA: Vienna, Austria, 2006. Available online: http://www-nds.iaea.org/RIPL-2/ (accessed on 20 January 2022).

31. Ermakov, A.N.; Ishkhanov, B.S.; Kamanin, A.N.; Pakhomov, N.I.; Khankin, V.V.; Shvedunov, V.I.; Shvedunov, N.V.; Zhuravlev, E.E.; Karev, A.I.; Sobenin, N.P. A multipurpose pulse race-track microtron with an energy of 55 MeV. *Instrum. Exp. Tech.* **2018**, *61*, 173–191. [CrossRef]
32. Belyshev, S.S.; Stopani, K.A. Automatic data acquisition and analysis in activation experiments. *Mosc. Univ. Phys. Bull.* **2013**, *68*, 88–91. [CrossRef]
33. Pourmand, A.; Dauphas, N. Distribution coefficients of 60 elements on TODGA resin: Application to Ca, Lu, Hf, U and Th isotope geochemistry. *Talanta* **2010**, *81*, 741–753. [CrossRef]

Article

Production and Semi-Automated Processing of ^{89}Zr Using a Commercially Available TRASIS MiniAiO Module

Vijay Gaja [1,2], **Jacqueline Cawthray** [3], **Clarence R. Geyer** [4] **and Humphrey Fonge** [1,5,*]

1   Department of Medical Imaging, University of Saskatchewan, College of Medicine, Saskatoon, SK S7N 0W8, Canada; vijay.gaja@lightsource.ca
2   Canadian Light Source, Saskatoon, SK S7N 2V3, Canada
3   Saskatchewan Centre for Cyclotron Sciences, Saskatoon, SK S7N 5C4, Canada; jacqueline.cawthray@fedorukcentre.ca
4   Department of Pathology and Laboratory Medicine, University of Saskatchewan, College of Medicine, Saskatoon, SK S7N 5E5, Canada; clg595@mail.usask.ca
5   Department of Medical Imaging, Royal University Hospital, Saskatoon, SK S7N 0W8, Canada
*   Correspondence: humphrey.fonge@usask.ca; Tel.: +1-306-655-3353; Fax: +1-306-655-1637

Academic Editors: Alessandra Boschi and Petra Martini
Received: 5 May 2020; Accepted: 3 June 2020; Published: 5 June 2020

Abstract: The increased interest in ^{89}Zr-labelled immunoPET imaging probes for use in preclinical and clinical studies has led to a rising demand for the isotope. The highly penetrating 511 and 909 keV photons emitted by ^{89}Zr deliver an undesirably high radiation dose, which makes it difficult to produce large amounts manually. Additionally, there is a growing demand for Good Manufacturing Practices (GMP)-grade radionuclides for clinical applications. In this study, we have adopted the commercially available TRASIS mini AllinOne (miniAiO) automated synthesis unit to achieve efficient and reproducible batches of ^{89}Zr. This automated module is used for the target dissolution and separation of ^{89}Zr from the yttrium target material. The ^{89}Zr is eluted with a very small volume of oxalic acid (1.5 mL) directly over the sterile filter into the final vial. Using this sophisticated automated purification method, we obtained satisfactory amount of ^{89}Zr in high radionuclidic and radiochemical purities in excess of 99.99%. The specific activity of three production batches were calculated and was found to be in the range of 1351–2323 MBq/μmol. ICP-MS analysis of final solutions showed impurity levels always below 1 ppm.

Keywords: Zirconium-89; automation; radiolabelling

---

1. Introduction

Zirconium-89 (^{89}Zr), a positron-emitting isotope, has emerged as an attractive radionuclide in the development of pre-clinical and clinical radiopharmaceuticals for positron-emission tomography (PET) imaging. This is due to the favourable physical characteristics of the isotope which decays via positron emission ($T_{1/2}$: 78.4 h, β^+ 22.7%, $E_{\beta+max}$ = 901 keV; average $E_{\beta+max}$ = 396 keV) [1,2] and electron capture (EC 77%, $E\gamma$ = 909 keV) to the stable yttrium-89 (^{89}Y). The longer half-life matches the need for large biomolecules such as antibodies, antibody fragments and nanoparticles which require prolonged circulation time in order to reach optimal target accumulation [3,4]. Furthermore, the β^+ branching ratio, average β^+ energy and short positron range (R^{ave} = 1.23 mm) provide high spatial resolution and image quality of the ^{89}Zr PET probes [5].

The combination of antibody-based targeting vectors and PET-based imaging, known as immuno-PET, is rapidly becoming a powerful tool for highly selective imaging agents [6,7]. Radiolabeling of the targeting vector requires a stable radiometal chelate that can be readily conjugated to an antibody or other proteins. At present, the bifunctional derivative of desferrioxamine (DFO), *p*-isothiocyanatobenzyl-desferoxamine (*p-SCN-Bz*-DFO) is considered the gold-standard and with facile reaction chemistry has had widespread application for preclinical research and clinical trials [8]. However, recently there have been some concerns surrounding the in vivo instability of the ^{89}Zr-DFO complex, presumably due to the unsaturated coordination sphere, which has led to the generation of new chelators with enhanced stability of the resulting ^{89}Zr complex [9–11].

The most common ^{89}Zr production method is via the ^{89}Y(p,n)^{89}Zr transmutation reaction. This route, employing ^{89}Y targets that is in 100% natural abundance and commercially available, is readily accessible by most medical cyclotrons. In general, the target is irradiated with incident proton beam energies of 13–16 MeV. At these low energies, there is a trade-off between radionuclidic purity and efficiency of the process owing to the competing reactions to produce ^{88}Zr via the ^{89}Y(p, 2n)^{88}Zr reaction and ^{88}Y via the ^{89}Y(p, pn)^{88}Y reaction at threshold energy of 13.3 MeV. The average maximum cross section value for ^{88}Zr is 786.9 ± 1.5 at 22.98 MeV and for ^{88}Y is 298.0 ± 5.7 mb at 28.30 MeV [1]. The production of ^{88}Zr and ^{88}Y is minimized at 12.8 MeV due to small cross sections and threshold energy. Following dissolution of the irradiated target in concentrated HCl the isolation and purification of ^{89}Zr proceeds via cation exchange chromatography using hydroxamate-modified resin [1,12]. Zirconium-89 is selectively eluted from the column using 1 M oxalic acid. With optimized conditions, this standardized method of separation can achieve high recovery, radionuclidic purity and effective specific activity (ESA) [1].

The increased interest in ^{89}Zr-immuno-PET imaging probes for use in preclinical and clinical studies has led to a rising demand for ^{89}Zr. Although ^{89}Zr is commercially available, the lengthy transportation time and high costs are a barrier for preclinical and routine clinical applications of ^{89}Zr immuno-PET. Furthermore, the highly penetrating 511 and 909 keV photons emitted by ^{89}Zr deliver an undesirably high radiation dose, which makes it difficult to produce large amounts manually. So far, the processing of ^{89}Zr from the target has mainly been achieved manually or has been based on semi-automated systems. These have not been widely adopted because they are based on in-house developed systems instead of commercially available equipment [13–15]. Moreover, regulatory demands are increasing and consequently the application of ^{89}Zr in clinical studies is becoming challenging. To maintain the high quality of the ^{89}Zr, a reliable process with quality control and good manufacturing practice-compliant automated module in the production is required. Additionally, there are several advantages of using automation in radiochemistry, such as lower radiation exposure to the production personnel, high reproducibility, more precise control of the parameters and ease of handling. Mostly, automated modules are placed inside dedicated lead shielded hot cells to protect the operator and environment from radiation exposure. Commercially available automated modules can be classified into two types, namely, those that consist of fixed tubing in which fluid flows are regulated by inert gas, and those that consist of sterile disposable cassettes where fluid flow is regulated by a syringe pump. The main disadvantage of using fixed tubing is bacterial contamination in the product if a validated cleaning method is not implemented. Usually this tubing is used for multiple productions and needs to be cleaned after each synthesis. Also, a single system cannot be used for multiple radioisotopes or for the production of other radiopharmaceuticals. These issues can be solved by using sterile, single-use cassettes manufactured in current Good Manufacturing Practices (cGMP)-compliant clean rooms.

The focus of this present work is the implementation of the semi-automated purification of ^{89}Zr using a commercially available automated synthesis unit that could fulfil regulatory requirements.

2. Materials and Methods

2.1. General

A commercially available sterile, single-use cassettes (produced according to GMP) used for ^{68}Ga-cationic prepurification (catalog number 10886) were purchased from TRASIS and adapted for the ^{89}Zr purification. PEEK tubing and PEEK connectors were purchased separately from IDEX Health & Science, LLC (Middleboro, MA, USA) and sanitized with hydrogen peroxide solution and dried using nitrogen gas. A customized glass target dissolution vial with a vented glass cover that also allowed the insertion of PEEK tubing for the addition and removal of solutions into the vial and a aluminium heating block for the vial were manufactured at the University of Saskatchewan (Saskatoon, SK, Canada). All commercially obtained chemicals were of the highest available purity grade and were used without further purification. Hydrochloric acid (99.999% trace metal basis), water (Omni *Trace* Ultra™), oxalic acid (99.999% trace metals basis), acetonitrile (99.999% trace metal basis), *N*-(3-dimethylaminopropyl)-*N*¢-ethyl carbodiimide hydrochloride (99.9%), hydroxylamine hydrochloride (99.999% trace metal basis), 2,3,5,6-tetrafluorophenol (97%), diethylenetriaminepentaacetic acid (99%), 1.5 mL Eppendorf® Safe-Lock microcentrifuge tubes, and empty reversible SPE tubes (0.5 mL) were purchased from Sigma Aldrich (Oakville, ON, Canada). Methanol (semiconductor grade) was purchased from Fisher Scientific (Ottawa, ON, Canada). *p-SCN-Bz*-DFO was purchased from Macrocyclics, Inc. (Dallas, TX, USA). PEEK tubing 1/16″ OD × .040″ ID, flangeless nut PEEK, Short, 1/4-28 flat-bottom, for 1/16″ OD, flangeless ferrule tefzel, (ETFE), 1/4-28 flat-bottom, Luer-adapter assembly 1/4-28 female–male PEEK, and Luer-adapter 1/4-28 female to female Luer PEEK, were obtained from IDEX Health & Science, LLC (Middleboro, MA, USA). Sep-Pak Accell Plus CM light cartridges containing weak cation-resin were purchased from Waters (Milford, MA, USA). A tube rotator was purchased from VWR International (Mississauga, ON, Canada). Instant thin-layer chromatography paper (iTLC) was purchased from Agilent Technology (SantaClara, CA, USA). The radio-TLC plates were scanned using a Bioscan AR-2000 radio-TLC scanner (Washington, DC, USA). Radioactivity was measured using a Capintec CRC-55tr dose calibrator (Florham Park, NJ, USA) or a 1470 Wallac Wizard automated gamma counter (Perkin Elmer, Ramsey MN, USA).

2.2. Cyclotron Irradiation

Yttrium-89-coated niobium coins were purchased from Advanced Cyclotron Systems Inc. (ACSI, Richmond BC, Canada). The niobium target body had a diameter of 24 mm and a thickness of 1 mm. The yttrium was adhered on the frontal face of niobium target body with a diameter of 10 mm and approximately 200 μm thickness. Aluminium degrader coins (99.0% 24 mm with 0.80 mm thickness) were purchased from Goodfellow (Coraopolis, PA, USA).

The target body along with 0.80 mm-thick aluminium degrader was mounted on TR-24 cyclotron (ACSI, Richmond, BC, Canada) at the Saskatchewan Centre for Cyclotron Sciences (SCCS) and bombarded for 2 h with the initial proton energy of 17.8 MeV and 40 μA current calculated as ≈12.8 MeV degraded beam energy at the target to produce ^{89}Zr via the ^{nat}Y(p, n)^{89}Zr reaction. During irradiation, the target was cooled on the frontal side by helium gas and on the back side by chilled water. After irradiation, the target was left on the target station for 2–3 h to allow for the decay of short-lived isotopes, in particular ^{88m}Zr ($t_{1/2}$ = 4.16 m). The target was released into a lead pig using target release valve located outside the cyclotron vault. The lead pig containing the irradiated target was manually retrieved from the vault and transported to the hot cells on a shielded cart.

2.3. Automation Adopted for ^{89}Zr Separation and Purification

A mini AlliOne (miniAiO) cassette-based automatic synthesis unit (ASU) from TRASIS (Ans, Belgium), with dimensions 21.5W × 41.2H × 40.8D cm, was used for the automated separation and purification of ^{89}Zr (Figure 1). The miniAiO consists of 12 rotary actuators and 2 linear actuators for operating syringes. All liquids were transferred through PEEK tubing. All tube connections were assembled using non-metallic connectors to reduce exposure to the metals which may affect radionuclidic purity and specific activity. Many of the liquid connection pieces were flangeless fittings designed for high pressure fluidic connection and ferrules are manufactured from tefzel™ (ETFE). The fluidic path was controlled using the built-in three-way valve actuators. All liquids and hydroxamate resin cartridge were preloaded onto the cassette prior to starting the sequence on the miniAiO.

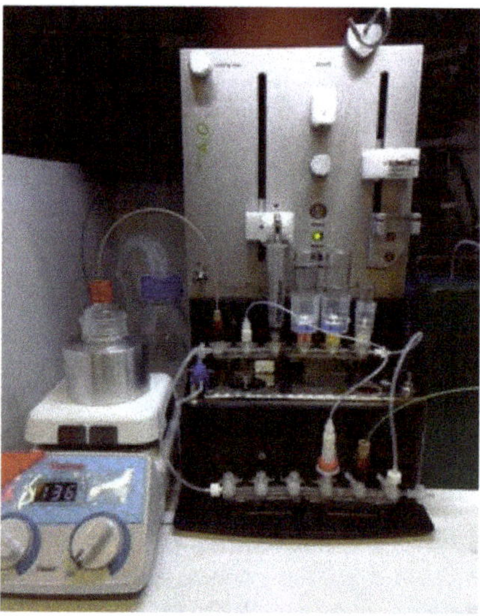

Figure 1. TRASIS mini AlliOne (miniAiO) setup for purification of zirconium-89.

The miniAiO ASU was located within a lead hot cell and remotely controlled through a graphical user interface (PC). A sequence was created on the TRASIS software using a graphical user interface which distributes commands and controls to the automation unit. The operator can initiate a pre-programmed ^{89}Zr target dissolution and purification process. Radioactive detectors are built-in at different locations within the ASU to monitor and record the location of the radioactivity. The radioactivity in the dissolution vial, hydroxamate cartridge and final vial was monitored.

2.4. Target Dissolution and Purification of ^{89}Zr

TRASIS MiniAiO Preparation

Automation of the process for separation and purification of ^{89}Zr, based on chemistry previously reported in the literature, was performed using a miniAiO ASU in a hot cell [1,12,16]. The miniAiO is a commonly used radiochemistry module for clinical GMP-grade radiopharmaceutical production. The ASU is designed for use with disposable kits and allows processing with full audit trail functionality for GMP production runs. Liquid transport was achieved using a syringe pump and transfer of liquids was controlled by three-way stopcock valves. For each liquid transfer step, the production protocols were adequately adapted and optimized. In brief, the following general steps were implemented in the production of ^{89}Zr. The hydroxamate resin (100 mg) was packed into an empty reversible SPE tube (0.5 mL). The hydroxamate resin was preconditioned with acetonitrile (MeCN) (8 mL, trace metal grade), water (15 mL, trace metal grade) and 2.0 M HCl (2 mL) and installed on the cassette (valve #10) of the miniAiO ASU (Figure 2). The following vials were installed on the cassette: 2 M HCl (20 mL) in a glass vial with septa installed at cassette valve #4, water (10 mL) in a glass vial with septa installed at cassette valve #5, 1 M oxalic acid (1.5 mL) in a glass vial with septa installed at cassette valve #6, a waste vial (40 mL) connected to cassette valve #7 and a [^{89}Zr]zirconium oxalate product vial (5 mL) connected to cassette valve #11.

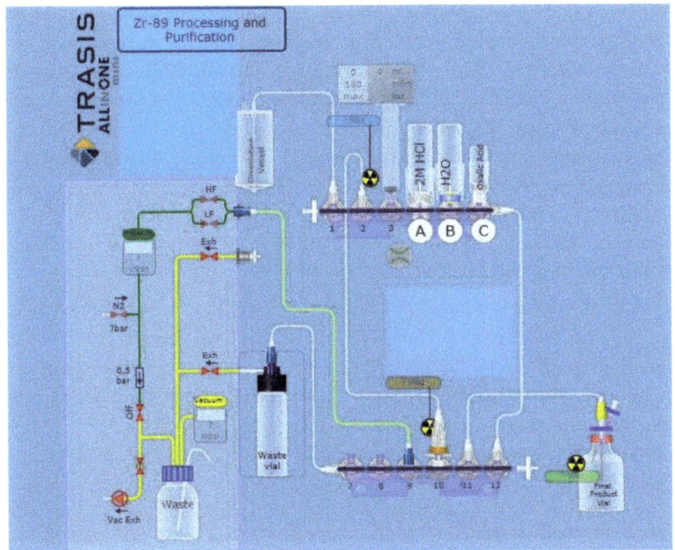

Figure 2. TRASIS miniAiO layout for automated purification and isolation of [89]Zr from irradiated yttrium coin. Valve 1 to dissolution vial; valve 3, 10 mL syringe; valve 4 (**A**) 2 M HCl (20 mL); valve 5 (**B**) water (10 mL); valve 6 (**C**) 1 M oxalic acid (1.5 mL); valve 10 hydroxamate resin; valve 11 to product vial.

The irradiated target was transferred to the custom-made dissolution vial and the vial was placed inside the aluminium block heater on a hot plate that had been pre-heated to 80 °C. After placing the vented glass cover with attached PEEK tubing on the dissolution vial, the hot cell was closed, and the purification sequence was initiated using the program interface for the miniAiO. Hydrochloric acid (2.0 M, 4 mL) was drawn into the syringe and pushed into the dissolution vial to dissolve the target. The target solution was heated at 80 °C for 20 min, and after cooling for 20 min, was passed through the hydroxamate resin using the syringe actuator at 1 mL/min flowrate to maximize trapping of [89]Zr on the resin. The dissolution vial was rinsed with 2 M HCl (2 mL) and passed through the hydroxamate resin again at 1 mL/min flowrate. The [89]Zr remained trapped on the resin as indicated by the in-process monitoring of the radiation detector at the column. The resin was washed with 2 M HCl (14 mL) at 3 mL/min flowrate followed by water (10 mL) at 3 mL/min flowrate. Finally, the [89]Zr was eluted with 1 M oxalic acid solution (1.5 mL) at 0.5 mL/min flow rate and collected in a sterile 5 mL product vial.

2.5. [89]Zr]Zirconium-Oxalate Characterization

2.5.1. Determination of Radionuclidic Purity

The radionuclidic purity of [89]Zr was determined by gamma-spectroscopy using a high purity germanium (HPGe) detector (Ortec, Oak Ridge, TN, USA). A 4-h scan of a diluted aliquot (50–80 kBq) was performed at the end of purification. To determine longer-lived contaminants, scanning was repeated after three half-lives.

2.5.2. Determination of Effective Specific Activity

The effective specific activity (MBq/µmol) of [89]Zr was determined by labeling *p-SCN-Bz*-DFO with purified [[89]Zr]zirconium oxalate. *p-SCN-Bz*-DFO was selected as it is a commonly used chelator for labeling antibodies with [89]Zr [17,18]. Specific activity was calculated for three consecutive batches of [[89]Zr]zirconium oxalate. A stock solution of *p-SCN-Bz*-DFO in DMSO (1 mg/mL) was prepared and 15 reaction tubes were prepared by 1:2 serial dilution in ultrapure water (500 µL) to give a final *p-SCN-Bz*-DFO concentration in the range of 2.07E-02 to 1.267E-06 µmol. A solution of [^{89}Zr]zirconium oxalate was neutralized with the 2 M NaCO$_3$ (pH 7 ± 0.2) and added to the *p-SCN-Bz*-DFO solutions. The resulting solutions were incubated at 37 °C on a shaker at 650 RPM for one hour. After incubation, 1 µL aliquots were spotted on iTLC and analysed by using citrate buffer (100 mM, pH 5.0) as a mobile phase. iTLC plates were measured using radio-TLC scanner. The results of the *p-SCN-Bz*-DFO titration with ^{89}Zr were plotted and fitted with a sigmoidal dose-response curve to generate an EC$_{50}$ value which was used to calculate specific activity in MBq/µmol.

2.5.3. Characterization of Radioactive Impurities Produced during Irradiation

Typical impurities in the production of ^{89}Zr are ^{88}Y, ^{88}Zr, ^{89}Y, ^{90}Zr and ^{56}Fe. Usually, impurities are derived from impure starting materials (^{89}Y), produced during proton bombardment due to inappropriate beam energy (^{88}Y and ^{88}Zr) or introduced via the buffers and recourses (^{90}Zr and ^{56}Fe). These impurities can be reduced using proper buffers and recourses and some of these impurities were removed by purifying the ^{89}Zr-solution using a hydroxamate column. All impurities were characterized using gamma-spectroscopy using a high purity germanium (HPGe) detector (Ortec, Oak Ridge, TN, USA).

2.5.4. Elemental Analysis by ICP-MS

The ICP-MS of three batches of final ^{89}Zr solution to determine metallic impurities were performed and validated by an ISO 9001:2015 certified Institution (Saskatchewan Research Council, Saskatoon SK, Canada). Niobium and yttrium were analyzed using an Agilent 7900 ICP-MS (Santa Clara, CA, USA) and all other elemental analysis were done using an Agilent 8800 ICP-MS (Santa Clara, CA, USA).

2.5.5. Preparation of p-SCN-Bz-DFO Conjugated Antibody

Conjugation of *p-SCN-Bz*-DFO to trastuzumab (DFO-trastuzumab) was performed following published procedures with slight modification [3,19]. Briefly, trastuzumab (5 mg/mL) in PBS was buffer exchanged in 0.1 M NaHCO$_3$ (pH 9) using centrifugal filters and concentrated to 10 mg/mL trastuzumab in the bicarbonate solution. A sixteen-fold mole excess of *p-SCN-Bz*-DFO (16 µg) in DMSO was added dropwise to the trastuzumab (2 mg) solution. The reaction mixture was incubated at 37 °C on a shaker at 650 RPM for an hour. The reaction mixture was cooled to room temperature and the unreacted DFO was removed by centrifugations using a spin-cap column (size 10–12 kDa). The buffer was exchanged with PBS using the same centrifugal filters.

2.5.6. Radiolabeling of DFO-Trastuzumab and Determination of Specific activity of [^{89}Zr]Zirconium Oxalate

Zirconium-89 in 1 M oxalic acid was neutralized by diluting with 1 M HEPES pH 7.4 followed by adding 2 M NaCO$_3$ (pH 11) dropwise until the solution was neutralized (pH 7 ± 0.2). Radiolabeling was performed using 15 different concentrations of DFO-trastuzumab and was prepared by 1:2 serial dilution in HEPES (100 µL) to give final DFO-trastuzumab masses in the range of 200 µg to 0.024 µg. Approximately 3.3 MBq of the neutralized [^{89}Zr] solution was added to each reaction tube. The reaction mixture was incubated at 37 °C in a shaker at 650 RPM for two hours. The reaction mixture was cooled to room temperature and 1 µL aliquots were analysed using iTLC with 0.15 M sodium citrate as a mobile phase. iTLC was measured using radioTLC scanner. The results of DFO-trastuzumab titration were plotted and fit with a sigmoidal dose–response curve to create an EC$_{50}$ value and used to calculate specific activity in MBq/ug.

3. Results and Discussion

3.1. Cyclotron Irradiation of ^{89}Y

Zirconium-89 was produced by proton irradiation of commercially available ^{89}Y sputtered on a niobium coin using the $^{89}Y(p, n)^{89}Zr$ reaction on a TR-24 cyclotron. Proton irradiation on ^{89}Y-targets has been investigated by several research groups [1,12,15,20,21]. Damage to the yttrium coins (yttrium solid and sputtered coins) was reported by Queern et al. [20]. In their production method, cyclotron irradiations were performed using proton beam energy 17.8 MeV degraded to 12.8 MeV using a 0.75 mm aluminium degrader. Beam currents of 40–45 µA were applied. Using the same irradiation conditions, we did not observe any damage to the yttrium target when irradiated for up to 4 h. The activity of the purified ^{89}Zr solutions was in the range of 866–1160 MBq after 2 h of irradiation using 45 µA beam current. Contamination of ^{88}Zr and ^{88}Y was not observed, as the formation of ^{88}Zr and ^{88}Y occurs via the $^{89}Y(p,2n)^{88}Zr$ reaction at higher incident proton energies >13 MeV.

3.2. Automation

With the vast development of ^{89}Zr radiolabeled antibodies for diagnostic application using PET imaging and increasing demand for their clinical use, there is a need for cGMP grade ^{89}Zr. In this study, we have adopted a commercially available cGMP-compliant TRASIS miniAiO to achieve efficient and reproducible batches of ^{89}Zr. This automated module was used for the target dissolution and separation of ^{89}Zr from the yttrium. In our method, the separation time including dissolution of the target was completed within 1 h. In addition, automated data acquisition was employed, allowing for full GMP-compliant documentation of the process. Furthermore, hydroxamate resin cartridge preconditioning and sterile-filter integrity testing were programmed in the same sequence file. The ^{89}Zr was eluted with a very small volume of oxalic acid (1.5 mL) directly over the sterile filter into the final vial. The automation of ^{89}Zr resulted in a significant reduction in the radiation dose and increased the GMP compliance.

Wooten et al. [14]. described an in-house automation system for ^{89}Zr processing constructed from extruded Al modular framing attached to walls made of ultra-high molecular weight polyethylene with components mounted on the plastic walls and a computer-controlled separation. This in-house automation method suffered from rather low and inconsistent recoveries (44–97%; average 74% ± 16%), while this work had a consistent recovery of 88.2–98.1% (average 93.6% ± 5%). Unlike the in-house module, the use of the GMP-certified AllinOne ASU produced under ISO 9001 offers an easier pathway to obtaining ^{89}Zr that meets cGMP needs for clinical studies.

3.3. Specific Activity and Radionuclidic Purity

The specific activity in the ^{89}Zr-immuno-PET imaging is vitally important because the amount of antibody injected in the preclinical or clinical subjects can alter the quality immuno-PET images and quantification. Particularly, tumour-specific antibodies or peptides for diagnostic and therapeutic applications where only small amounts of antibodies are injected to ensure site-specific uptake. Metallic impurities can compete with the radioisotope for binding sites or ligands during the radiolabelling process with a resulting decrease in radiolabelling efficiency. In order to achieve higher radiolabelling efficiency, radionuclides need to be free from metallic impurities and this can be achieved by selecting high purity target material, energy window for irradiation and trace metal grade solutions for processing. The conventional method to determine effective specific activity (ESA) of ^{89}Zr is titration with commonly used chelator deferoxamine (DFO). In this method, the known quantities of *p-SCN-Bz-DFO* is titrated with ^{89}Zr. The specific activity of three production batches was calculated and was found to be in the range of 1351–2323 MBq/µmol. Our effective specific activities by titration method were low compared to reported data by Queern et al. [20], but similar to those of Wooten et al. [14]. Our ICP-MS analysis showed significantly lower concentration of metallic impurities compared with those of Queern et al. [20].

The radionuclidic purity of ^{89}Zr mainly depends on the purity of the target material and irradiation conditions. The [^{89}Zr]zirconium oxalate solution was tested for radionuclidic identity and purity using a high purity germanium (HPGe) detector (Figure 3A). The analysis showed the presence of 909 and 511 keV peaks. No additional radioactive contamination was detected. In these studies, the radionuclidic purity of the isolated ^{89}Zr fractions was found to be >99.99%. Similar values for radionuclidic purity have been reported by other authors [1,14,20].

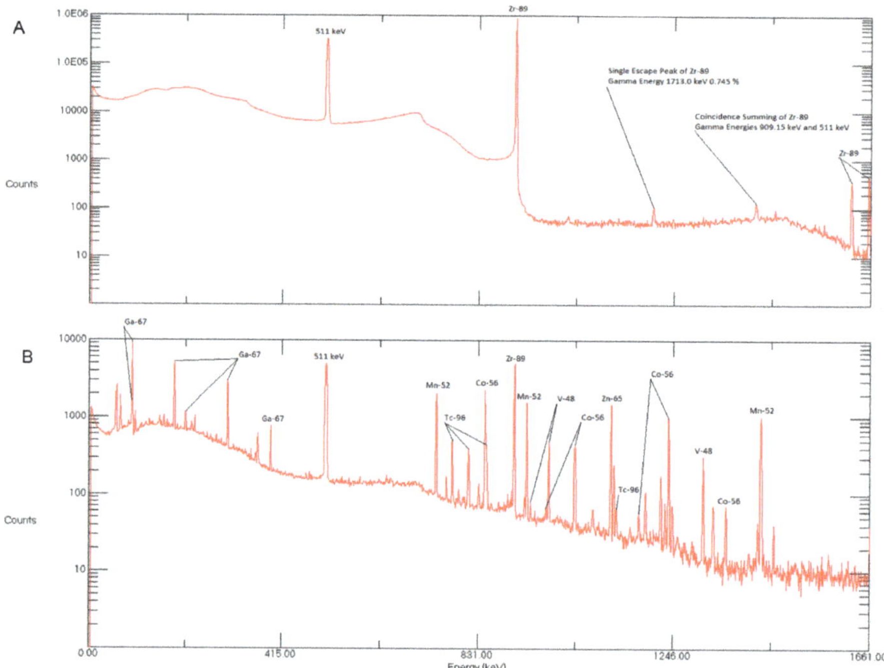

Figure 3. (**A**) Gamma spectrum of purified sample of ^{89}Zr taken 8h after the end-of-bombardment (EOB). (**B**) Gamma spectrum of impurities found in waste vial following ^{89}Zr production.

Gamma spectrometry of waste was also performed to determine radioactive impurities produced during irradiation. These impurities were successfully removed using hydroxamate resin as these impurities were not adhered on hydroxamate resin cartridge. The radioactive impurities produced during production were ^{52}Mn, ^{54}Mn, ^{56}Co, ^{65}Zn, ^{67}Ga, ^{96}Tc, ^{48}V (Figure 3B).

3.4. Radiolabeling and Characterization of ^{89}Zr-DFO-Trastuzumab

Radiolabeling experiments were performed as described in the literature [3,19]. To obtain the quantitative radiolabeling yield and to determine specific activity, several reactions with different concentrations of DFO-trastuzumab were performed and analysed on iTLC (Figure 4A,B). Reported specific activities of ^{89}Zr-DFO-trastuzumab typically range from 0.067 to 0.296 MBq/µg [20,22,23].

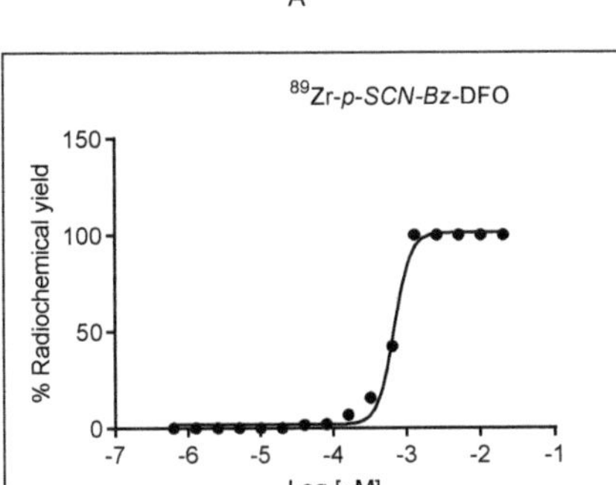

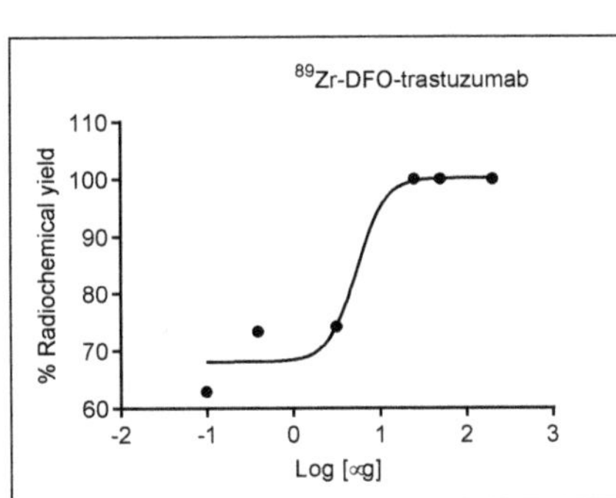

Figure 4. (**A**). Sigmoidal dose–response curve of *p*-isothiocyanatobenzyl-desferoxamine (*p-SCN-Bz*-DFO)–^{89}Zr titration, showing effective specific activity of 2323 MBq/μmol. (**B**) Sigmoidal dose–response curve of ^{89}Zr-DFO-trastuzumab titration showing effective specific activity of 0.308 MBq/μg.

The labeling efficiency of ^{89}Zr-DFO-trastuzumab was 100% ± 5% when 25 μg of DFO-trastuzumab and 3.3 MBq ^{89}Zr was used. The specific activity was 0.308 MBq/μg. which is comparable to Queern et al. [20] (specific activity 0.296 MBq/μg with DFO-trastuzumab). The results of radiolabelling efficiencies are summarized in Table 1.

Table 1. The labelling efficiency of ^{89}Zr with DFO conjugated trastuzumab were evaluated at 37 °C with varying concentrations of conjugate in HEPES buffer, pH 7. A constant ~3.3 MBq of ^{89}Zr-oxalate was added to the DFO-trastuzumab solution for labeling.

Mass of DFO-Trastuzumab (μg)	Labeling Efficiency (%)
200	100
50	100
25	100
3.125	74.24
0.39	73.41
0.048	62.43

3.5. Elemental Analysis

The ICP-MS analyses of three batches of the final ^{89}Zr solution (1.0 mL) were performed after decay of ^{89}Zr. The concentrations of impurities such as yttrium, zirconium, zinc, aluminium, copper, nickel, iron, chromium, niobium and magnesium were quantified by ICP-MS (Table 2). All impurity levels were significantly lower than previously reported values [15,20] and always below <1.0 ppm

Table 2. ICP-MS analysis of three consecutive ^{89}Zr production runs.

Sample #	Zr (ppm)	Al (ppm)	Y (ppm)	Fe (ppm)	Cu (ppm)	Cr (ppm)	Ni (ppm)	Zn (ppm)	Mg (ppm)	Nb (ppm)
PV1 ZR8920180905	0.006	0.019	0.001	0.0084	0.0003	0.0018	0.0013	0.0099	<0.1	0.045
PV2 ZR8920180907	0.006	0.014	<0.001	0.0044	0.0005	0.0016	0.0024	0.019	<0.1	0.022
PV3 ZR8920180910	0.007	0.016	0.009	0.0042	0.0004	0.0009	0.0007	0.0094	<0.1	0.018

4. Conclusions

We report a simplified and efficient automated purification method for [^{89}Zr]zirconium oxalate with high radionuclidic purity using the commercially available TRASIS miniAiO automation module. This method showed reproducible results for the production of [^{89}Zr]zirconium oxalate. Additionally, the automated process with disposable cassettes provides sterile [^{89}Zr]zirconium oxalate and documentation of the manufacturing process that can be used to fulfil cGMP requirements. This automated process can also be adopted for [^{89}Zr]Cl$_2$ production. With adequate shielding, the module ensures safe user operation and environmental radiation protection. This study strongly reinforces the utility of commercially available automated modules for the production of cGMP grade ^{89}Zr.

Author Contributions: V.G. and J.C. performed the experiments and wrote the manuscript. H.F. designed experiments, supervised the study and edited the manuscript. C.R.G. edited the manuscript and helped obtain grant for the study. All authors have read and agreed to the published version of the manuscript.

Funding: This study was funded by a grant from Sylvia Fedoruk Center (grant # J2016-0036).

Acknowledgments: The authors like to acknowledge the contribution and support from the staffs of the Saskatchewan Centre for Cyclotron Science (SCCS). Authors also like to acknowledge the contributions from the staffs of Advanced Cyclotron Systems Inc. (ACSI).

Conflicts of Interest: All authors declare they have no conflicts of interest.

References

1. Holland, J.P.; Sheh, Y.; Lewis, J.S. Standardized methods for the production of high speci fic-activity zirconium-89. *Nucl. Med. Biol.* **2009**, *36*, 729–739. [CrossRef]
2. Nuclear Structure & Decay Data. Available online: https://www.nndc.bnl.gov/nudat2/ (accessed on 5 May 2020).
3. Chekol, R.; Solomon, V.R.; Alizadeh, E.; Bernhard, W.; Fisher, D.; Hill, W.; Barreto, K.; DeCoteau, J.F.; Parada, A.C.; Geyer, R.C.; et al. ^{89}Zr-nimotuzumab for immunoPET imaging of epidermal growth factor receptor I. *Oncotarget* **2018**, *9*, 17117–17132. [CrossRef]
4. Makhlouf, A.; Hajdu, I.; Hartimath, S.V.; Alizadeh, E.; Wharton, K.; Wasan, K.M.; Badea, I.; Fonge, H. ^{111}In-labeled glycoprotein non-metastatic b (GPNMB) targeted gemini surfactant-based nanoparticles against melanoma: In vitro characterization and in vivo evaluation in melanoma mouse xenograft model. *Mol. Pharm.* **2019**, *16*, 542–551. [CrossRef]
5. Lee, Y.S.; Kim, J.S.; Kim, J.Y.; Kim, B., II; Lim, S.M.; Kim, H.J. Spatial resolution and image qualities of Zr-89 on siemens biograph truePoint PET/CT. *Cancer Biother. Radiopharm.* **2015**, *30*, 27–32. [CrossRef] [PubMed]
6. Jauw, Y.W.S.; Menke-van der Houven van Oordt, C.W.; Hoekstra, O.S.; Hendrikse, H.N.; Vugts, D.J.; Zijlstra, J.M.; Huisman, M.C.; van Dongen, G.A.M.S. Immuno-positron emission tomography with zirconium-89-labeled monoclonal antibodies in oncology: What can we learn from initial clinical trials? *Front. Pharmacol.* **2016**, *7*, 131. [CrossRef] [PubMed]
7. Deri, M.A.; Zeglis, B.M.; Francesconi, L.C.; Lewis, J.S. PET imaging with ^{89}Zr: From radiochemistry to the clinic. *Nucl. Med. Biol.* **2013**, *40*, 3–14. [CrossRef]
8. Perk, L.R.; Vosjan, M.J.W.D.W.D.; Visser, G.W.M.M.; Budde, M.; Jurek, P.; Kiefer, G.E.; Van Dongen, G.A.M.S.M.S. P-Isothiocyanatobenzyl-desferrioxamine: A new bifunctional chelate for facile radiolabeling of monoclonal antibodies with zirconium-89 for immuno-PET imaging. *Eur. J. Nucl. Med. Mol. Imaging* **2010**, *37*, 250–259. [CrossRef]
9. Bhatt, N.; Pandya, D.; Wadas, T. Recent Advances in Zirconium-89 Chelator Development. *Molecules* **2018**, *23*, 638. [CrossRef] [PubMed]
10. Price, E.W.; Zeglis, B.M.; Lewis, J.S.; Adam, M.J.; Orvig, C. H$_6$phospa-trastuzumab: Bifunctional methylenephosphonate-based chelator with ^{89}Zr, ^{111}In and ^{177}Lu. *Dalton Trans.* **2014**, *43*, 119–131. [CrossRef]
11. Buchwalder, C.; Rodríguez-Rodríguez, C.; Schaffer, P.; Karagiozov, S.K.; Saatchi, K.; Häfeli, U.O. A new tetrapodal 3-hydroxy-4-pyridinone ligand for complexation of ^{89}zirconium for positron emission tomography (PET) imaging. *Dalton Trans.* **2017**, *46*, 9654–9663. [CrossRef]

12. Meijs, W.E.; Herscheid, J.D.M.; Haisma, H.J.; Wijbrandts, R.; van Langevelde, F.; Van Leuffen, P.J.; Mooy, R.; Pinedo, H.M. Production of highly pure no-carrier added [89]Zr for the labelling of antibodies with a positron emitter. *Appl. Radiat. Isot.* **1994**, *45*, 1143–1147. [CrossRef]

13. Lin, M.; Mukhopadhyay, U.; Waligorski, G.J.; Balatoni, J.A.; González-Lepera, C. Semi-automated production of [89]Zr-oxalate/[89]Zr-chloride and the potential of [89]Zr-chloride in radiopharmaceutical compounding. *Appl. Radiat. Isot.* **2016**, *107*, 317–322. [CrossRef] [PubMed]

14. Wooten, A.; Madrid, E.; Schweitzer, G.; Lawrence, L.; Mebrahtu, E.; Lewis, B.; Lapi, S. Routine Production of [89]Zr Using an Automated Module. *Appl. Sci.* **2013**, *3*, 593–613. [CrossRef]

15. Alnahwi, A.; Tremblay, S.; Guérin, B. Comparative Study with [89]Y-foil and [89]Y-pressed Targets for the Production of [89]Zr. *Appl. Sci.* **2018**, *8*, 1579. [CrossRef]

16. Verel, I.; Visser, G.W.M.; Boellaard, R.; Stigter-van Walsum, M.; Snow, G.B.; Van Dongen, G.A.M.S. [89]Zr immuno-PET: Comprehensive procedures for the production of [89]Zr-labeled monoclonal antibodies. *J. Nucl. Med.* **2003**, *44*, 1271–1281.

17. Meijs, W.E.; Herscheid, J.D.M.; Haisma, H.J.; Pinedo, H.M. Evaluation of desferal as a bifunctional chelating agent for labeling antibodies with Zr-89. *Int. J. Radiat. Appl. Instrum. Part.* **1992**, *43*, 1443–1447. [CrossRef]

18. Zeglis, B.M.; Lewis, J.S. The bioconjugation and radiosynthesis of [89]Zr-DFO-labeled antibodies. *J. Vis. Exp.* **2015**, *96*, 1–8.

19. Vosjan, M.J.W.D.; Perk, L.R.; Visser, G.W.M.; Budde, M.; Jurek, P.; Kiefer, G.E.; Van Dongen, G.A.M.S. Conjugation and radiolabeling of monoclonal antibodies with zirconium-89 for PET imaging using the bifunctional chelate p-isothiocyanatobenzyl-desferrioxamine. *Nat. Protoc.* **2010**, *5*, 739–743. [CrossRef]

20. Queern, S.L.; Aweda, T.A.; Vidal, A.; Massicano, F.; Clanton, N.A.; El Sayed, R.; Sader, J.A.; Zyuzin, A.; Lapi, S.E. Production of Zr-89 using sputtered yttrium coin targets. *Nucl. Med. Biol.* **2017**, *50*, 11–16. [CrossRef]

21. Dejesus, O.T.; Nickles, R.J. Production and purification of [89]Zr, a potential PET antibody label. *Int. J. Radiat. Appl. Instrum. Part.* **1990**, *41*, 789–790. [CrossRef]

22. Dijkers, E.C.F.; Kosterink, J.G.W.; Rademaker, A.P.; Perk, L.R.; Van Dongen, G.A.M.S.; Bart, J.; De Jong, J.R.; De Vries, E.G.E.; Lub-de Hooge, M.N. Development and characterization of clinical-grade[89]Zr-trastuzumab for HER2/neu immunoPET imaging. *J. Nucl. Med.* **2009**, *50*, 974–981. [CrossRef] [PubMed]

23. Kristensen, L.K.; Christensen, C.; Jensen, M.M.; Agnew, B.J.; Schjöth-Frydendahl, C.; Kjaer, A.; Nielsen, C.H. Site-specifically labeled [89]Zr-DFO-trastuzumab improves immuno-reactivity and tumor uptake for immuno-PET in a subcutaneous HER2-positive xenograft mouse model. *Theranostics* **2019**, *9*, 4409–4420. [CrossRef] [PubMed]

Sample Availability: Samples of the compounds [89]Y sputtered niobium coins are available from the authors.

Article

A Universal Cassette-Based System for the Dissolution of Solid Targets

Gabriele Sciacca [1,2,*,†], Petra Martini [3,†], Sara Cisternino [1,2], Liliana Mou [1], Jonathan Amico [4], Juan Esposito [1], Giancarlo Gorgoni [4] and Emiliano Cazzola [4]

1   Legnaro National Laboratories, National Institute for Nuclear Physics, 35020 Legnaro, Italy; sara.cisternino@lnl.infn.it (S.C.); liliana.mou@lnl.infn.it (L.M.); juan.esposito@lnl.infn.it (J.E.)
2   Department of Industrial Engineering, University of Padova, 35131 Padova, Italy
3   Department of Translational Medicine, University of Ferrara, 44121 Ferrara, Italy; mrtptr1@unife.it
4   Cyclotron & Radiopharmacy Department, Sacro Cuore Hospital, 37024 Negrar, Italy; jonathan.amico@sacrocuore.it (J.A.); giancarlo.gorgoni@sacrocuore.it (G.G.); emiliano.cazzola@sacrocuore.it (E.C.)
*   Correspondence: gabriele.sciacca@lnl.infn.it
†   These authors contributed equally to this work.

Abstract: Cyclotron-based radionuclides production by using solid targets has become important in the last years due to the growing demand of radiometals, e.g., ^{68}Ga, ^{89}Zr, $^{43/47}Sc$, and $^{52/54}Mn$. This shifted the focus on solid target management, where the first fundamental step of the radiochemical processing is the target dissolution. Currently, this step is generally performed with commercial or home-made modules separated from the following purification/radiolabelling modules. The aim of this work is the realization of a flexible solid target dissolution system to be easily installed on commercial cassette-based synthesis modules. This would offer a complete target processing and radiopharmaceutical synthesis performable in a single module continuously. The presented solid target dissolution system concept relies on an open-bottomed vial positioned upon a target coin. In particular, the idea is to use the movement mechanism of a syringe pump to position the vial up and down on the target, and to exploit the heater/cooler reactor of the module as a target holder. All the steps can be remotely controlled and are incorporated in the cassette manifold together with the purification and radiolabelling steps. The performance of the device was tested by processing three different irradiated targets under different dissolution conditions.

Keywords: dissolution system; radiopharmaceutical; solid targets; automation; radiometals production; cyclotron production; radiochemistry; manganese-52; zirconium-89; technetium-99m

Citation: Sciacca, G.; Martini, P.; Cisternino, S.; Mou, L.; Amico, J.; Esposito, J.; Gorgoni, G.; Cazzola, E. A Universal Cassette-Based System for the Dissolution of Solid Targets. *Molecules* **2021**, *26*, 6255. https://doi.org/10.3390/molecules26206255

Academic Editor: Anne Roivainen

Received: 17 September 2021
Accepted: 14 October 2021
Published: 16 October 2021

1. Introduction

Radioisotopes (RIs), largely used worldwide in diagnostic imaging procedures in the fields of oncology, neurology and cardiology, are currently produced by medical cyclotron accelerators, starting from the irradiation of a specific target [1]. The growing number of cyclotrons of different energies installed worldwide has given a strong impulse to the production of conventional and emerging radionuclides for medical applications [2–14]. In particular, the great advantage of using medical cyclotrons is the possibility to produce the medical radionuclide of interest on site and on demand. Recently, the technological advancement in the radionuclides cyclotron-based production sector has encouraged the use of novel radioisotopes (mainly radiometals) in medical applications, for implementing the so-called personalized medicine approach. In particular, the strength of this approach relies on the possibility of selecting patients responding positively to the targeted treatment by performing a preliminary diagnostic imaging using the same radiopharmaceutical (theranostic approach) [15–18].

The most used RIs for PET (Positron Emission Tomography), such as ^{18}F and ^{11}C, are produced by liquid or gas targets, whereas the availability of radiometals through a cyclotron-based production requires the use of solid targets. This kind of target is generally composed by a pellet of the desired material, usually costly isotopically enriched material either in metal or oxide form, bonded to a backing plate or encapsulated in a holding shell [19–21]. Liquid target is also an alternative for radiometal production, certainly preferred in case of short half-life radionuclides, necessitating fast irradiation and processing chemistry [22]. However, limitations in the target concentration dissolved in the acidic solution lead to low production yield. In addition, this process is not exempt from issues related to the corrosiveness of the acidic target solution and possible gas formation [23–25]. On the other hand, despite the higher production yield achievable in choosing a solid target-based production, it requires important technological and structural investments for target manufacturing, specific target station, complex automated delivery systems and a dedicated post-irradiation target dissolution system [15,26]. This is in addition to the need to recover the costly enriched material.

Research and development on solid target technologies as well as commercial interest is very dynamic and is evolving very rapidly. Nevertheless, regarding target dissolution systems, to our knowledge mainly three devices are available to date on the market for solid target treatment (IBA-Pinctada® metal [27], ARTMS_ QIS® [28], and Comecer-ALCEO [29]). These are all self-standing systems independent from the synthesis module. Each system is suited to the characteristics of the target provided by the same company and not adaptable to alternative solutions. In some cases (e.g., ALCEO retrofit [29]) the target can be processed directly at the irradiation site to avoid the transfer of the solid target to the radiochemistry lab. From this point of view, other research prototypes have been reported in the literature. Gelbart and Johnson have recently proposed a solid target system with in situ target dissolution featuring heating up to 100 °C, and gas bubbling agitation [30]. Similarly, Beaudoin et al. also developed an in-vault system solid target dissolution system [31].

Having the opportunity to work with a medical cyclotron already equipped with a solid target station and a pneumatic transfer system to the radiochemistry lab, the purpose of this work was the development of a simple and efficient solid target dissolution system compatible with commercial cassette-based synthesis modules. In this way it would be possible to perform the radiochemical processing, from the dissolution to the labelling, all at once using a single remotely controlled device. Keeping the system compact allows containing all the process in a single hot cell, lowering the probability of external and operator contamination. At the same time, this reduces the processing time and maximizes the recovery yield thanks to the absence of wasteful transfers from one system to another. The entire process, starting from dissolution up to radiopharmaceutical formulation, can be applied continuously.

In this regard, a specific solid target dissolution system has been developed in a collaboration between the LARAMED (LAboratory of RAdionuclides for MEDicine) group of the Italian National Institute for Nuclear Physics (INFN), at the Legnaro National Laboratories, and the Sacro Cuore Don Calabria Hospital (SCDCH), located at Negrar di Valpolicella, VR, Italy [32]. The idea of this new reactor originates from the INFN WO/2019/053570 patent [33], describing the technology for manufacturing solid targets by Magnetron Sputtering technique. Indeed, this patent also includes a dissolution reactor system based on open-bottomed vial, which was originally applied on a semi-automatic prototype system used for ^{99m}Tc and ^{64}Cu dissolution and recovery [8,34].

In this paper we describe the developed system and tests performed with three different irradiated targets to demonstrate the performance of the system under different dissolution conditions.

2. Materials and Methods

2.1. Irradiated Targets

Yttrium, chromium, and molybdenum metal targets have been prepared by Spark Plasma Sintering (SPS) [21]. This technique allows the sintering of a pellet starting from powder, and bonding it to a different material without the need of a filler [35,36]. In this work, Y targets were manufactured in one step: ^{Nat}Y disc (Ø 12 mm, thickness 150 μm, purity 99%, Goodfellow) was bonded to a Nb backing disc (Ø 23.5 mm, thickness 1.7 mm, purity 99.99%, Goodfellow). Chromium and Molybdenum targets were prepared in 3 steps: First, green pellets of Cr or Mo (Ø 10 mm, thickness 400 μm and 280 μm, respectively) were prepared starting from the powder form; then, an inert Au foil (Ø 20 mm, thickness 25 μm, purity 99.95%, Goodfellow) was bonded to an Oxygen-Free High thermal Conductivity (OFHC) Cu backing disc (Ø 23.5 mm, thickness 1.7 mm, purity 99.95%) by SPS; finally, Mo or Cr pellets were press-bonded to the backing plate system (Au/Cu) by SPS. SPS machine prototype at University of Pavia (Italy) and a commercial machine (Dr. SINTER® SPS1050, Sumitomo Coal & Mining Co. Ltd., now SPS Syntex Inc., Tokyo, Japan) were used. Manufactured targets are shown in Figure 1.

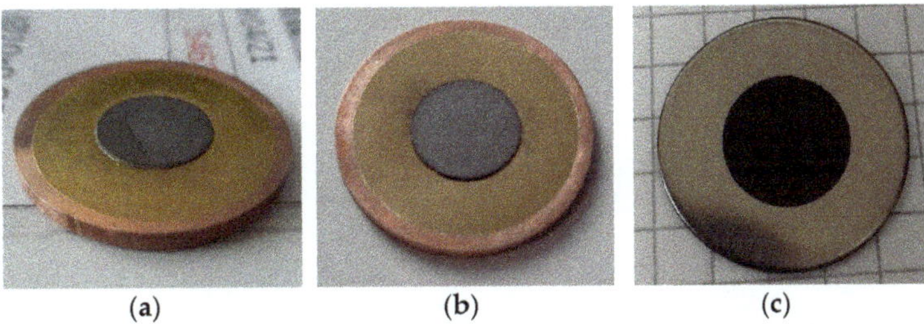

Figure 1. Mo (**a**), Cr (**b**), and Y (**c**) targets prepared with SPS technique, prior to irradiation.

The target sizes fit the target station of ACSI TR-19 Cyclotron, whose details can be found in [37] and is installed at SCDCH where the studies were performed.

2.2. Design of the Reactor Components

The reactor was devised so as to be compatible with the electro-mechanics of commercial synthesis modules. The developed system fundamentally consists of a series of adaptors able to exploit the original purpose of some components of commercial modules.

The concept was tested on the Eckert & Ziegler (E&Z) Modular Lab and Trasis AllinOne commercial modules. These devices are based on disposable cassettes and have a suitable number of valves and components to implement the dissolution process together with the following purification and synthesis protocol. In particular, stepped motors and heated reactors are present. In the original modules' configuration, they are intended for the movement of syringe drivers/activity plunger and the kinetics regulation of the chemical process, respectively. In both modules these components are conveniently located to implement a dissolution process of solid targets: in the proposed configuration, the stepped motor is used for the movement of a dissolution vial by means of a vial-holder and mounting rods, while the reactor's heater is used to plug a target holder suitable for coin-shaped targets.

Hence, the design of the components was performed with the Computer Aided Design tool Solidworks®, after the geometry of the available modules had been reconstructed with the same software. All the parts were made of materials easily washable and as much as possible inert to the strong dissolution condition required for the radiometal isotope recovery. Since the system components are principally made of metallic materials, they were manufactured with Computerized Numerical Control machines. The pieces realized for this purpose are listed in Table 1, and a description of the various parts is hereafter reported.

Table 1. List of the dissolution system components.

Component	Material
Dissolution Vial	quartz; borosilicate glass; PEEK [1]
Vial-Holder and mounting rods	aluminium alloy 6082; stainless steel
Target holder	aluminium; copper
O-ring	NBR [2] 70

[1] PolyEther-Ether-Ketone. [2] Nitrile Butadien Rubber.

2.2.1. Reactor Dissolution Vial

This component, where the dissolution reaction process occurs, is most of the time in contact with strong acid solvents. For that reason, it must be made of inert material. In particular, three different materials were chosen for the proof of concept:

- Borosilicate glass;
- Quartz;
- PEEK.

Indeed, these materials offer different inertness to acidic solutions. While PEEK vials could be machined in house, the ones in glass and quartz were manufactured by the French company Intellion S.a.r.l (Paris). The top of the vial was shaped to screw with the Vial Head for 3 connectors 1/4", shown in Figure 2b, provided as accessories of E & Z modules. This allows the reaction vial to be connected with the cassette by standard tubing and connectors used in these kinds of applications. By contrast, the vial bottom part has a cavity dedicated to installation of a 15.08 mm ID (Internal Diameter) × 2.62 mm thick O-ring, used to seal the reactor with the backing disc. In this way, only the target material will be dissolved, thanks to the inertness of the backing plate.

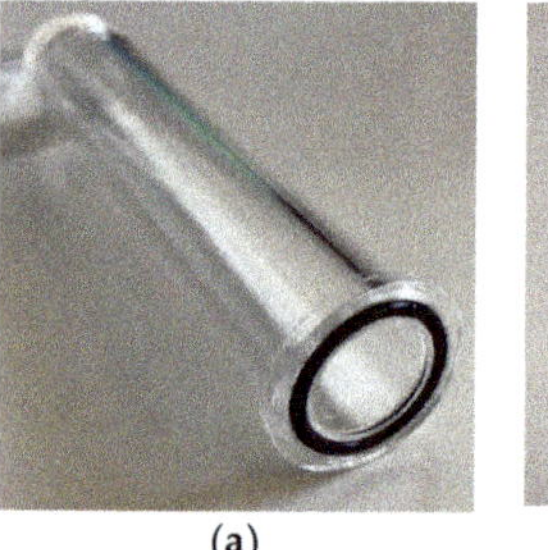
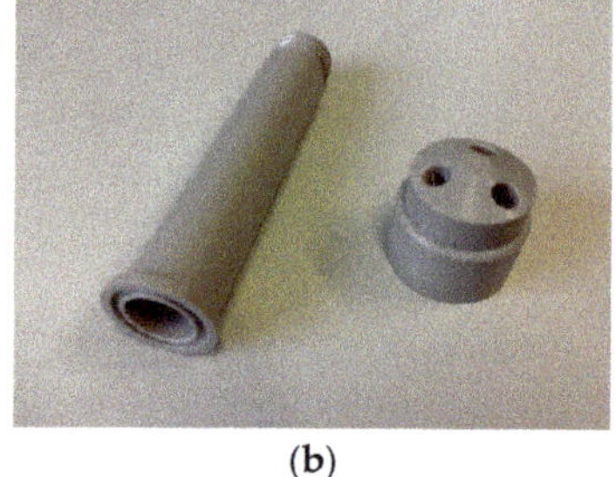

(a)   (b)

Figure 2. (**a**) Reaction vial made of borosilicate glass with mounted O-ring; (**b**) PEEK vial, with the PEEK Vial Head.

2.2.2. Vial-Holder and Mounting Rods

These components, made either in machined aluminium alloy 6082 or stainless steel in relation to the desired rigidity, were needed to connect the reactor vial to the stepping motor in order to perform the vial movement. Indeed, the vial could be kept suspended without slipping down owing to the contact of its holder with the cap, while it also allowed the vial to be held in contact with the target when pushed against it. Therefore, the component is structured in such manner as to facilitate the vial insertion and, in case of a transparent vessel, to allow some visibility inside it during the chemical attack.

Since the moving elements of the modules are not well aligned with the target holder platform, connecting elements have to be adopted. Thus, an assembly of two rods was realized to mount the vessel holder to the syringe actuator. The images in Figure 3 illustrate the parts manufactured for the E&Z module. Slots were applied in correspondence to the bolted joints for the correction of possible misalignments. Enough thickness was then provided in the design of these elements to avoid their bending during the reactor sealing, which can cause solvent losses during operation. At the same time, their dimensions were adjusted to not cause collision with the modules' components.

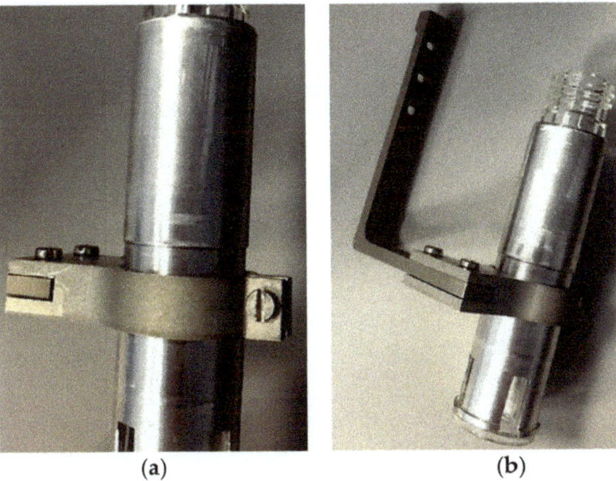

(a)   (b)

Figure 3. (**a**) Assembly of the vial holder; (**b**) mounting rods used for the E&Z module.

2.2.3. Target Holder

A dedicated target holder was developed in order to keep in position the target coin during the dissolution process. This component should have a good thermal conductivity, to efficiently heat the target when control on the reaction kinetics is desired. Its shape was devised as an extension of the reactor's heater to the outside, while a dedicated cavity on the holder's top allows the accommodation of the target. Moreover, as show in Figure 4, a groove surrounding the target slot is intended to confine possible losses from the vessel.

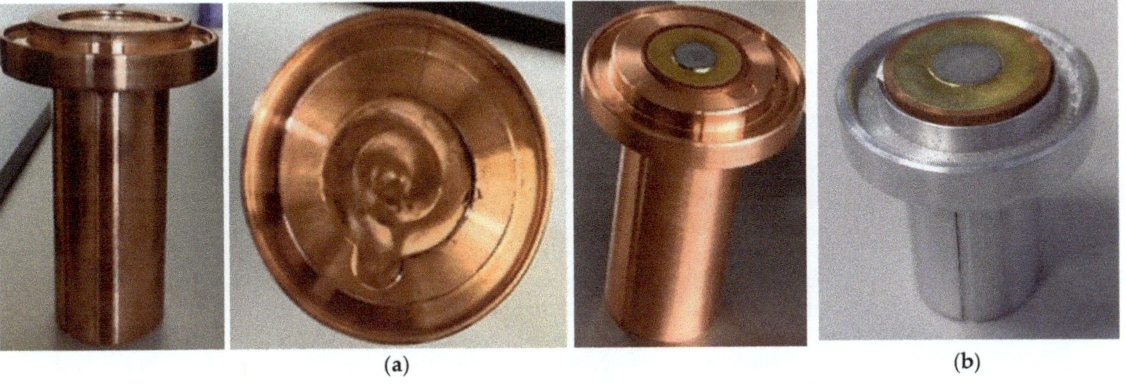

(a)   (b)

Figure 4. (**a**) Target holder made for the E&Z module shown in lateral view (left), upside view (middle), and with a sample target inserted (right). (**b**) Target holder made for the Trasis AllinOne, with the sample target inserted.

2.3. Dissolution Tests

Three different dissolution procedures were performed with the two reactors, installed on E&Z and Trasis systems, on three different irradiated targets (Y, Cr, and Mo) for the production of ^{89}Zr, ^{52}Mn, and ^{99m}Tc, respectively.

Prior to each test, the sealing capability of the system was assessed to prevent possible losses of reactants during the dissolution. After passing the leakage test, the two synthesizers were then used for dissolution experiments of irradiated target at the conditions reported in Table 2. All chemicals and reagents involved in the purification process were of analytical grade, unless otherwise specified. Hydrogen peroxide 30% *w/w* and HCl 37% ACS reagents were purchased from Merck (Darmstadt, Germany).

Table 2. Summary of the performed tests. Time refers to the total time which the process lasted.

Target	Pellet Mass/Thickness	Irradiation Parameters	Module	Vial Material	Process Parameters
^{Nat}Y on Nb	0.0758 g 150 µm	12.5 MeV 10 µA 5 min	Trasis	PEEK	2 mL HCl 2 M RT [1] Time 1 h
^{Nat}Cr on Au/Cu	0.02 g 400 µm	16 MeV 10 µA 15 min	E&Z	Borosilicate glass	3 mL HCl 8 M Heating up to 70 °C Time 1 h
^{Nat}Mo on Au/Cu	0.02 g 280 µm	19 MeV 1 µA 2 min	E&Z	Borosilicate glass	1.5 mL (×3 times) H_2O_2 30% Heating up to 90 °C Time 30 min

[1] Room temperature.

All the dissolution steps were remotely controlled by the module's dedicated software. To evaluate the dissolution efficiency, the target coins were always weighed before and after dissolution. Once dissolution was completed, visual inspection, gamma spectrometry, and activity measurement of the target solutions were performed for each production run. Gamma spectrometric measurements were performed using a High-Purity Germanium (HPGe) detector (Sw GENIE II Canberra, Meriden, CT, USA). The efficiency calibration was carried out in the energy window 17 keV to 1923.1 keV for three different geometries (Eppendorf 1ml; vial 5ml; vial 1ml) by using a multi-peak certified liquid source (containing the reference radionuclides ^{241}Am, ^{109}Cd, ^{139}Ce, ^{57}Co, ^{60}Co, ^{137}Cs, ^{113}Sn, ^{85}Sr, ^{88}Y, and ^{51}Cr), with Genie 2000 software. Activity measurement was carried out with a Capintec, Inc. dose calibrator, model CRC-25PET, periodically subjected to a quality control program. An appropriate factor was set for each isotope measured.

3. Results

3.1. Dissolution Reactor Assembly

The two dissolution reactors have been assembled, respectively, on the E&Z Modular-Lab and the Trasis' AllinOne modules in a hot cell. All the parts were installed without particular difficulties on the commercial modules, and the slots available on the mounting rods allowed to align the vial with the target holder's baseplate with the required precision. Connection of the reactor with the cassette is made easy thanks to the compatibility of the vial with the PEEK head.

Figure 5 shows the developed systems implemented on the tested commercial modules, hosted inside the hot cell. In the dissolution process, after target placement, the reactor vial can be positioned on top of the coin target by controlling the movement of the syringe actuator. The pressure applied by the stepper motor adequately seals, by means of the O-ring, the vial during the chemical attack. Once the dissolution is completed, the solution can be pumped through the outlet channel connected to the top of the vial to the subsequent radiochemistry steps. All the operations can be remotely controlled with the respective modules' software, and the system allows for the incorporation of the purification and radiolabelling steps within a single cassette manifold. Furthermore, the proposed reactor offers an intrinsic flexibility in processing targets having different thicknesses of the deposited target pellet.

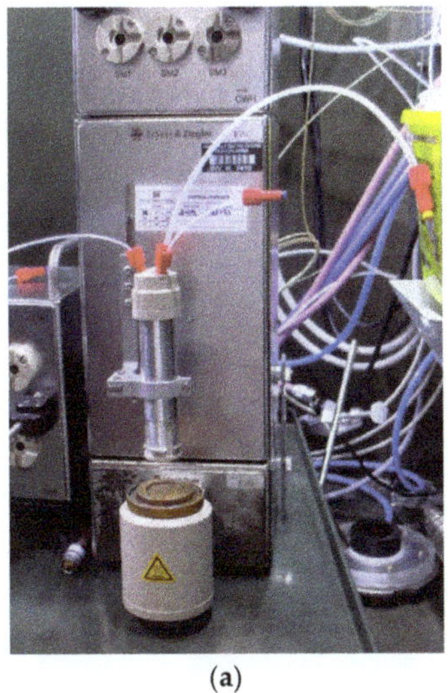
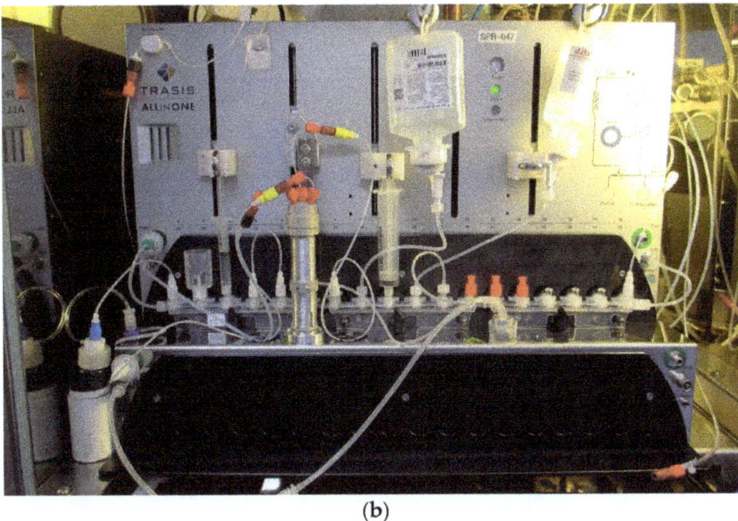

(a) (b)

Figure 5. Pictures of the solid target dissolution system mounted on: (**a**) an E&Z module; (**b**) a TRASIS AllinOne module.

3.2. Operational Testing

All the devices successfully passed the preliminary leakage tests without losses, thus proving to have enough rigidity to prevent possible release of radioactive liquid.

Following the automated transportation of the irradiated target from the cyclotron target station to the hot cell docking station, the target is transferred to the dissolution unit of the automatic system with tongs or telemanipulators. The target can be easily accommodated on the reactor heater baseplate with the target material pointing upwards in alignment with the bottom of the vial. The sealing of the vial on the target by the O-ring, during the chemical attack, allows for the selective dissolution of the target material minimizing the contact of the solution with the backing and avoiding liquid leakage. Liquid leakage never occurred during our dissolution tests. The dissolution of the irradiated target can be activated or hastened by heating the conductive baseplate. The use of a transparent glass vial allowed for the monitoring of the process, as was useful in the case of non-well-established procedures, like with the dissolution of Cr and Mo, whereas with Y target it was possible to use PEEK vial since the procedure was well-known.

Average weights of the target/backing ensembles and of the backing after target material dissolution are listed in Table 3. The amount of dissolved weight corresponds to the amount of the original target material. Therefore, in all three cases, all the target material was efficiently dissolved and removed from the backing in a reproducible manner.

Table 3. Mean weight with standard deviation before and after dissolution of the tested targets. Accuracy interval of $\pm 10^{-4}$ g.

Target	N Test	Target Weight		Difference (g)	Pellet's Initial Weight (g)
		Before Dissolution (g)	After Dissolution (g)		
^{Nat}Y on Nb	6	6.50 ± 0.04	6.43 ± 0.04	0.0758 ± 0.0007	0.0758
^{Nat}Cr on Au/Cu	5	6.90 ± 0.02	6.70 ± 0.02	0.200 ± 0.002	0.2
^{Nat}Mo on Au/Cu	3	6.86 ± 0.02	6.66 ± 0.02	0.200 ± 0.002	0.2

Table 4 reports the measured activity at the end of dissolution (EOD) and rescaled to end of bombardment (EOB). These measurements can be compared with the theoretical predicted activity calculated at EOB by using the on-line tool ISOTOPIA [38]. The reported values are in agreement with the predicted activity and the gamma-spectrometry analysis.

Furthermore, as additional confirmation of complete target dissolution, an additional chemical attack was performed with fresh solvents after dissolved target removal from the reactor in order to detect any residual activity remaining undissolved on the backing. From the following activity measurement, no relevant activity was found in any of the three solutions.

Table 4. Activity measured with dose calibrator at EOD and rescaled to EOB compared to the predicted activity at EOB calculated by using ISOTOPIA tool.

Isotope	Activity @ EOD (MBq)	Activity @ EOB (MBq)	
	Measured	Rescaled	Predicted
^{89}Zr	8.8	8.9	9.7
^{52}Mn	21	22	23
^{99m}Tc	0.87	0.88	0.92

As an example, Figure 6 shows a picture of Au/Cu backing after dissolution of not irradiated ^{nat}Cr target analysed by SEM-EDS. The grooves on the right side of the SEM image correspond to the part where Cr pellet was attached. It is clearly visible that the Cr pellet was completely dissolved. Indeed, the EDS analysis detected traces of Cr (about 30% at.) that could be due to the bonding of Cr pellet to the Au layer. Considering that the SEM-EDS electron beam penetration depth is about 2 μm, the 30% at. of Cr corresponds to about 0.24 mg, approximately 0.1% of the Cr pellet mass, over the 1 cm diameter Cr spot, supporting the data shown in Table 3.

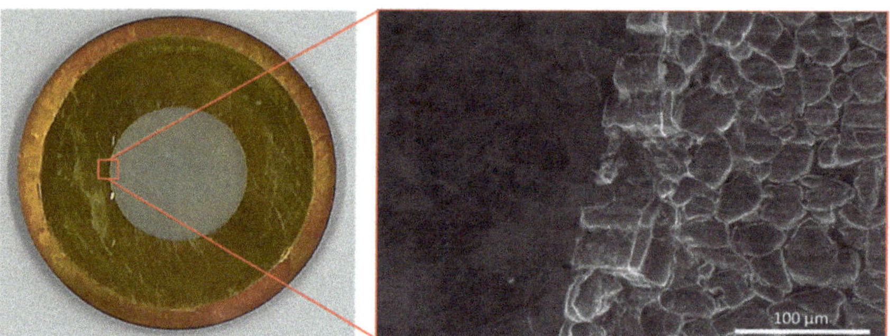

Figure 6. Left: image of the Au/Cu backing after the dissolution of Cr pellet; Right: SEM image of the Au layer surface at the boundary of the area where Cr was bonded.

Below we report a spectrum of the ^{nat}Cr dissolved target (Figure 7). In the spectrum only energy peaks corresponding to manganese and chromium isotopes are identifiable. In all the three performed dissolution studies, no radioactive contaminants coming from the backing material were detected. Minimum detectable activity (MDA) of the most prominent gamma lines for the contaminants $^{197m/g}$Hg and ^{93m}Mo (potentially coming from the activation of Au and Nb backing materials, respectively) calculated from the spectra of the dissolved target solution for ^{52}Mn, ^{99m}Tc, and ^{89}Zr are reported in Table 5.

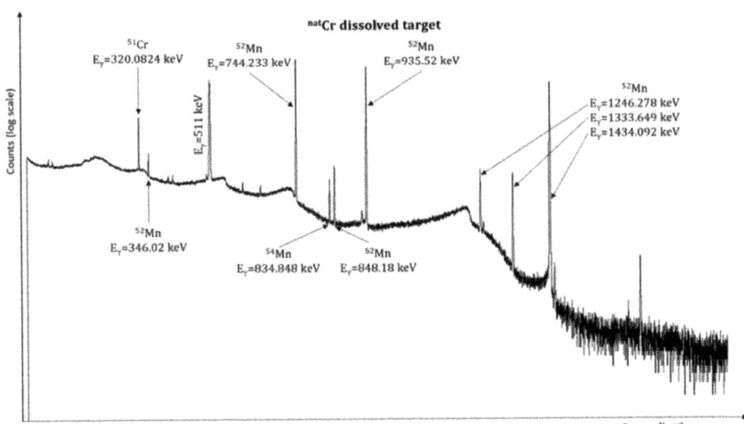

Figure 7. γ-spectrum of the dissolved ^{nat}Cr target.

Table 5. MDA of the most prominent gamma lines for the contaminants $^{197m/g}$Hg and ^{93m}Mo.

Backing	Pellet	Desired Isotope	Isotope Produced in the Backing	Half-Life (h)	E (keV)	MDA (Bq)
Nb	Y	^{89}Zr	^{93m}Mo	6.85	684.693	17.6
Cu/Au	Cr	^{52}Mn	^{197m}Hg	23.8	133.98	23.4
			^{197g}Hg	64.14	191.364	1843.8
Cu/Au	Mo	^{99m}Tc	^{197m}Hg	23.8	133.98	10.4
			^{197g}Hg	64.14	191.364	785.0

4. Discussion

In this work the realization of a flexible solid target dissolution system to be easily installed on commercial cassette-based synthesis modules is described. The developed system allows for the dissolution of targets characterized by various diameters/thicknesses of the target material attached to the backing, regardless of their manufacturing protocol. The target material can be selectively dissolved and radiochemically processed in order to achieve an injectable radiopharmaceutical product of high purity. Both Cr and Mo targets were processed with the E&Z module by using transparent borosilicate glass vials enabling the visualization inside the vessel during the chemical attack to monitor the reaction. At the same time, dissolution of Y irradiated target was performed with a Trasis module using a PEEK vial since the dissolution procedure was standard and already established. Both vials correctly fitted with the cap and reactor ensuring a proper system operation. This demonstrates the possibility of manufacturing this particular design of open-bottomed vial with different materials (from PEEK to quartz). Moreover, the system properly works also changing the components' material, both vial or reactor block, which can be selected according to the needs of thermal conductivity, resistance, and chemical inertness to the solvents involved in the process.

The dissolution system configuration has been tested on both commercial modules considered in this study and was successfully used to perform the dissolution of chromium, molybdenum and yttrium targets used for the ^{52}Mn, ^{99m}Tc, and ^{89}Zr production, respectively. The system is versatile, since it can be used with targets made with different manufacturing techniques and may be adapted with different cassette-based commercial automatic modules (e.g., Eckert & Ziegler Modular-Lab and the Trasis' AllinOne). Thanks to this system, the dissolution of the target can be remotely controlled, directly connected, and totally integrated with the separation and purification processes, keeping the operations safe and clean.

The use of a unique automated system for dissolution, separation, purification, and labelling will sharply decrease the radiation exposure of operators in handling high radioactive materials and, secondly, will definitely contribute to making the whole process much more reproducible, faster, and traceable, minimizing environmental contamination. That is indeed a key prerequisite for attaining a clinical-grade quality for the recovered radioisotope and radiopharmaceutical.

It is thus possible, for those who already own synthesis modules like E&Z modular-lab or Trasis AllinOne, to simply implement their modules with the component described in this paper to get an integrated solid target dissolution system. These components may also be compatible with commercial modules of other brands, making only small design/structural refinements (e.g., GE, IBA radiopharma solutions, ORA-NEPTIS, SCINTOMICS, IPHASE).

Author Contributions: Conceptualization, G.S., P.M., J.E., G.G. and E.C.; methodology, G.S., P.M., S.C., J.E. and E.C.; validation, G.S., P.M., J.A. and E.C.; formal analysis, P.M., L.M. and E.C.; investigation, G.S., P.M., S.C., J.A. and E.C.; resources, G.G. and J.E.; writing—original draft preparation, G.S. and P.M.; writing—review and editing, G.S., P.M., S.C., L.M., J.E. and E.C.; visualization, G.S., P.M., S.C., L.M. and E.C.; supervision, P.M., J.E., G.G. and E.C.; project administration, J.E., G.G. and E.C.; funding acquisition, J.E. All authors have read and agreed to the published version of the manuscript.

Funding: This research was funded by the Istituto Nazionale di Fisica Nucleare, Italy, in the framework of the "METRICS" CSN5 project.

Data Availability Statement: Technical drawings of the reactor components are available from the corresponding author.

Acknowledgments: This work was performed in the framework of the ongoing METRICS project (Multimodal pET/mRI imaging with Cyclotron-produced $^{51/52}$Mn iSotopes) in collaboration with the Sacro Cuore Don Calabria Hospital Radiopharmacy, within the LARAMED (LAboratory of RAdioisotopes for MEDicine) research program running at LNL-INFN. We thank the LNL mechanical workshop for the manufacturing of the reactor's components.

Conflicts of Interest: The authors declare no conflict of interest. The funders had no role in the design of the study, in the collection, analyses, or interpretation of data, in the writing of the manuscript, or in the decision to publish the results.

Sample Availability: Samples are not available from the authors.

References

1. *Cyclotron Produced Radionuclides: Principles and Practice*; International Atomic Energy Agency: Vienna, Austria, 2008; ISBN 978-92-0-100208-2.
2. McCarthy, D.W.; Shefer, R.E.; Klinkowstein, R.E.; Bass, L.A.; Margeneau, W.H.; Cutler, C.S.; Anderson, C.J.; Welch, M.J. Efficient Production of High Specific Activity 64Cu Using a Biomedical Cyclotron. *Nucl. Med. Biol.* **1997**, *24*, 35–43. [CrossRef]
3. Obata, A.; Kasamatsu, S.; McCarthy, D.W.; Welch, M.J.; Saji, H.; Yonekura, Y.; Fujibayashi, Y. Production of Therapeutic Quantities of 64Cu Using a 12 MeV Cyclotron. *Nucl. Med. Biol.* **2003**, *30*, 535–539. [CrossRef]
4. Fonslet, J.; Tietze, S.; Jensen, A.I.; Graves, S.A.; Severin, G.W. Optimized Procedures for Manganese-52: Production, Separation and Radiolabeling. *Appl. Radiat. Isot.* **2017**, *121*, 38–43. [CrossRef]
5. Pyles, J.M.; Massicano, A.V.; Appiah, J.-P.; Bartels, J.L.; Alford, A.; Lapi, S.E. Production of 52Mn Using a Semi-Automated Module. *Appl. Radiat. Isot.* **2021**, *174*, 109741. [CrossRef]
6. Kasbollah, A.; Eu, P.; Cowell, S.; Deb, P. Review on Production of 89Zr in a Medical Cyclotron for PET Radiopharmaceuticals. *J. Nucl. Med. Technol.* **2013**, *41*, 35–41. [CrossRef] [PubMed]
7. Solbach, C.; Bertram, J.; Scheel, W.; Baur, B.; Machulla, H.; Reske, S. Production of Zr-89 on a PETtrace Cyclotron 2013. *J. Nucl. Med.* **2013**, *54*, 1098.
8. Martini, P.; Boschi, A.; Cicoria, G.; Zagni, F.; Corazza, A.; Uccelli, L.; Pasquali, M.; Pupillo, G.; Marengo, M.; Loriggiola, M. In-House Cyclotron Production of High-Purity Tc-99m and Tc-99m Radiopharmaceuticals. *Appl. Radiat. Isot.* **2018**, *139*, 325–331. [CrossRef]

9.   Martini, P.; Boschi, A.; Cicoria, G.; Uccelli, L.; Pasquali, M.; Duatti, A.; Pupillo, G.; Marengo, M.; Loriggiola, M.; Esposito, J. A Solvent-Extraction Module for Cyclotron Production of High-Purity Technetium-99m. *Appl. Radiat. Isot.* **2016**, *118*, 302–307. [CrossRef] [PubMed]

10. Bénard, F.; Buckley, K.R.; Ruth, T.J.; Zeisler, S.K.; Klug, J.; Hanemaayer, V.; Vuckovic, M.; Hou, X.; Celler, A.; Appiah, J.-P. Implementation of Multi-Curie Production of 99mTc by Conventional Medical Cyclotrons. *J. Nucl. Med.* **2014**, *55*, 1017–1022. [CrossRef] [PubMed]

11. Synowiecki, M.A.; Perk, L.R.; Nijsen, J.F.W. Production of Novel Diagnostic Radionuclides in Small Medical Cyclotrons. *EJNMMI Radiopharm. Chem.* **2018**, *3*, 1–25. [CrossRef] [PubMed]

12. Uccelli, L.; Martini, P.; Cittanti, C.; Carnevale, A.; Missiroli, L.; Giganti, M.; Bartolomei, M.; Boschi, A. Therapeutic Radiometals: Worldwide Scientific Literature Trend Analysis (2008–2018). *Molecules* **2019**, *24*, 640. [CrossRef] [PubMed]

13. Pupillo, G.; Mou, L.; Boschi, A.; Calzaferri, S.; Canton, L.; Cisternino, S.; De Dominicis, L.; Duatti, A.; Fontana, A.; Haddad, F. Production of 47 Sc with Natural Vanadium Targets: Results of the PASTA Project. *J. Radioanal. Nucl. Chem.* **2019**, *322*, 1711–1718. [CrossRef]

14. Pupillo, G.; Mou, L.; Martini, P.; Pasquali, M.; Boschi, A.; Cicoria, G.; Duatti, A.; Haddad, F.; Esposito, J. Production of 67Cu by Enriched 70Zn Targets: First Measurements of Formation Cross Sections of 67Cu, 64Cu, 67Ga, 66Ga, 69mZn and 65Zn in Interactions of 70Zn with Protons above 45 MeV. *Radiochim. Acta* **2020**, *108*, 593–602. [CrossRef]

15. Boschi, A.; Martini, P.; Costa, V.; Pagnoni, A.; Uccelli, L. Interdisciplinary Tasks in the Cyclotron Production of Radiometals for Medical Applications. The Case of 47Sc as Example. *Molecules* **2019**, *24*, 444. [CrossRef] [PubMed]

16. Smilkov, K.; Janevik, E.; Guerrini, R.; Pasquali, M.; Boschi, A.; Uccelli, L.; Di Domenico, G.; Duatti, A. Preparation and First Biological Evaluation of Novel Re-188/Tc-99m Peptide Conjugates with Substance-P. *Appl. Radiat. Isot.* **2014**, *92*, 25–31. [CrossRef]

17. Srivastava, S.C. A Bridge Not Too Far: Personalized Medicine with the Use of Theragnostic Radiopharmaceuticals. *J. Postgrad. Med. Educ. Res.* **2013**, *47*, 31–46. [CrossRef]

18. Qaim, S.M. *Medical Radionuclide Production*; De Gruyter: Berlin, Germany, 2019.

19. Gagnon, K.; Wilson, J.S.; Holt, C.M.B.; Abrams, D.N.; McEwan, A.J.B.; Mitlin, D.; McQuarrie, S.A. Cyclotron Production of 99mTc: Recycling of Enriched 100Mo Metal Targets. *Appl. Radiat. Isot.* **2012**, *70*, 1685–1690. [CrossRef]

20. Skliarova, H.; Cisternino, S.; Cicoria, G.; Marengo, M.; Palmieri, V. Innovative Target for Production of Technetium-99m by Biomedical Cyclotron. *Molecules* **2019**, *24*, 25. [CrossRef]

21. Skliarova, H.; Cisternino, S.; Cicoria, G.; Cazzola, E.; Gorgoni, G.; Marengo, M.; Esposito, J. Cyclotron Solid Targets Preparation for Medical Radionuclides Production in the Framework of LARAMED Project. In *Journal of Physics: Conference Series*; IOP Publishing: Bristol, UK, 2020; Volume 1548, p. 012022.

22. Pandey, M.K.; DeGrado, T.R. Cyclotron Production of PET Radiometals in Liquid Targets: Aspects and Prospects. *Curr. Radiopharm.* **2020**, *14*, 325–339. [CrossRef]

23. Riga, S.; Cicoria, G.; Pancaldi, D.; Zagni, F.; Vichi, S.; Dassenno, M.; Mora, L.; Lodi, F.; Morigi, M.P.; Marengo, M. Production of Ga-68 with a General Electric PETtrace Cyclotron by Liquid Target. *Phys. Med.* **2018**, *55*, 116–126. [CrossRef]

24. Oehlke, E.; Hoehr, C.; Hou, X.; Hanemaayer, V.; Zeisler, S.; Adam, M.J.; Ruth, T.J.; Celler, A.; Buckley, K.; Benard, F. Production of Y-86 and Other Radiometals for Research Purposes Using a Solution Target System. *Nucl. Med. Biol.* **2015**, *42*, 842–849. [CrossRef] [PubMed]

25. Pandey, M.K.; Engelbrecht, H.P.; Byrne, J.F.; Packard, A.B.; DeGrado, T.R. Production of 89Zr via the 89Y (p, n) 89Zr Reaction in Aqueous Solution: Effect of Solution Composition on in-Target Chemistry. *Nucl. Med. Biol.* **2014**, *41*, 309–316. [CrossRef] [PubMed]

26. Boschi, A.; Martini, P.; Pasquali, M.; Uccelli, L. Recent Achievements in Tc-99m Radiopharmaceutical Direct Production by Medical Cyclotrons. *Drug Dev. Ind. Pharm.* **2017**, *43*, 1402–1412. [CrossRef] [PubMed]

27. IBA Pinctada Metal. Available online: https://www.iba-radiopharmasolutions.com/more-chemistry (accessed on 15 September 2021).

28. ARTMS QIS. Available online: http://artms.ca/hardware-and-consumables (accessed on 15 September 2021).

29. Comecer ALCEO. Available online: https://www.comecer.com/it/scopri-le-nuove-caratteristiche-alceo/ (accessed on 15 September 2021).

30. Gelbart, W.Z.; Johnson, R.R. Solid Target System with In-Situ Target Dissolution. *Instruments* **2019**, *3*, 14. [CrossRef]

31. Beaudoin, J.-F.; Tremblay, S.; Alnahwi, A.; Guerin, B. Target Carrier and Dissolution System for Facilitating End-to-End in-Vault Cyclotron Production of Radiometals in Large Quantities. *J. Nucl. Med.* **2019**, *60*, 1162.

32. Esposito, J.; Bettoni, D.; Boschi, A.; Calderolla, M.; Cisternino, S.; Fiorentini, G.; Keppel, G.; Martini, P.; Maggiore, M.; Mou, L.; et al. LARAMED: A Laboratory for Radioisotopes of Medical Interest. *Molecules* **2018**, *24*, 20. [CrossRef] [PubMed]

33. Palmieri, V.; Skliarova, H.; Cisternino, S.; Marengo, M.; Cicoria, G. Method for Obtaining a Solid Target for Radiopharmaceuticals Production. International Patent Application PCT/IB2018/056826, 7 September 2018.

34. Cicoria, G.; Pancaldi, D.; Lodi, F.; Malizia, C.; Costa, S.; Lucconi, G.; Lnfantino, A.; Zagni, F.; Fanti, S.; Boschi, S. A Complete Compact Automatic Modular System for 64CuCl2 Production and Labelling of 64Cu-Tracers. In *European Journal of Nuclear Medicine and Molecular Imaging*; 233 Spring St: New York, NY, USA, 2014; Volume 41, p. 348.

35. Anselmi-Tamburini, U. Spark Plasma Sintering. In *Reference Module in Materials Science and Materials Engineering*; Elsevier: Amsterdam, The Netherlands, 2019.

36. Hu, Z.-Y.; Zhang, Z.-H.; Cheng, X.-W.; Wang, F.-C.; Zhang, Y.-F.; Li, S.-L. A Review of Multi-Physical Fields Induced Phenomena and Effects in Spark Plasma Sintering: Fundamentals and Applications. *Mater. Des.* **2020**, *191*, 108662. [CrossRef]
37. Skliarova, H.; Cisternino, S.; Cicoria, G.; Marengo, M.; Cazzola, E.; Gorgoni, G.; Palmieri, V. Medical Cyclotron Solid Target Preparation by Ultrathick Film Magnetron Sputtering Deposition. *Instruments* **2019**, *3*, 21. [CrossRef]
38. IAEA ISOTOPIA. Available online: https://www-nds.iaea.org/relnsd/isotopia/isotopia.html (accessed on 10 October 2020).

Article

Chemical Conversion of Hardly Ionizable Rhenium Aryl Chlorocomplexes with *p*-Substituted Anilines

Martin Štícha [1,*] , Ivan Jelínek [2] and Mikuláš Vlk [2]

[1] Department of Chemistry, Faculty of Science, Charles University, 12000 Prague 2, Czech Republic
[2] Department of Analytical Chemistry, Faculty of Science, Charles University, 12000 Prague 2, Czech Republic; ijelinek@natur.cuni.cz (I.J.); mikulas.vlk@natur.cuni.cz (M.V.)
*   Correspondence: sticha@natur.cuni.cz

Abstract: Fast and selective analytical methods help to ensure the chemical identity and desired purity of the prepared complexes before their medical application, and play an indispensable role in clinical practice. Mass spectrometry, despite some limitations, is an integral part of these methods. In the context of mass spectrometry, specific problems arise with the low ionization efficiency of particular analytes. Chemical derivatization was used as one of the most effective methods to improve the analyte's response and separation characteristics. The Schotten–Baumann reaction was successfully adapted for the derivatization of ESI hardly ionizable Re(VII) bis(catechol) oxochlorocomplex. Various alkyl and halogen *p*-substituted anilines as possible derivatization agents were tested. Unlike the starting complex, the reaction products were easily ionizable in electrospray, providing structurally characteristic molecular and fragment anions. DFT computer modeling, which proposed significant conformation changes of prepared complexes within their deprotonation, proved to have a close link to MS spectra. High-resolution MS and MS/MS measurements complemented with collision-induced dissociation experiments for detailed specification of prepared complexes' fragmentation pathways were used. The specified fragmentation schemes were analogous for all studied derivatives, with an exception for [Re(O)(Cat)$_2$PIPA].

Keywords: high-resolution mass spectrometry; rhenium complexes; chemical derivatization; coordination chemistry; DFT

Citation: Štícha, M.; Jelínek, I.; Vlk, M. Chemical Conversion of Hardly Ionizable Rhenium Aryl Chlorocomplexes with *p*-Substituted Anilines. *Molecules* **2021**, *26*, 3427. https://doi.org/10.3390/molecules26113427

Academic Editors: Alessandra Boschi and Petra Martini

Received: 29 April 2021
Accepted: 1 June 2021
Published: 5 June 2021

Publisher's Note: MDPI stays neutral with regard to jurisdictional claims in published maps and institutional affiliations.

1. Introduction

In analytical chemistry, derivatization is primarily used to modify an analyte that cannot be analyzed by a particular analytical method or to improve selectivity. Chemical derivatization helps to improve separation characteristics and the sensitivity of detection [1]. Chemical derivatization has played an important role in GC/MS analysis, where derivatization is used to increase volatility, change the analyte's ionization properties, or affect analyte fragmentation [2]. The most commonly used derivatives in GC/MS are methyl, ethyl, acetyl, or silyl esters of fatty acids. Mostly in situ derivatization methods are used, where the sample preparation and chemical modification take place in one step. Derivatization in liquid chromatography has a different rationale and, therefore, different rules apply when selecting reagents [3]

The goal of chemical derivatization in ESI/MS is to convert the poorly ionizable or non-ionizable substance into an easily detectable one by changing its chemical and physical properties. ESI is considered very sensitive toward polar compounds. However, for low polarity or non-polar compounds it has been considered less satisfactory than APCI. Substances that form ions in a solution are generally well ionizable by ESI. In contrast, APCI is more suitable for ionizing low to medium polar substances containing atoms with high proton affinity [4]. Although ESI/MS is one of the most efficient analytical methods, its use in detecting some low polar rhenium complexes is limited. Problems with the ionization of these molecules are caused by the absence of acidic or basic groups in the structure. The analyte can be chemically modified to increase ionization efficiency by introducing groups capable of protonation or deprotonation [5–17].

The derivatization potential for ESI-MS in analytical chemistry has been presented in numerous publications [7,8,10,16–22]. One of the disadvantages of derivatization is the possibility of affecting not only the target analyte but also other components of the sample. The Schotten–Baumann reaction (SB reaction) is a commonly used procedure for the derivatization of primary, secondary, and tertiary amines in GC/NPD, GC/FPD, and GC/MS. Various alkyl chloroformates as derivatizing agents have been tested for these purposes; their utilization in aqueous solutions and two-phase solvent solutions have been reported and reviewed [23–25]. Aniline and its substituents are among the simplest weak bases that are highly susceptible to electrophilic and nucleophilic substitutions as the basis of SB reactions. For chloro- and bromo-substituted derivatives, a characteristic isotope pattern can be successfully used to identify fragments. This is why these substances have been chosen as potential derivatizing agents to analyze non-ionizable rhenium complexes. The resulting derivative contains a nitrogen atom as an easily ionizable group and improves the ionization efficiency of the analytes by ESI.

Recently, we showed the suitability of MS for structure characterization of $Re^{V, VI, VII}$ complexes with aromatic bidentate ligands [26]. Although MS provided useful structural information in a series of observed reaction and degradation products, we have occasionally met hardly ionizable structural types [27]. Namely, Re^{VII} oxochloro catechol complex resisted against ionization under ESI, APCI, and APPI conditions. We found an immediate solution in a reaction with *p*-bromoaniline and assumed this reaction worthwhile for further investigation and optimization. This contribution aims to systematically investigate the possibility of chemical conversion of bis(1,2-dihydroxybenzen)-chloro-oxorhenium complex to ESI ionizable products via reactions with aniline and its *p*-substituted derivatives (*p*-chloroaniline, *p*-bromoaniline, *p*-isopropylaniline, *p*-toluidine. The time-course of the derivatization reaction was followed both by ESI/MS and UV-Vis kinetic measurements. Collision-induced dissociation experiments revealed typical fragmentation pathways of formed molecular anions.

2. Results and Discussion

2.1. Chemistry

The hardly ionizable Chlorocomplexes $[Re^{VII}(O)Cl(Cat)_2]$ were prepared using a procedure that was adapted from [26]. One equivalent (4.4 mg) of tetrabutylammonium tetrachlorooxorhenate(V) $[(n\text{-}Bu_4\text{-}N)(ReOCl_4)]$ was dissolved in 1.5 mL of acetonitrile. Two equivalents of ligand 1,2-dihydroxybenzene and two equivalents of triethylamine (10% (*v*/*v*) solution in acetonitrile) were added and the reaction mixture was stirred for 3 days at laboratory temperature. Since catechol ligand lacks the free dissociable group, the yielding compound remains uncharged and its structural characterization by MS is impossible. Its presence in the reaction mixture was presumed entirely from a similarity between absorption spectra describing the formation of deprotonated pyrogallol analog. However, its ESI-MS structure identification is possible after the reaction with aniline (Figure 1), yielding an ESI ionizable reaction product [27].

Derivatives were prepared by mixing 10 µL of the reaction mixture described above with 50 µL of *p*- substituted aniline (5% (*v*/*v*) solution in acetonitrile). The reaction scheme of derivatization is shown in Figure 1. All prepared complexes are described in Table 1. For full chemical names see Table S6 in the Supplementary Materials.

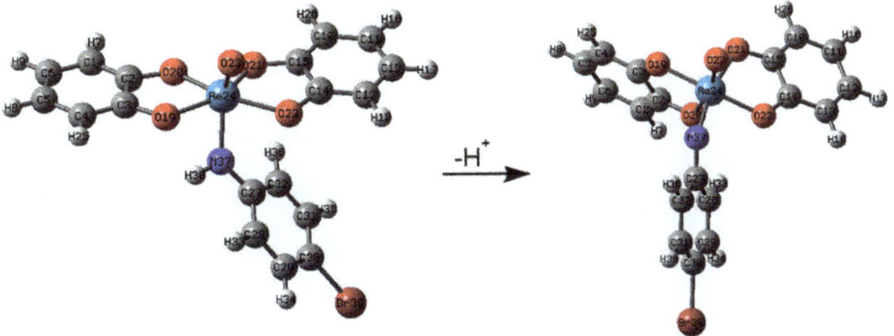

$X = -Cl, -Br, -H, -CH_3, -C(CH_3)_2$

p-chloroaniline   p-bromoroaniline   aniline   p-toluidine   p-isopropylaniline

Figure 1. Derivatization reaction of uncharged rhenium(VII) chlorocomplexes with *p*-substituted aniline.

Table 1. Labels and formulas of prepared rhenium complexes.

Entry	X	Complex	Formula
1	Cl	$[Re^{VII}(O)(Cat)_2PClA]^{-}$ [a]	$C_{18}H_{12}ClNO_5Re$
2	Br	$[Re^{VII}(O)(Cat)_2PBrA]^{-}$ [a]	$C_{18}H_{12}BrNO_5Re$
3	H	$[Re^{VII}(O)(Cat)_2An]^{-}$ [a]	$C_{18}H_{13}NO_5Re$
4	CH_3	$[Re^{VII}(O)(Cat)_2PT]^{-}$ [a]	$C_{19}H_{15}NO_5Re$
5	$C(CH_3)_2$	$[Re^{VII}(O)(Cat)_2PIPA]^{-}$ [a]	$C_{21}H_{19}NO_5Re$

[a] Deprotonated ion.

2.2. Molecular Modeling

Density functional theory could be a possible MS/MS prediction tool useful in structure elucidating. Molecular modeling proposed the significant change of molecular conformation of studied aniline derivatives before and after ionization. We illustrate the structures of neutral and deprotonated [Re(O)(Cat)$_2$PBrA] in Figure 2.

Figure 2. The DFT calculated structures of neutral and deprotonated [Re(O)(Cat)$_2$PBrA].

Evident is the shortening of the Re-N bond in the course of deprotonation. As it results from more detailed modeling, the shortening of the Re-N bond is accompanied by the prolongation of the Re-O(20) bond in a ligand, the increase in the Re-N-C(22) angle, and the decrease in the dihedral angle C(14)-O(22)-Re-O(19). For details, see the corresponding computation data in the Supplementary Materials. The extent of such structural changes depends on the aromatic substituent linked via derivatization. The decisive element here seems to be the basicity of the aniline derivative entering the SB reaction. The bar graph showing the correlation between Re-N bond shortening and the basicity of the used aniline derivative is shown in Figure 3. The pKa values of used aniline derivatives were obtained from [28].

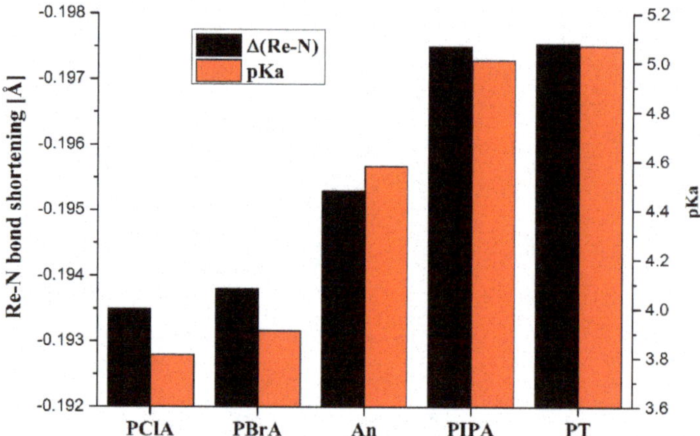

Figure 3. Correlation between Re-N bond shortening Δ(Re-N) and pKa of *p*-substituted anilines; PClA-*p*-chloroaniline, PBrA-*p*-bromoaniline, An-aniline, PIPA-*p*-isopropylaniline, PT-*p*-toluidine.

2.3. High-Resolution Mass Spectrometry Characterization

All molecular ions of prepared derivatives exhibit characteristic isotopic distributions. We compared the similarity between the calculated and experimental molecular ion isotopic patterns using the similarity index (SI) [29]. The [(1-SI) 100] values given in Table 2 indicate an apparent coincidence between the calculated and experimental spectra and the correct assignment of the elemental composition to the m/z values of the observed ions.

Table 2. Theoretical and experimental exact masses of molecular anions of studied derivatives. The error values express the difference between theoretical and experimental masses; the similarity index expresses the resemblance between theoretical and experimental isotope patterns of molecular anions.

Entry	Molecular Formula	Theoretical m/z	Measured m/z	Error (mDa)	Error (ppm)	(1-SI) 100 (%)
[Re(O)(Cat)$_2$PClA]$^-$	C$_{18}$H$_{12}$ClNO$_5$Re	543.9958	543.9969	−1.1	−2.0	98.6
[Re(O)(Cat)$_2$PBrA]$^-$	C$_{18}$H$_{12}$BrNO$_5$Re	587.9445	587.9440	0.4	0.8	94.2
[Re(O)(Cat)$_2$An]$^-$	C$_{18}$H$_{13}$NO$_5$Re	510.0357	510.0360	−0.3	−0.6	98.1
[Re(O)(Cat)$_2$PT]$^-$	C$_{19}$H$_{15}$NO$_5$Re	524.0514	524.0507	0.7	1.2	89.2
[Re(O)(Cat)$_2$PIPA]$^-$	C$_{21}$H$_{19}$NO$_5$Re	552.0827	552.0828	−0.2	−0.3	98.1

Excellent agreement between the isotope pattern calculated and that obtained by MS of the [Re(O)(Cat)$_2$PBrA]- complex is documented in Figure 4. An analogous satisfactory agreement of exact mass and isotopic distribution for all other derivatives was observed.

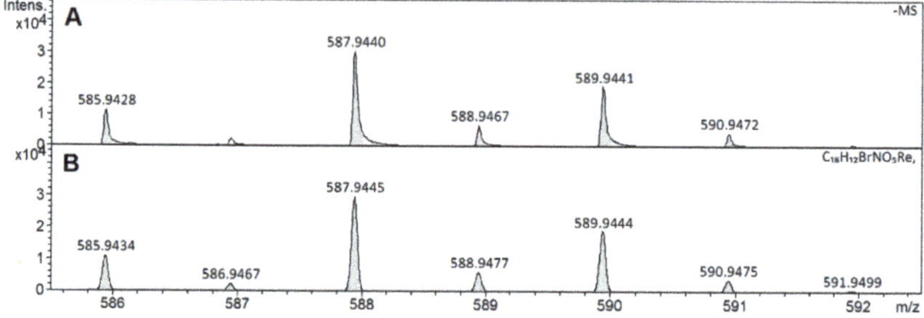

Figure 4. Experimental (**A**) and calculated (**B**) isotope pattern of [Re(O)(Cat)$_2$PBrA].

2.4. Reaction Time-Course

A significant color change accompanies the reaction of the $[Re^{VII}(O)Cl(Cat)_2]$ complex with the aniline derivative. Therefore, UV-Vis absorption spectrophotometry can be applied for the evaluation of its reaction rate. The major absorption band at λ_{max} = 630 nm and the minor one at λ_{max} = 390 nm characterize the intensive blue-green coloration of the $[Re^{VII}(O)Cl(Cat)_2]$ complex. The product of the reaction with the aniline derivative is pale yellow, showing an absorption maximum at 355 nm. The rate of the derivatization reaction with *p*-bromoaniline was followed by the UV/Vis absorption measurement. The corresponding spectra recorded at defined time intervals are shown in Figure 5. A decrease of more than 90% over the 15 min interval was observed for the absorption band at λ_{max} = 630 nm. The proportional decrease in the height of the related absorption band at λ_{max} = 390 nm was observed. After 50 min, both absorption maxima disappear in favor of a minor absorption band at λ_{max} = 355 nm (see inset).

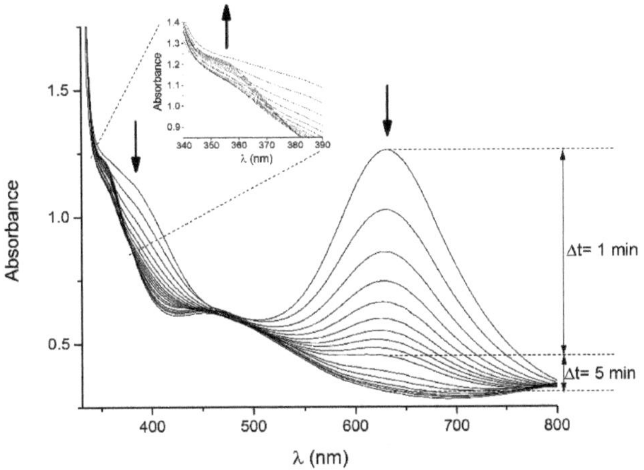

Figure 5. UV/Vis kinetics of derivatization reaction of uncharged $[Re^{VII}(O)Cl(Cat)_2]$ chlorocomplex with *p*-bromoaniline.

An alternative insight into the mechanism of derivatization reaction has been provided by complementary ESI/MS kinetic measurement displayed in Figure 6. The intensity of ion m/z 588 (red line), as the desired derivatization product, achieves a maximum at approximately 15 min. This is consistent with the observed rate of the decrease in the $[Re^{VII}(O)Cl(Cat)_2]$ absorption band (λ_{max} = 630 nm). The appearance of ion m/z 454 (blue line) revealed another function of aniline derivative acting as a weakly basic accelerator of the Re^{V} complex oxidation to a higher Re^{VI} form. Therefore, we can observe a decrease in the intensity of peak m/z 419 (black line) due to the slow transformation of complex $[Re^{V}(O)(Cat)_2]^-$ present in the reaction mixture. As the Re^{VI} complex species are prone to further oxidation, the intensity of ion m/z 454, reaching a maximum within 15 min, decreases at a moderate rate.

The time-course of ionic intensities helps to reveal the actual structure of ion m/z 454. Although the high-resolution MS data are available, there is still uncertainty about the exact structure of ion m/z 454, where both $[Re^{VI}(O)Cl(Cat)_2]^-$ and the adduct with chlorine $[Re^{VII}(O)(Cat)_2]Cl^-$ fit in the same mass. We believe that the shape of the time dependence of the ionic intensities presented in Figure 6 precludes the presence of a chlorine adduct.

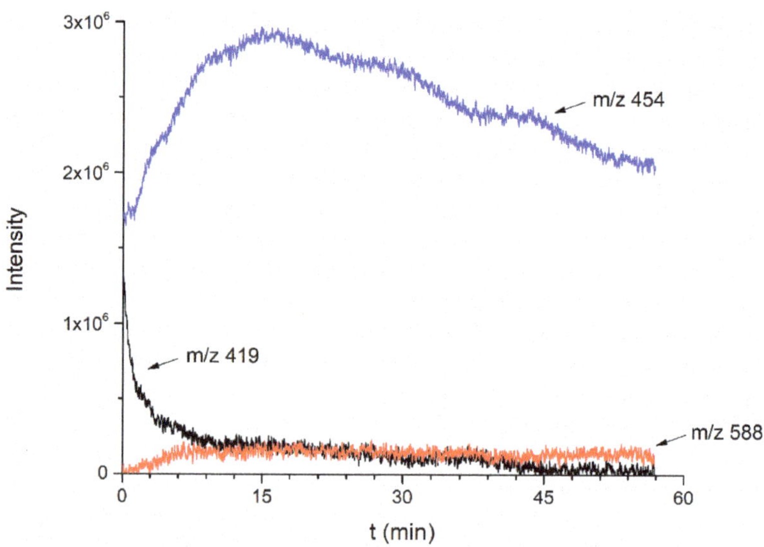

Figure 6. ESI-MS kinetics of derivatization reaction of uncharged [Re^{VII}(O)Cl(Cat)$_2$] chlorocomplex with *p*-bromoaniline.

2.5. Collision Induced Dissociation (CID)

The tandem mass spectrometry at 35eV collision energy in negative ionization mode was used to determine the structure of the prepared derivatives. The obtained MS/MS spectrum in Table 3 is very simple. The accurate mass measurement indicates that the main peak m/z 480 is formed by the loss of catechol moiety from the precursor marked by the blue diamond. It is evident that the isotopic profile retains the distribution confirming the presence of rhenium and bromine isotopes. It is even possible to observe the loss of the aromatic ring from the other catechol moiety to form ion M-2L m/z 404. Fragmentation behavior suggested the presence of a remarkably strong Re-N bond. On the other hand, the ion m/z 327 is formed by the loss of substituted aniline and the whole process ends in ReO$_4^-$ and ReO$_3^-$ ions resp. Since the same behavior was observed for almost all prepared complexes, the fragmentation pattern was demonstrated in this example only. The proposed fragmentation pathway is presented in Figure 7. High mass-accuracy measurements according to Table 3 and fragmentation patterns allowed us to identify the structure of all prepared derivatives.

Table 3. Theoretical and experimental masses of [Re(O)(Cat)$_2$PBrA] CID fragment ions measured at collision energy 35 eV. The error values express the difference between theoretical and experimental masses.

Nom. m/z	Ion Formula	Theoretical m/z	Measured m/z	Error (mDa)	Error (ppm)	Rel. Abundance (%)
588	C$_{18}$H$_{12}$BrNO$_5$Re$^-$	587.9462	587.9459	0.5	0.3	0.3
480	C$_{12}$H$_8$BrNO$_3$Re$^-$	479.9233	479.9271	−7.9	−3.8	−3.8
404	C$_6$H$_4$BrNO$_3$Re$^-$	403.892	403.8931	−2.8	−1.1	−1.1
327	C$_6$H$_4$O$_4$Re$^-$	326.9673	326.9681	−2.5	−0.8	−0.8
251	ReO$_4^-$	250.936	250.9354	2.3	0.6	0.6
235	ReO$_3^-$	234.9411	234.9383	11.7	2.8	2.8

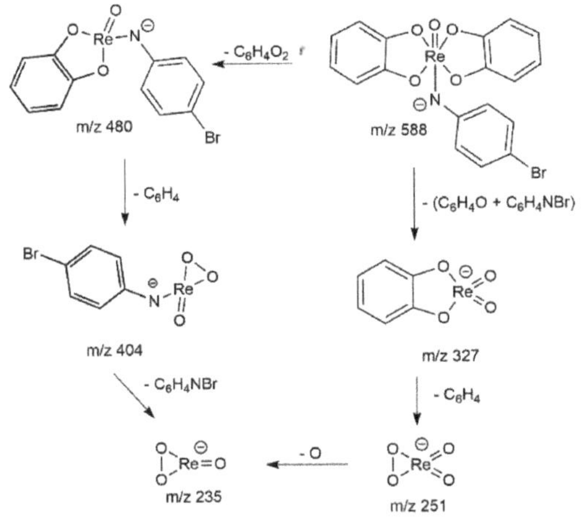

Figure 7. Fragmentation scheme of [Re(O)(Cat)$_2$PBrA].

According to the collision-induced dissociation results in Figure 8, it can be seen that the intensity of the product ion formed by the loss of one catechol ligand M-L (m/z 480) reaches a maximum at the same collision energy as the ion formed by simultaneous fragmentation of substituted aniline (m/z 327), but the intensity of this ion is only around 10% relative to the major peak. This is consistent with the observed shortening of the Re-N bond.

The green curve on the CID diagram describing the formation of the ion m/z 251 does not exhibit a significant maximum, and it is evident that the formation of that ion corresponds to different processes. The spotting of this ion already at zero collision energy can be attributed to the decomposition of the complex; for example, by air humidity. Another mechanism of m/z 251 ion formation is the loss of the aromatic ring from m/z 327. Dissociation of the Re-N bond and cleavage of aniline from the m/z 404 ion occurs only at high CE.

Calculated significant deformation of the molecule, which should be linked both with the shortening of the Re-N bond and the prolongation of a Re-O(20) bond (approximately 0.2 Å) in a single ligand, has an experimental consequence in CID experiments. Such a bond is going to be the most easily fragmented, yielding ion m/z 480. The intensity of the simultaneously arising ion m/z 327 is then significantly lower due to the lower energetical convenience of such fragmentation pathway. As expected, the ion on the favorable fragmentation pathway has the highest negative energy. The difference in energies is about 177 kcal mol^{-1} in favor of ion m/z 480. The energy differences were converted from Hartrees into kcal mol^{-1} using a conversion factor of 627.5.

The fragmentation of all prepared complexes was analogous. The data are available in the Supplementary Materials (Figures S1–S12). We observed the only exception for [Re(O)(Cat)$_2$PIPA]$^-$. Here, the different behavior is not related to the change of bond length but the product stability of arising ions. As is evident from the corresponding CID diagram (Figure 8B), a significant shift to higher collision energies has been observed for the ion ReO$_3^-$. The collision energy where the ion m/z 235 reaches a maximum is almost 40 eV or 3.2 eV regarding E$_{CM}$ higher. Unlike the other complexes, the ReO$_3^-$ ion (m/z 235) in a [Re(O)(Cat)$_2$PIPA]$^-$ fragmentation pathway is not formed directly from ion M-2L but through the ion m/z 325 as an intermediate. Ion m/z 325 is formed by a loss of the methane molecule (Figure 9), where aryl-vinyl stabilization due to π electrons conjugation takes part. Such types of stabilization are unique for [Re(O)(Cat)$_2$PIPA]$^-$ and not possible for other prepared complexes. We calculated that the difference in energies of the fragments with and without aryl-vinyl stabilization is around 48 kcal mol^{-1}. The elemental composition of ion m/z 325 was verified using HRMS. We have obtained the exact mass of 351.9993 Da by measuring, while the calculated value for $C_8H_7NO_3Re^-$ is 351.9989 Da. It means the error is -1.0 ppm.

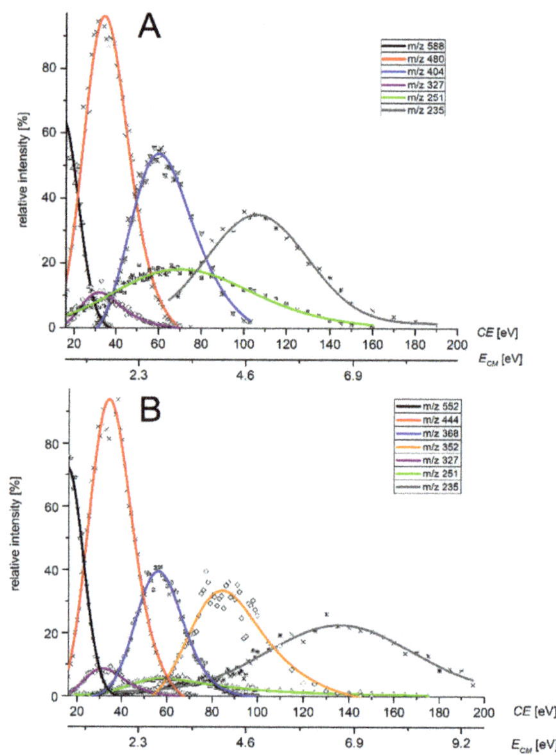

Figure 8. CID diagram of the dependence of the relative intensity of fragmented ions on the collision energy: (**A**) $[Re(O)(Cat)_2PBrA]^-$ (**B**) $[Re(O)(Cat)_2PIPA]^-$.

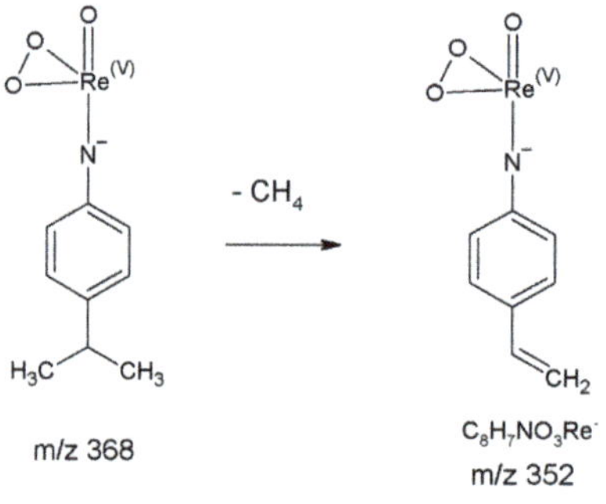

Figure 9. Proposed fragmentation mechanism of ion M-2L yielding from $[Re(O)(Cat)_2PIPA]$ complex.

3. Materials and Methods

3.1. Materials and Reagents

Tetrabutylammonium tetrachlorooxorhenate(V), 4-Chloroaniline, 4-Bromoaniline, 4-Methylaniline, 4-isopropylaniline, aniline, 4-Methylcatechol, and 1,2-dihydroxybenzene were purchased from Sigma-Aldrich. Acetonitrile (HPLC grade) and triethylamine were purchased from Fisher Scientific. Nitrogen used as nebulizing and drying gas was generated by MS-NGM 11 (Bruker Daltonics, Bremen, Germany) nitrogen generator.

3.2. Instrumentation and Software

ESI/MS experiments were conducted on a Bruker QqTOF compact instrument operated using Compass otofControl 4.0 (Bruker Daltonics, Bremen, Germany) software. Compass DataAnalysis 4.4 (Build 200.55.2969) (Bruker Daltonics, Bremen, Germany) software was used for data processing. Molecular structures and fragmentation schemes were drawn using ChemDraw (PerkinElmer Informatics, Waltham, MA, USA). Isotope patterns and exact masses of ions were calculated using IsotopePattern 3.0 (Build 201.9.27) (Bruker Daltonics, Bremen, Germany) utility. Analytical scales Kern ALJ 220-4 (Kern & Sohn, Balingen, Germany) were used to weigh solids. Stirring procedures were performed using a Stuard SA8 (Cole Parmer, UK) stirrer.

ESI/MS data were collected in negative ion mode at scan range from m/z 50 to m/z 1000. The temperature of the drying gas was set to 220 °C at 3.0 L min^{-1} flow rate. Cone voltage was 2800 V. Samples were injected into the nebulizer by a syringe pump (Cole Parmer, USA) at a flow rate 3 μL min^{-1}.

Time-based ESI/MS measurement was performed by mixing the reactants in concentrated form and diluting the reaction mixture right before the ESI ion source using the second syringe pump with acetonitrile.

The isolation width of parent ions in CID experiments was set to 5 Da, the pressure of collision gas (nitrogen) in the collision cell was 2.5×10^{-3} mbar. Measurements were conducted in the range from 10 eV to 200 eV collision energy (ELAB) with a step of 1 eV. Mass spectrometer was calibrated using clusters of ammonium formate. OriginPro 9.0 was used for fitting CID dependences.

Agilent 8453 spectrophotometer (Agilent, Santa Clara, CA, USA) was used for UV/Vis kinetic measurements. Spectra were recorded at 0.1 nm resolution from 330 to 750 nm and processed with UV-Visible Chemstation (Agilent, Santa Clara, CA, USA).

3.3. DFT Calculation

Optimizations of molecular geometries and other theoretical calculations were performed at the density functional level of theory (DFT B3LYP) using Gaussian 16 [30] and LanL2DZ basis set. Both non-ionized and ionized forms of studied complexes were optimized, and frequency calculations were performed with optimized structures at the same level of theory.

4. Conclusions

The Schotten–Baumann (SB) reaction has been successfully adapted for the derivatization of MS hardly ionizable Re(VII) chlorocomplexes. We systematically studied the reaction of the Re(VII) bis(catechol) chlorocomplex with the set of halogen and alkyl anilines as derivatization agents. The SB reaction products are easily ionizable under common ESI conditions providing structurally characteristic molecular and fragment anions. Based on DFT computation, the effect of Re-N bond shortening in the course of complex deprotonation was simulated and also correlated with the basicity of aniline derivative used as a derivatization agent. Our conclusions follow the known relation between the basicity of the reaction environment and the yield of the SB reaction. However, an attempt to increase the yield of the derivatization reaction by adding triethylamine (TEA) to the reaction mixture was unsuccessful. Such a conclusion probably refers to the fast reaction providing the dioxorhenium complex as a competition to the SB reaction itself.

The shortening of the Re-N bond throughout neutral molecule deprotonation and concurrent prolongation of Re-O(20) makes the cleavage of one or both ligands the most probable initial fragmentation pathway of studied complexes, leading to the formation of abundant M-L and M-2L anions.

Although the fragmentation of all studied complexes was analogous, we observed a notable difference concerning the formation of ReO_3^- (m/z 235) ion in a fragmentation scheme of [Re(O)(Cat)$_2$PIPA]$^-$. Unlike the other complexes, this ion is not formed directly from an M-2L fragment but through an aryl-vinyl stabilized unique m/z 325, unseen in other studied complexes' fragmentation schemes.

Supplementary Materials: The following are available online, Figures S1–S13 and Tables S1–S7. Figure S1. HR ESI–MS/MS spectra of complex 1; collision energy, CE, was 40 eV, Figure S2. HR ESI–MS/MS spectra of complex 2; collision energy, CE, was 40 eV, Figure S3. HR ESI–MS/MS spectra of complex 3; collision energy, CE, was 40 eV, Figure S4. HR ESI–MS/MS spectra of complex 4; collision energy, CE, was 40 eV, Figure S5. HR ESI–MS/MS spectra of complex 5; collision energy, CE, was 40 eV, Figure S6. Pro-posed fragmentation scheme of complex 1, Figure S7. Proposed fragmentation scheme of complex 3, Figure S8. Proposed fragmentation scheme of complex 4, Figure S9. Proposed fragmentation scheme of complex 5, Figure S10. CID diagram of complex 1, Figure S11. CID diagram of complex 3, Figure S12. CID diagram of complex 4. Figure S13. Graph of calculated bond elongation of prepared complexes. Table S1. Cartesian coordinates calculated for the optimized neutral structure of complex 1, Table S2. Cartesian coordinates calculated for the optimized neutral structure of complex 2, Table S3. Cartesian coordinates calculated for the optimized neutral structure of complex 3, Table S4. Cartesian coordinates calculated for the optimized neutral structure of complex 4, Table S5. Cartesian coordinates calculated for the optimized neutral structure of complex 5, Table S6. Chemical names, labels, and formulas of prepared rhenium complexes, Table S7. Calculated bond elongation of prepared complexes.

Author Contributions: M.Š. conceived, designed and performed the experiments; M.Š. and I.J. wrote the paper; M.V. was responsible for data analysis and visualization. All authors have read and agreed to the published version of the manuscript.

Funding: Financial support from Charles University Centre of Advanced Materials (CUCAM) (OP VVV Excellent Research Teams, project number CZ.02.1.01/0.0/0.0/15_003/ 0000417) is greatly acknowledged.

Institutional Review Board Statement: Not applicable.

Informed Consent Statement: Not applicable.

Data Availability Statement: Data is contained within the article or Supplementary Materials.

Conflicts of Interest: The authors declare no conflict of interest.

Sample Availability: Samples of the compounds mentioned in the materials section are available from the authors.

Abbreviations

electrospray ionization (ESI); atmospheric pressure photoionization (APPI); atmospheric pressure chemical ionization (APCI); high-resolution mass spectrometry (HRMS); tandem mass spectrometry (MS/MS); gas chromatography/mass spectrometry (GC/MS); gas chromatography/nitrogen phosphorous detector (GC/NPD); gas chromatography/flame photometric detector (GC/FPD); collision-induced dissociation (CID); density functional theory (DFT).

References

1. Anderegg, R.J. Derivatization in mass spectrometry: Strategies for controlling fragmentation. *Mass Spectrom. Rev.* **1988**, *7*, 395–424. [CrossRef]
2. Eggink, M.; Wijtmans, M.; Ekkebus, R.; Lingeman, H.; De Esch, I.J.P.; Kool, J.; Niessen, W.M.A.; Irth, H. Development of a Selective ESI-MS Derivatization Reagent: Synthesis and optimization for the analysis of aldehydes in biological mixtures. *Anal. Chem.* **2008**, *80*, 9042–9051. [CrossRef]
3. Blau, K.; King, G.S. Chapter 2–3. In *Handbook of Derivatives for Chromatography*; Heyden & Son Ltd.: London, UK, 1977.
4. Niessen, W.M.A. State-of-the-art in liquid chromatography–mass spectrometry. *J. Chromatogr. A* **1999**, *856*, 179–197. [CrossRef]
5. Zaikin, V.G.; Halket, J.M. Derivatization in mass spectrometry—7. On-line derivatization/degradation. *Eur. J. Mass Spectrom.* **2006**, *12*, 79–115. [CrossRef] [PubMed]

6. Van Berkel, G.J.; Asano, K.G. Electrospray as a controlled current electrolytic cell—Electrochemical ionization of neutral analytes for detection by electrospray mass-spectrometry. *Anal. Chem.* **1994**, *66*, 2096–2102. [CrossRef]
7. Quirke, J.M.E.; Van Berkel, G.J.; Adams, C.L. Chemical Derivatization for Electrospray Ionization Mass Spectrometry. 1. Alkyl Halides, Alcohols, Phenols, Thiols, and Amines. *Anal. Chem.* **1994**, *66*, 1302–1315. [CrossRef]
8. Van Berkel, G.J.; Quirke, J.M.E.; Tigani, R.A.; Dilley, A.S.; Covey, T.R. Derivatization for electrospray ionization mass spectrometry. 3. Electrochemically ionizable derivatives. *Anal. Chem.* **1998**, *70*, 1544–1554. [CrossRef] [PubMed]
9. Niessen, W.M.A. *Liquid Chromatography-Mass Spectrometry*, 2nd ed.; Marcel Dekker: New York, NY, USA, 1999.
10. Gao, S.; Zhang, Z.P.; Karnes, H.T. Sensitivity enhancement in liquid chromatography/atmospheric pressure ionization mass spectrometry using derivatization and mobile phase additives. *J. Chromatogr. B Anal. Technol. Biomed. Life Sci.* **2005**, *825*, 98–110. [CrossRef]
11. Ming Ng, K.; Ling Ma, N.; Wai Tsang, C. Differentiation of isomeric polyaromatic hydrocarbons by electrospray Ag(I) cationization mass spectrometry. *Rapid Commun. Mass Spectrom.* **2003**, *17*, 2082–2088. [CrossRef]
12. Frenking, G.; Fröhlich, N. The Nature of the Bonding in Transition-Metal Compounds. *Chem. Rev.* **2000**, *100*, 717–774. [CrossRef]
13. Nikolova-Damyanova, B. Retention of lipids in silver ion high-performance liquid chromatography: Facts and assumptions. *J. Chromatogr. A* **2009**, *1216*, 1815–1824. [CrossRef] [PubMed]
14. Nikolova-Damyanova, B.; Momchilova, S. Silver ion HPLC for the analysis of positionally isomeric fatty acids. *J. Liq. Chromatogr. Relat. Technol.* **2002**, *25*, 1947–1965. [CrossRef]
15. Momchilova, S.; Nikolova-Damyanova, B. Stationary phases for silver ion chromatography of lipids: Preparation and properties. *J. Sep. Sci.* **2003**, *26*, 261–270. [CrossRef]
16. Bayer, E.; Gfrörer, P.; Rentel, C. Coordination-Ionspray-MS (CIS-MS), a Universal Detection and Characterization Method for Direct Coupling with Separation Techniques. *Angew. Chem. Int. Ed.* **1999**, *38*, 992–995. [CrossRef]
17. Moriwaki, H. Electrospray ionization mass spectrometric detection of low polar compounds by adding NaAuCl4. *J. Mass Spectrom.* **2016**, *51*, 1096–1102. [CrossRef]
18. Van Berkel, G.J.; Quirke, J.M.E.; Adams, C.L. Derivatization for electrospray ionization-mass spectrometry. 4. Alkenes and alkynes. *Rapid Commun. Mass Spectrom.* **2000**, *14*, 849–858. [CrossRef]
19. Johnson, D.W. Contemporary clinical usage of LC/MS: Analysis of biologically important carboxylic acids. *Clin. Biochem.* **2005**, *38*, 351–361. [CrossRef] [PubMed]
20. Higashi, T.; Shimada, K. Derivatization of neutral steroids to enhance their detection characteristics in liquid chromatography-mass spectrometry. *Anal. Bioanal. Chem.* **2004**, *378*, 875–882. [CrossRef] [PubMed]
21. Honda, A.; Hayashi, S.; Hifumi, H.; Honma, Y.; Tanji, N.; Iwasawa, N.; Suzuki, Y.; Suzuki, K. MPAI (Mass Probes Aided Ionization) Method for Total Analysis of Biomolecules by Mass Spectrometry. *Anal. Sci.* **2007**, *23*, 11–15. [CrossRef]
22. Van Berkel, G.J.; McLuckey, S.A.; Glish, G.L. Charge determination of product ions formed from collision-induced dissociation of multiply protonated molecules via ion molecule reactions. *Anal. Chem.* **1991**, *63*, 2064–2068. [CrossRef]
23. Zaikin, V.G.; Halket, J.M. Derivatization in mass spectrometry—2. Acylation. *Eur. J. Mass Spectrom.* **2003**, *9*, 421–434. [CrossRef] [PubMed]
24. Hušek, P. Chloroformates in gas chromatography as general purpose derivatizing agents. *J. Chromatogr. B* **1998**, *717*, 57–91. [CrossRef]
25. Kataoka, H. 2.1.2—Gas Chromatography of Amines as Various Derivatives. *Journal of Chromatography Library* **2005**, *70*, 364–404. [CrossRef]
26. Štícha, M.; Jelínek, I.; Poláková, J.; Kaliba, D. Characterization of Rhenium(V) Complexes with Phenols Using Mass Spectrometry with Selected Soft Ionization Techniques. *Anal. Lett.* **2015**, *48*, 2329–2342. [CrossRef]
27. Sticha, M.; Jelinek, I.; Kaliba, D.; Polakova, J. Analytical study of rhenium complexes with pyrogallol and catechol. *Chem. Pap.* **2017**, *71*, 819–830. [CrossRef]
28. Williams, R. pKa Data Compiled by R. Williams. Available online: https://organicchemistrydata.org/hansreich/resources/pka/pka_data/pka-compilation-williams.pdf (accessed on 28 April 2021).
29. Wan, K.X.; Vidavsky, I.; Gross, M.L. From similarity index to spectral contrast angle. *J. Am. Soc. Mass Spectrom.* **2002**, *13*, 85–88. [CrossRef]
30. Frisch, M.J.; Trucks, G.W.; Schlegel, H.B.; Scuseria, G.E.; Robb, M.A.; Cheeseman, J.R.; Scalmani, G.; Barone, V.; Mennucci, B.; Petersson, G.A.; et al. *Gaussian 16, Revision C.01*; Gaussian Inc.: Wallingford, CT, USA, 2016.

Article

Highly Efficient Micro-Scale Liquid-Liquid In-Flow Extraction of ^{99m}Tc from Molybdenum

Petra Martini [1,*], Licia Uccelli [1], Adriano Duatti [2], Lorenza Marvelli [2], Juan Esposito [3] and Alessandra Boschi [2]

1 Department of Translational Medicine, University of Ferrara, Via Fossato di Mortara, 70 c/o viale Eliporto, 44121 Ferrara, Italy; ccl@unife.it
2 Department of Chemical, Pharmaceutical and Agricultural Sciences, University of Ferrara, Via L. Borsari, 46, 44121 Ferrara, Italy; dta@unife.it (A.D.); lorenza.marvelli@unife.it (L.M.); bsclsn@unife.it (A.B.)
3 Legnaro National Laboratories, National Institute of Nuclear Physics, Viale dell'Università, 2, 35020 Legnaro, Italy; juan.esposito@lnl.infn.it
* Correspondence: mrtptr1@unife.it

Abstract: The trend to achieve even more compact-sized systems is leading to the development of micro-scale reactors (lab-on-chip) in the field of radiochemical separation and radiopharmaceutical production. Technetium-99m extraction from both high and low specific activity molybdenum could be simply performed by MEK-driven solvent extraction if it were not for unpractical automation. The aim of this work is to develop a solvent extraction and separation process of technetium from molybdenum in a micro-scale in-flow chemistry regime with the aid of a capillary loop and a membrane-based separator, respectively. The developed system is able to extract and separate quantitatively and selectively (91.0 ± 1.8% decay corrected) the [^{99m}Tc]TcO$_4$Na in about 20 min, by using a ZAIPUT separator device. In conclusion, we demonstrated for the first time in our knowledge the high efficiency of a MEK-based solvent extraction process of ^{99m}Tc from a molybdenum-based liquid phased in an in-flow micro-scale regime.

Keywords: microfluidics; liquid–liquid extraction; technetium-99m; in-flow chemistry

Citation: Martini, P.; Uccelli, L.; Duatti, A.; Marvelli, L.; Esposito, J.; Boschi, A. Highly Efficient Micro-Scale Liquid-Liquid In-Flow Extraction of ^{99m}Tc from Molybdenum. *Molecules* **2021**, *26*, 5699. https://doi.org/10.3390/molecules26185699

Academic Editor: Makoto Tsunoda

Received: 7 September 2021
Accepted: 17 September 2021
Published: 21 September 2021

Publisher's Note: MDPI stays neutral with regard to jurisdictional claims in published maps and institutional affiliations.

1. Introduction

Liver, kidney, brain, thyroid scans, imaging of bone lesions, and localization of myocardial infarctions are just some of the basic diagnostic tests performed daily with technetium-99m (^{99m}Tc), a gamma ray emitting radioisotope (Eγ = 140 keV, $t_{1/2}$ = 6 h) which covers over 85% of diagnostic applications in Nuclear Medicine. Technetium-99m is an everlasting radionuclide which has seen the birth of Nuclear Medicine, and the recent advances in technetium chemistry and detector technologies make it still modern and competitive with trendy PET radionuclides [1]. In the late 2000s, a recurrent lack of availability of ^{99m}Tc in the hospitals due to a global shortage of molybdenum-99 (^{99}Mo) because of frequent shutdowns of the ageing reactor-based ^{235}U fission production chain, revealed the fragility of the traditional supply chain and, consequently, prompted the research community to look for alternative production routes.

Strong commitment has been devoted to the cyclotron-based direct production of ^{99m}Tc, through the ^{100}Mo(p, 2n)^{99m}Tc nuclear reaction, as a valuable alternative [2–12]. Accurate studies on the cross section of this reaction and of collateral nuclear reactions have determined the optimal energy range (15–24 MeV) to maximize the production of ^{99m}Tc, ensuring a radionuclidic purity level suitable for clinical applications [13]. This energy range is covered by most of the conventional medical cyclotrons already in operation in many hospital radiopharmacies, which can therefore be used for the production of ^{99m}Tc on site ensuring a constant and on demand supply, without the aid of nuclear reactors.

Other reactor-based routes have been investigated based on the indirect production of ^{99m}Tc induced by thermal or fast neutron beams, respectively through the ^{98}Mo(n,γ)^{99}Mo and ^{100}Mo(n,2n)^{99}Mo nuclear reactions, or gamma-ray beam ^{100}Mo(γ,n)^{99}Mo [11,14,15]. These routes are all affected by low specific activity molybdenum, being products and targets made of the same element material, thus complicating the generator-like extraction and separation process required for the isolation of high purity ^{99m}Tc, since it would require large columns to adsorb molybdenum, decreasing the radioactive concentration of ^{99m}Tc obtained to unacceptably low levels [16–18].

The extraction and purification of ^{99m}Tc from the irradiated metal target is a key step in the production cycle, necessary to make the radioisotope suitable for further radiopharmaceutical processing and patient injection. This process must be simple, reproducible, and efficient, in order to rapidly supply a radioisotope with high purity. The automation of this process is essential to ensure these requirements and to minimize operators' radiation exposure.

In this regard and in the framework of the LARAMED project of the Legnaro National Laboratories of the INFN, we developed an automatic module for the extraction, separation, and purification of cyclotron-produced ^{99m}Tc from the molybdenum metal target, based on the solvent extraction technique [8,9,19]. This system exploits a helium-bubbling into a separation column to boost solvent extraction of technetium into a biphasic system, composed of an alkaline aqueous solution (containing Tc, Mo and contaminants) and an organic phase of methyl ethyl ketone (MEK). The developed system, although very efficient, may however be improved in terms of processing times and costs. Indeed, both dissolution and extraction-separation are the time-consuming steps of all procedures (about 20 and 30 min respectively, over 60 min total). Moreover, the developed module consists of an assembly of commercially-available modular units which are overall quite expensive (i.e., in the range 50–80 k€).

The current trend in technology aiming to achieve even more compact systems is leading to the development of micro-scale reactors (lab-on-chip) in the field of radiochemical separation and radiopharmaceutical production, in order to improve performance and minimize chemical and radiological risks [20–31]. In this view, a latest generation device, the membrane-based Liquid–Liquid separator, 10 September (Figure 1), patented and produced by ZAIPUT Flow Technologies company (Cambridge, MA, USA), has been recently used for the radiochemical separation of radioisotopes of nuclear medical interest [23,29,30,32], for the miniaturization of liquid–liquid extraction processes in an in-flow chemistry regime. This device allows two immiscible phases to be effectively separated by exploiting interfacial tension and the affinity of one of the two phases for a microporous hydrophilic or hydrophobic polytetrafluoroethylene membrane (PTFE). Finally, thanks to a self-regulating differential pressure applied inside the device by a diaphragm, the separation can take place continuously [32].

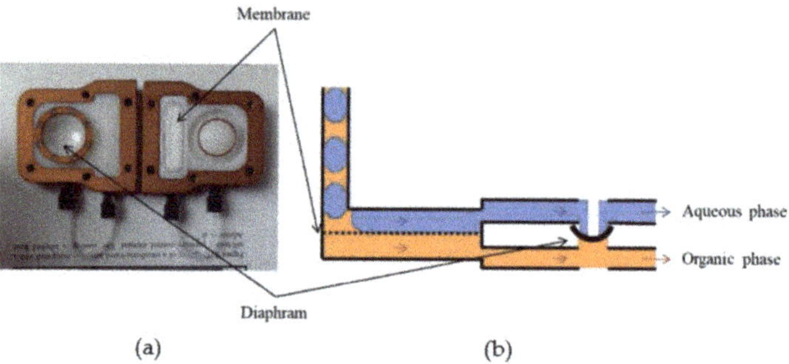

Figure 1. Picture (**a**) and scheme (**b**) of the ZAIPUT separation device.

The in-flow extraction process carried out at the micro-scale level creates some advantages when compared to similar processes performed on a macro-scale, including: shorter extraction times, an increase in the mass transfer coefficient, a better surface-volume interface ratio (S/V), etc. [31,33–36].

The aim of this work is to test the efficiency of the solvent extraction and separation process of technetium from molybdenum in an in-flow chemistry regime with the aid of the Zaiput separator. A versatile separation system that would allow for minimization of costs, times, dimensions, and volumes involved in the process, as well as being applicable also to ^{99m}Tc indirect production methods, such as from low specific activity reactor-produced ^{99}Mo, where the solvent extraction remains the best extraction-separation method [16,37].

2. Results and Discussion

In the production of radioisotopes for nuclear medicine, automation of the radiochemical separation process of the radioisotope of interest from the target and contaminants is necessary to make the results reproducible, minimize losses of radioactive material, decrease process times, and maximize recovery yields. Not least, automation allows for drastic reduction of radiation exposure to the operator conducting the separation.

Recently, we developed a separation and purification module based on the liquid–liquid extraction method applied to the cyclotron-production of ^{99m}Tc from target of metallic molybdenum. Despite this system being extremely efficient, it requires rather long process times for the extraction of technetium with MEK with gas bubbling (helium or argon) within a glass separation column, and subsequent separation of the phases.

The purpose of this work was to develop a different, more compact, and efficient, automatic system for the extraction of ^{99m}Tc from molybdenum, based on the liquid–liquid in-flow extraction process and separation with a membrane separator device, which would allow for minimization of costs, times, dimensions and involved volumes, while keeping high yield and quality.

The Zaiput device (Figure 1), a membrane separator, was therefore selected as the heart of the automatic system that allows two immiscible phases to be separated, taking advantage of the interfacial tension between them and the affinity of one of the two phases for a microporous membrane. Thanks to a differential pressure applied inside the device from a simple diaphragm, the separation can take place continuously [32]. This system allows the macro process of solvent extraction to be reduced to a micro scale, and therefore follows flow chemistry laws.

While the device takes care of the phase's separation, the liquid–liquid extraction, in which the two phases are put in intimate contact allowing for solute transfer, takes place inside a capillary according to flow chemistry [38–41]. The capillary, optimized in terms of length and diameter and wrapped in a loop to minimize space, allows the two co-injected phases to alternate, forming a sort of train technically called the slug-flow regime. This typical micro-arrangement of phases is optimal for liquid–liquid extraction, as it allows the surface/volume ratio of the phases to be maximized, and therefore the ability to transfer solutes according to the chemical affinities involved [34,36,40].

In a preliminary phase, a prototype of the automatic system was assembled and tested with the aim of optimizing the flow rates of the biphasic system (NaOH 6 M/MEK) to achieve a proper separation. Subsequently, "cold" tests (without radioactivity) were performed to optimize the process conditions (volumes, transfer speed, slug-flow regime, loop dimensions, residence times of the sample in the loop, speed, and separation times). Capillary dimensions (ETFE capillary loop 1/8 inch, length = 106 cm; internal diameter 1.59 mm; internal volume about 2.1 mL) and flow rate (0.5 mL/min) are calibrated to maximize the contact between the two phases by achieving a slug-flow regime), the most suitable regime for liquid–liquid extraction on a micro-scale [40]. A schematic diagram of the overall system as it was assembled and installed is reported in Figure 2.

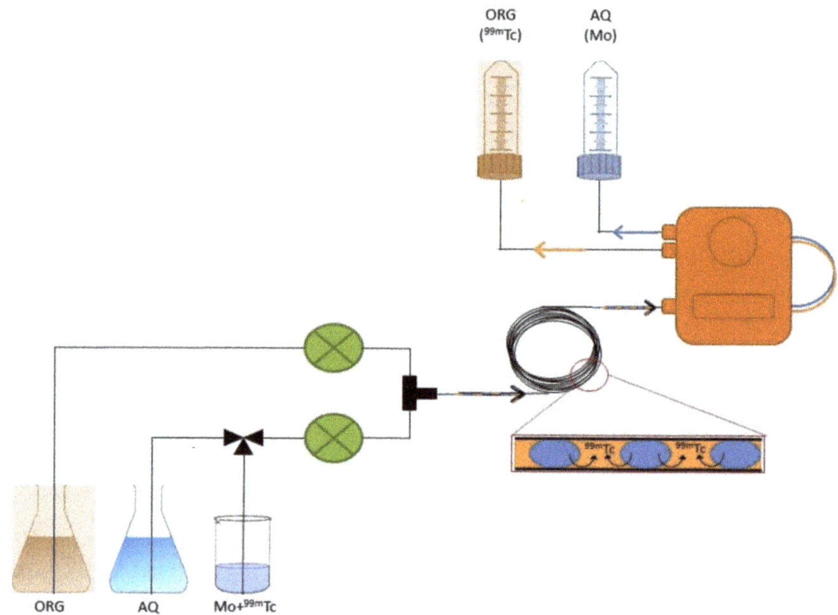

Figure 2. Diagram of the extraction and separation system.

In order to assess how the efficiency of the extraction and separation system concerned, three "hot" tests (with radioactivity) were performed. The separation tests were carried out by adding ^{99m}Tc-sodium pertechnetate eluted by a ^{99}Mo/^{99m}Tc generator in order to trace the activity in the process phases and to determine its efficiency. The collected results, summarized in Table 1, show that the system is able to quantitatively and selectively extract and separate (91.0 ± 1.8% decay corrected) the [^{99m}Tc]TcO$_4$Na in about 20 min.

Table 1. Average percentage of extraction ^{99m}Tc (n = 3). (ORG/AQ OUT 1: preconditioning, ORG/AQ OUT 2: sample collection, ORG/AQ OUT 3: residual washing).

Sample	Volume [mL]	^{99m}Tc %
Sample vial	empty	3.3 ± 0.7
ORG OUT 1	4	-
AQ OUT 1	5	-
ORG OUT 2	8.5	91.0 ± 1.8
AQ OUT 2	7	2.1 ± 0.6
ORG OUT 3	2.5	2.2 ± 1.4
AQ OUT 3	3	1.5 ± 2.1
H$_2$O washing	5	-
MEK washing	5	-

After the phase separation, the ^{99m}TcO$_4{}^-$ contained in the organic solution was purified through a silica and an alumina column, in order to remove molybdate traces and the MEK solvent, respectively. Finally, [^{99m}Tc]TcO$_4{}^-$ was collected from the alumina column with saline. In Table 2, chemical (CP) and radiochemical (RCP) purity values of the [^{99m}Tc]TcO$_4{}^-$ solution obtained at the end of the overall procedure is reported. Post-separation purification with column cartridges allows a pharmaceutical product compliant with the Pharmacopoeia standards to be obtained.

Table 2. CP and RCP of $[^{99m}Tc]TcO_4{}^-$ obtained at the end of the overall procedure of this work compared with the European pharmacopeia requirements for injection, other solvent-extraction-based techniques from the literature, and the quality of $[^{99m}Tc]TcO_4{}^-$ eluted from a UTK generator (Curium). * EU Phar. = European Pharmacopoeia.

Parameters		This Work	EU Phar. *[42–45]	Other SE-Based Techniques [16,37]	Generator Eluate
CP	pH	5	4–8	6–7	6
	Mo	<5 ppm		<10 ppm	-
	Al	<5 ppm	<5 ppm	<10 ppm	<5 ppm
	MEK	<0.0006% (*v/v*)	<0.5% (*v/v*)	<0.1% (*v/v*)	-
RCP	$^{99m}TcO_4{}^-$	>98%	≥95%	>99%	>99%

Given the high ^{99m}Tc extraction yield and purity achieved, this system can be efficiently involved in the separation of technetium from both low and high specific activity molybdenum. In the case of indirect technetium production, the generator-like extraction of ^{99m}Tc from low specific activity ^{99}Mo can be performed running the protocol described here. Moreover, following further decay of ^{99}Mo into ^{99m}Tc, the molybdenum-rich aqueous phase coming out from the separator system, can be recirculated to the capillary loop and to the separator with fresh organic solvent for the extraction and separation of ^{99m}Tc in continuous. Finally, the extraction process, starting with an alkaline aqueous solution of molybdenum, allows extension of the application of the system to not only the treatment of molybdenum metal targets, but also to oxides.

3. Material and Methods

3.1. Materials

The automatic system for the technetium/molybdenum liquid–liquid flow extraction on a microscale is composed of:

- a Zaiput SEP-10 device (purchased by ZAIPUT Flow Technologies company, Cambridge, MA, USA) mounting an OB-100 hydrophobic membrane, supplied with the device;
- two WPX1-U1/16S2-J8-CP peristaltic pumps (WELCO, Tokyo, Japan), flow rate of 0.5 mL/min at 6 V;
- two power supplies;
- a teflon T-junction;
- an ETFE capillary loop (1/8 ", length = 106 cm; internal diameter 1.59 mm; internal volume about 2.1 mL);
- ETFE tubes (1/8 ", internal diameter = 1.59 mm) and connectors.

Other materials involved in the process are: H_2O_2 30% *w/w* (Titolchimica, Rovigo, Italy), methyl ethyl ketone (Carlo Erba, Milano, Italy), NaOH and sodium chloride 0.9% (Fresenius Kabi, Verona, Italy); $[^{99m}Tc]NaTcO_4$, in physiological solution eluted by a $^{99}Mo/^{99m}Tc$ generator (UTK, Curium, London). Silica and acidic alumina SepPak Cartridges were purchased from Waters Corporation (Milford, MA, USA). Natural molybdenum metal target (^{nat}Mo, 99.95% purity, 150 mg) were produced by Spark Plasma Sintering technique (SPS) [46].

3.2. Procedure

A molybdenum metal target, not irradiated, was dissolved in hot concentrated H_2O_2 2 mL, and subsequently 3.5 mL of NaOH 6 M was added to convert MoO_3 into Na_2MoO_4.

In order to "mimic" the irradiated target conditions, a few microliters of generator eluted $[^{99m}Tc]$-sodium pertechnetate were added to the target solution. Subsequently, the sample containing the mixture of Mo and Tc (about 5 mL), together with the organic phase MEK (5 mL), was pumped through the extraction-separation system, previously conditioned with the biphasic NaOH/MEK system (5 mL/phase).

The two phases converge, by means of a Teflon T-junction, in the capillary loop by forming a kind of train technically called the "slug-flow" regime (Figure 3b). The residence time in the loop is approximately 2.5 min and, at this stage, the transfer of $[^{99m}Tc]TcO_4^-$, from the aqueous to the organic phase occurs. Exiting from the loop, the two phases enter the ZAIPUT separator device, from which the technetium-rich organic and molybdenum-rich aqueous phases come out separated (ORG/AQ OUT 1 for the preconditioning, ORG/AQ OUT 2 for the sample).

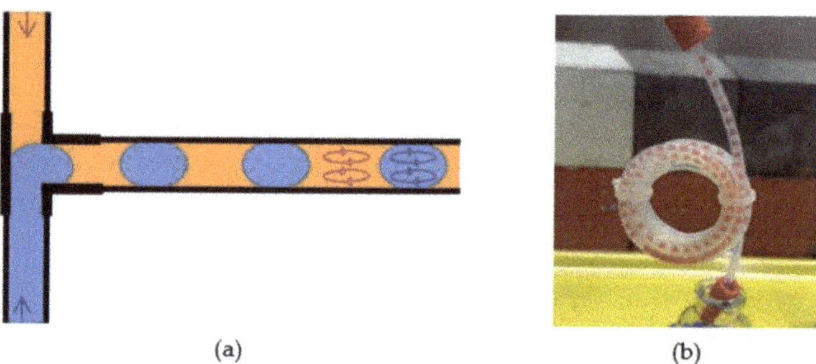

(a)   (b)

Figure 3. (**a**) Operating principle with schematic representation of internal circulation within the two phases and (**b**) photo of a slug-flow regime in the loop in which the organic phase has been colored red by adding a dye.

Once drained of all the sample, a washing with the biphasic NaOH/MEK system (5 mL/phase) was performed to collect any residual sample in the system (ORG/AQ OUT 3). Between one experiment and another, the system was washed with water and MEK. All the samples collected at the output of the ZAIPUT device were then measured with an activity calibrator to detect the amount of ^{99m}Tc activity present in each sample, thus determining the extraction yield. The technetium was then extracted and purified from the ^{99m}Tc-rich organic phase with a silica and alumina column in series, preconditioned with 5 mL of MEK and 10 mL of deionized water, respectively. The technetium was then eluted from the alumina column with 3 mL of saline solution.

3.3. Quality Controls

Paper chromatography was carried out on Whatman 1 strips using methanol/water (8:2) and Physiological solution as mobile phases, Chromatograph Cyclone Plus Storage Phosphor System and Software Optiquant ™.

Kit Merckoquant Molibdeno 5–250 MG/L was purchased from Merck (Darmstadt, Germany) and Kit Tec-Control Aluminium Breakthru from Biodex (Shirley, NY, USA).

Dose Calibrator Capintec CRC-15R (Ramsey, NJ, USA) was used for activity measurements.

Gas-chromatography (GC) was performed using Agilent-Pal H6500-CTC injector, gas-chromatograph Agilent Gc 6850 Series II Network, mass spectrometry detector (MS) Agilent Mass Selective Detector 5973 Network and column Agilent J&W DB-624 UI (20 m, 0.18 mm, 1.00 u) was used for chromatographic separation.

The operating conditions of the headspace were: incubation time: 50 min; incubation temperature: 80 °C; magnetic stirring speed: 250 rpm. The column oven temperature was set at 35 °C and remained constant for 4 min. After this time, the temperature was raised up to 240 °C at 15 °C/min. A constant column flow of 0.7 mL/min of helium was used.

4. Conclusions

In this work we demonstrated for the first time the efficiency of the MEK-based solvent extraction process of ^{99m}Tc from a molybdenum-based liquid phased in an in-flow micro-scale regime, using a compact ZAIPUT separator device. Further developments will be carried out aimed at completing automation of the system thereof, including the dissolution and post-separation purification phases.

This system allows the extraction time and separation of technetium from the organic phase to be drastically reduced and can be used both to purify technetium from molybdenum metal targets in the direct cyclotron ^{99m}Tc production, as well as in indirect ^{99m}Tc production such as the ^{99}Mo production, by irradiating a natural molybdenum using a 14-MeV accelerator-driven neutron source.

Author Contributions: Conceptualization, P.M. and A.B.; data curation, P.M.; methodology, P.M.; investigation, P.M., L.U., L.M. and A.B.; writing—original draft preparation, P.M.; writing—review and editing, A.B.; supervision, A.D.; funding acquisition, J.E. All authors have read and agreed to the published version of the manuscript.

Funding: This research was funded by National Institute for Nuclear Physics, grant number METRICS-CSN5.

Institutional Review Board Statement: Not applicable.

Informed Consent Statement: Not applicable.

Data Availability Statement: Not applicable.

Conflicts of Interest: The authors declare no conflict of interest.

References

1. Boschi, A.; Uccelli, L.; Martini, P. A Picture of Modern Tc-99m Radiopharmaceuticals: Production, Chemistry, and Applications in Molecular Imaging. *Appl. Sci.* **2019**, *9*, 2526. [CrossRef]
2. Qaim, S.M. Nuclear data for medical radionuclides. *J. Radioanal. Nucl. Chem.* **2015**, *305*, 233–245. [CrossRef]
3. Scholten, B.; Lambrecht, R.M.; Cogneau, M.; Ruiz, H.V.; Qaim, S.M. Excitation functions for the cyclotron production of 99mTc and 99Mo. *Appl. Radiat. Isot.* **1999**, *51*, 69–80. [CrossRef]
4. Uzunov, N.M.; Melendez-Alafort, L.; Bello, M.; Cicoria, G.; Zagni, F.; De Nardo, L.; Selva, A.; Mou, L.; Rossi-Alvarez, C.; Pupillo, G.; et al. Radioisotopic purity and imaging properties of cyclotron-produced ^{99m}Tc using direct ^{100}Mo(p,2n) reaction. *Phys. Med. Biol.* **2018**, *63*, 1–30. [CrossRef]
5. Bénard, F.; Buckley, K.R.; Ruth, T.J.; Zeisler, S.K.; Klug, J.; Hanemaayer, V.; Vuckovic, M.; Hou, X.; Celler, A.; Appiah, J.-P.; et al. Implementation of Multi-Curie Production of (99m)Tc by Conventional Medical Cyclotrons. *J. Nucl. Med.* **2014**, *55*, 1017–1022. [CrossRef]
6. Schaffer, P.; Buckley, K.; Celler, A.; Klug, J.; Kovacs, M.; Prato, F.; Valliant, J.; Zeisler, S.; Ruth, T.; Benard, F. A tale of three targets: Direct, multi-Curie production of ^{99m}Tc on three different cyclotrons. *J. Nucl. Med.* **2015**, *56*, 164.
7. Schaffer, P.; Bénard, F.; Bernstein, A.; Buckley, K.; Celler, A.; Cockburn, N.; Corsaut, J.; Dodd, M.; Economou, C.; Eriksson, T.; et al. Direct Production of ^{99m}Tc via ^{100}Mo(p,2n) on Small Medical Cyclotrons. *Phys. Procedia* **2015**, *66*, 383–395. [CrossRef]
8. Martini, P.; Boschi, A.; Cicoria, G.; Uccelli, L.; Pasquali, M.; Duatti, A.; Pupillo, G.; Marengo, M.; Loriggiola, M.; Esposito, J. A solvent-extraction module for cyclotron production of high-purity technetium-99m. *Appl. Radiat. Isot.* **2016**, *118*, 302–307. [CrossRef]
9. Martini, P.; Boschi, A.; Cicoria, G.; Zagni, F.; Corazza, A.; Uccelli, L.; Pasquali, M.; Pupillo, G.; Marengo, M.; Loriggiola, M.; et al. In-house cyclotron production of high-purity Tc-99m and Tc-99m radiopharmaceuticals. *Appl. Radiat. Isot.* **2018**, *139*, 325–331. [CrossRef]
10. Beaver, J.E.; Hupf, H.B. Production of ^{99m}Tc on a medical cyclotron: A feasibility study. *J. Nucl. Med.* **1971**, *12*, 739–741.
11. Pillai, M.R.A.; Dash, A.; Knapp, F.F.R.J. Sustained availability of ^{99m}Tc: Possible paths forward. *J. Nucl. Med.* **2013**, *54*, 313–323. [CrossRef]
12. Boschi, A.; Martini, P.; Pasquali, M.; Uccelli, L. Recent achievements in Tc-99m radiopharmaceutical direct production by medical cyclotrons. *Drug Dev. Ind. Pharm.* **2017**, *43*, 1402–1412. [CrossRef]
13. Esposito, J.; Vecchi, G.; Pupillo, G.; Taibi, A.; Uccelli, L.; Boschi, A.; Gambaccini, M. Evaluation of 99Mo and 99m Tc productions based on a high-performance cyclotron. *Sci. Technol. Nucl. Install.* **2013**. [CrossRef]
14. IAEA. IAEA Nuclear Energy Series Non-HEU Production Technologies for Molybdenum-99 and Technetium-99m. *IAEA Nucl. Energy Ser.* **2013**, 1–75. Available online: https://www-pub.iaea.org/MTCD/Publications/PDF/Pub1589_web.pdf (accessed on 19 September 2021).

15. Capogni, M.; Pietropaolo, A.; Quintieri, L.; Angelone, M.; Boschi, A.; Capone, M.; Cherubini, N.; De Felice, P.; Dodaro, A.; Duatti, A.; et al. 14 MeV neutrons for ^{99}Mo/^{99m}Tc production: Experiments, simulations and perspectives. *Molecules* **2018**, *23*, 1872. [CrossRef]
16. Chattopadhyay, S.; Saha Das, S.; Barua, L.; Pal, A.K.; Kumar, U.; Madhusmita; Alam, M.N.; Hudait, A.K.; Banerjee, S. A compact solvent extraction based ^{99}Mo/^{99m}Tc generator for hospital radiopharmacy. *Appl. Radiat. Isot.* **2019**, *143*, 41–46. [CrossRef]
17. Chattopadhyay, S.; Das, S.S.; Das, M.K.; Goomer, N.C. Recovery of ^{99m}Tc from Na$_2$[^{99}Mo]MoO$_4$ solution obtained from reactor-produced (n,γ) ^{99}Mo using a tiny Dowex-1 column in tandem with a small alumina column. *Appl. Radiat. Isot.* **2008**, *66*, 1814–1817. [CrossRef]
18. Uccelli, L.; Boschi, A.; Pasquali, M.; Duatti, A.; Di Domenico, G.; Pupillo, G.; Esposito, J.; Giganti, M.; Taibi, A.; Gambaccini, M. Influence of the generator in-growth time on the final radiochemical purity and stability of ^{99m}Tc radiopharmaceuticals. *Sci. Technol. Nucl. Install.* **2013**. [CrossRef]
19. Esposito, J.; Bettoni, D.; Boschi, A.; Calderolla, M.; Cisternino, S.; Fiorentini, G.; Keppel, G.; Martini, P.; Maggiore, M.; Mou, L.; et al. Laramed: A laboratory for radioisotopes of medical interest. *Molecules* **2019**, *24*, 20. [CrossRef]
20. Kralj, J.G.; Sahoo, H.R.; Jensen, K.F. Integrated continuous microfluidic liquid–liquid extraction. *Lab. Chip* **2007**, *7*, 256–263. [CrossRef]
21. Mariet, C.; Vansteene, A.; Losno, M.; Pellé, J.; Jasmin, J.P.; Bruchet, A.; Hellé, G. Microfluidics devices applied to radionuclides separation in acidic media for the nuclear fuel cycle. *Micro Nano Eng.* **2019**, *3*, 7–14. [CrossRef]
22. Wang, J.; van Dam, R.M. High-Efficiency Production of Radiopharmaceuticals via Droplet Radiochemistry: A Review of Recent Progress. *Mol. Imaging* **2020**, *19*, 1536012120973099. [CrossRef]
23. Pedersen, K.S.; Imbrogno, J.; Fonslet, J.; Lusardi, M.; Jensen, K.F.; Zhuravlev, F. Liquid–liquid extraction in flow of the radioisotope titanium-45 for positron emission tomography applications. *React. Chem. Eng.* **2018**, *3*, 898–904. [CrossRef]
24. Rensch, C.; Jackson, A.; Lindner, S.; Salvamoser, R.; Samper, V.; Riese, S.; Bartenstein, P.; Wängler, C.; Wängler, B. Microfluidics: A groundbreaking technology for PET tracer production? *Molecules* **2013**, *18*, 7930–7956. [CrossRef]
25. Maurice, A.; Theisen, J.; Gabriel, J.C.P. Microfluidic lab-on-chip advances for liquid–liquid extraction process studies. *Curr. Opin. Colloid Interface Sci.* **2020**, *46*, 20–35. [CrossRef]
26. Farahani, A.; Rahbar-Kelishami, A.; Shayesteh, H. Microfluidic solvent extraction of Cd(II) in parallel flow pattern: Optimization, ion exchange, and mass transfer study. *Sep. Purif. Technol.* **2021**, *258*, 118031. [CrossRef]
27. Priest, C.; Zhou, J.; Sedev, R.; Ralston, J.; Aota, A.; Mawatari, K.; Kitamori, T. Microfluidic extraction of copper from particle-laden solutions. *Int. J. Miner. Process.* **2011**, *98*, 168–173. [CrossRef]
28. Zhang, L.; Hessel, V.; Peng, J.; Wang, Q.; Zhang, L. Co and Ni extraction and separation in segmented micro-flow using a coiled flow inverter. *Chem. Eng. J.* **2017**, *307*, 1–8. [CrossRef]
29. Hellé, G.; Mariet, C.; Cote, G. Microfluidic Tools for the Liquid-liquid Extraction of Radionuclides in Analytical Procedures. *Procedia Chem.* **2012**, *7*, 679–684. [CrossRef]
30. Ban, Y.; Kikutani, Y.; Tokeshi, M.; Morita, Y. Extraction of Am(III) at the Interface of Organic-Aqueous Two-Layer Flow in a Microchannel. *J. Nucl. Sci. Technol.* **2011**, *48*, 1313–1318. [CrossRef]
31. Martini, P.; Adamo, A.; Syna, N.; Boschi, A.; Uccelli, L.; Weeranoppanant, N.; Markham, J.; Pascali, G. Perspectives on the use of liquid extraction for radioisotope purification. *Molecules* **2019**, *24*, 334. [CrossRef] [PubMed]
32. Adamo, A.; Heider, P.L.; Weeranoppanant, N.; Jensen, K.F. Membrane-based, liquid-liquid separator with integrated pressure control. *Ind. Eng. Chem. Res.* **2013**, *52*, 10802–10808. [CrossRef]
33. Kashid, M.N.; Harshe, Y.M.; Agar, D.W. Liquid-liquid slug flow in a capillary: An alternative to suspended drop or film contactors. *Ind. Eng. Chem. Res.* **2007**, *46*, 8420–8430. [CrossRef]
34. Ghaini, A.; Kashid, M.N.; Agar, D.W. Effective interfacial area for mass transfer in the liquid–liquid slug flow capillary microreactors. *Chem. Eng. Process. Process. Intensif.* **2010**, *49*, 358–366. [CrossRef]
35. Woitalka, A.; Kuhn, S.; Jensen, K.F. Scalability of mass transfer in liquid–liquid flow. *Chem. Eng. Sci.* **2014**, *116*, 1–8. [CrossRef]
36. Holbach, A.; Kockmann, N. Counter-current arrangement of microfluidic liquid-liquid droplet flow contactors. *Green Process. Synth.* **2013**, *2*, 157–167. [CrossRef]
37. Chattopadhyay, S.; Das, S.S.; Barua, L. A simple and rapid technique for recovery of ^{99m}Tc from low specific activity (n,γ)^{99}Mo based on solvent extraction and column chromatography. *Appl. Radiat. Isot.* **2010**, *68*, 1–4. [CrossRef]
38. Dessimoz, A.-L.; Cavin, L.; Renken, A.; Kiwi-Minsker, L. Liquid–liquid two-phase flow patterns and mass transfer characteristics in rectangular glass microreactors. *Chem. Eng. Sci.* **2008**, *63*, 4035–4044. [CrossRef]
39. Jovanović, J.; Rebrov, E.V.; Nijhuis, T.A.; Kreutzer, M.T.; Hessel, V.; Schouten, J.C. Liquid-liquid flow in a capillary microreactor: Hydrodynamic flow patterns and extraction performance. *Ind. Eng. Chem. Res.* **2012**, *51*, 1015–1026. [CrossRef]
40. Kashid, M.N.; Gerlach, I.; Goetz, S.; Franzke, J.; Acker, J.F.; Platte, F.; Agar, D.W.; Turek, S. Internal circulation within the liquid slugs of a liquid-liquid slug-flow capillary microreactor. *Ind. Eng. Chem. Res.* **2005**, *44*, 5003–5010. [CrossRef]
41. Tsaoulidis, D.; Angeli, P. Effect of channel size on mass transfer during liquid–liquid plug flow in small scale extractors. *Chem. Eng. J.* **2015**, *262*, 785–793. [CrossRef]
42. Sodium pertechnetate (99mTc) injection (Accelator- Produced). In *EU Pharmacopoeia 9.3*; EDQM, Ed.; European Directorate for the Quality of Medicines & HealthCare Council of Europe: Strasbourg, France, 2018; pp. 4801–4803.

43. Sodium pertechnetate (99mTc) injection (FISSION). In *EU Pharmacopoeia 8.0*; EDQM, Ed.; European Directorate for the Quality of Medicines & HealthCare Council of Europe: Strasbourg, France, 2014; pp. 1090–1091.

44. Sodium pertechnetate (99mTc) injection (NON-FISSION). In *EU Pharmacopoeia 8.0*; EDQM, Ed.; European Directorate for the Quality of Medicines & HealthCare Council of Europe: Strasbourg, France, 2014.

45. Solvents with low toxic potential. In *EU Pharmacopoeia 8.0*; EDQM, Ed.; European Directorate for the Quality of Medicines & HealthCare Council of Europe: Strasbourg, France, 2014; p. 642.

46. Skliarova, H.; Cisternino, S.; Cicoria, G.; Cazzola, E.; Gorgoni, G.; Marengo, M.; Esposito, J. Cyclotron solid targets preparation for medical radionuclides production in the framework of LARAMED project. *J. Phys. Conf. Ser.* **2020**, *1548*, 012022. [CrossRef]

molecules

Article

Preliminary Study of a 1,5-Benzodiazepine-Derivative Labelled with Indium-111 for CCK-2 Receptor Targeting

Marco Verona [1], Sara Rubagotti [2], Stefania Croci [3], Sophia Sarpaki [4], Francesca Borgna [1], Marianna Tosato [5], Elisa Vettorato [1], Giovanni Marzaro [1,6], Francesca Mastrotto [1,6] and Mattia Asti [2,*]

[1] Department of Pharmaceutical Sciences, University of Padova, via Marzolo 5, 35131 Padova, Italy; marco.verona@phd.unipd.it (M.V.); francesca.borgna@psi.ch (F.B.); elisa.vettorato@unipd.it (E.V.); giovanni.marzaro@unipd.it (G.M.); francesca.mastrotto@unipd.it (F.M.)

[2] Radiopharmaceutical Chemistry Section, Nuclear Medicine Unit, AUSL-IRCCS di Reggio Emilia, Viale Risorgimento 80, 42122 Reggio Emilia, Italy; sara.rubagotti@libero.it

[3] Clinical Immunology, Allergy, and Advanced Biotechnologies Unit, Diagnostic Imaging and Laboratory Medicine Department, AUSL-IRCCS di Reggio Emilia, Viale Risorgimento 80, 42122 Reggio Emilia, Italy; Stefania.Croci@ausl.re.it

[4] Bioemtech, Lefkippos Attica Technology Park N.C.S.R. "DEMOKRITOS" Patr. Gregoriou E & 27 Neapoleos Str. Ag. Paraskevi, 15341 Athens, Greece; ssarpaki@bioemtech.com

[5] Department of Chemical Sciences, University of Padova, via Marzolo 1, 35131 Padova, Italy; marianna.tosato@phd.unipd.it

[6] Legnaro National Laboratories, Italian Institute of Nuclear Physics, Viale dell'Università 2, 35020 Legnaro (Padova), Italy

* Correspondence: asti.mattia@ausl.re.it; Tel.: +39-0522-296766

Citation: Verona, M.; Rubagotti, S.; Croci, S.; Sarpaki, S.; Borgna, F.; Tosato, M.; Vettorato, E.; Marzaro, G.; Mastrotto, F.; Asti, M. Preliminary Study of a 1,5-Benzodiazepine-Derivative Labelled with Indium-111 for CCK-2 Receptor Targeting. *Molecules* **2021**, *26*, 918. https://doi.org/10.3390/molecules26040918

Academic Editors: Alessandra Boschi and Petra Martini

Received: 30 December 2020
Accepted: 3 February 2021
Published: 9 February 2021

Abstract: The cholecystokinin-2 receptor (CCK-2R) is overexpressed in several human cancers but displays limited expression in normal tissues. For this reason, it is a suitable target for developing specific radiotracers. In this study, a nastorazepide-based ligand functionalized with a 1,4,7,10-tetraazacyclododecane-1,4,7,10-tetraacetic acid (DOTA) chelator (IP-001) was synthesized and labelled with indium-111. The radiolabeling process yielded >95% with a molar activity of 10 MBq/nmol and a radiochemical purity of >98%. Stability studies have shown a remarkable resistance to degradation (>93%) within 120 h of incubation in human blood. The in vitro uptake of [^{111}In]In-IP-001 was assessed for up to 24 h on a high CCK-2R-expressing tumor cell line (A549) showing maximal accumulation after 4 h of incubation. Biodistribution and single photon emission tomography (SPECT)/CT imaging were evaluated on BALB/c nude mice bearing A549 xenograft tumors. Implanted tumors could be clearly visualized after only 4 h post injection (2.36 ± 0.26% ID/cc), although a high amount of radiotracer was also found in the liver, kidneys, and spleen (8.25 ± 2.21%, 6.99 ± 0.97%, and 3.88 ± 0.36% ID/cc, respectively). Clearance was slow by both hepatobiliary and renal excretion. Tumor retention persisted for up to 24 h, with the tumor to organs ratio increasing over-time and ending with a tumor uptake (1.52 ± 0.71% ID/cc) comparable to liver and kidneys.

Keywords: cholecystokinin-2 receptor; indium-111 labelling; nastorazepide; radiopharmaceuticals

1. Introduction

Receptor-specific targeting ligands were recently harnessed to transport and selectively deliver cytotoxic payloads to cancer tissues in vivo. Payloads may vary from chemotherapeutics to radionuclides and are usually strongly tethered to a chemically modified endogenous ligand or a synthetic agent that expresses a high affinity for the selected receptors [1,2]. Focusing on radioactive cargos, the use of metal radionuclides allows for the achievement of both diagnostic and therapeutic purposes since the same targeting vector can be labelled with metals exhibiting different emissions but similar or equal chemical features. A stable bond between a radiometal and ligand is usually provided by using proper bifunctional chelators, i.e., chemical compounds covalently bind to the backbone of the targeting vector and are able to form thermodynamically and kinetical stable complexes with the metal [3,4].

Among target receptors, cholecystokinin-2 receptor (CCK-2R) is a G-protein-coupled receptor that is normally expressed in the central nervous system and gastric mucosa, where it conveys a regulatory function. Endogenous ligands for CCK-2R are low molecular-weight peptides mainly synthesized in the central nervous system and the gastrointestinal tract such as gastrin and cholecystokinin. CCK-2R is also overexpressed in various human cancers (e.g., lung, medullary thyroid, pancreatic, colon, and gastrointestinal stromal tumors), where it stimulates cell growth, migration, and tumor metastasis, but it displays limited expression in other normal tissues [5–8]. For this reason, CCK-2R is a suitable target for functional imaging and therapy with radiopharmaceuticals, and its potential has been largely explored in the last few years by using various radiolabeled derivatives of its endogenous activators [9]. However, the use of peptide-based ligands still raises some concerns because derivatives exhibiting the highest tumor uptake are also characterized by high retention in the kidneys. Conversely, radiolabeled peptides that display low accumulation in kidneys show little retention in tumor tissues as well. Secondly, being the peptide chains prone to degradation by endogenous peptidases and physiological oxidation on methionine residue, the affinity for receptors can be thwarted during circulation, thus resulting in a suboptimal distribution with a consequent accumulation of high radioactive doses in healthy tissues. Finally, the use of agonist such as CCK- or gastrin-derivatives may activate the receptor signal stimulating the growth, proliferation, and survival of cancer cells [10,11]. As a consequence, the development of a peptide-based CCK-2R-targeting molecule for radiotherapeutic applications has been hampered so far, although many improvements have been obtained over the years through several interesting approaches [12–14].

An alternative path to outperform these issues could be achieved by labelling antagonist ligands based on small organic molecules rather than amino acid sequences. As opposed to agonists, antagonists do not activate the signaling cascade when bound to the receptor, and a non-peptide-based structure should provide a higher resistance to enzymatic degradation. Nastorazepide (Z-360) is a selective 1,5-benzodiazepine-derivative CCK-2 receptor antagonist with potential antineoplastic activity. Z-360 binds to CCK-2R, leading to the avoidance of its activation with sub-nanomolar affinity and high selectivity, in contrast to CCK-1R (K_d = 0.47 nmol/L; selectivity relative to CCK-1R = 672) [15].

The usefulness of Z-360 as directing moiety for the preparation of radiopharmaceutical has already been demonstrated in the literature. For example, a series of nastorazepide-based derivatives have been synthesized and tethered to a N_3S- or N_4-system bifunctional chelator through different spacers [16,17]. The introduction of these chelators provided a suitable moiety for technetium-99m ($t_{1/2}$ = 6 h; E_γ = 140 keV) complexation, while appropriate spacers were introduced to optimize affinity and water solubility for CCK-2R. As a result, it has been demonstrated that Z360-based radiopharmaceuticals are valuable tools that are able to yield high-resolution images of CCK-2R-expressing tumors, especially at longer acquisition times.

Based on these groundbreaking results, in the present study, a 1,5-benzodiazepine-based ligand functionalized with a 1,4,7,10-tetraazacyclododecane-1,4,7,10-tetraacetic acid (DOTA) chelator was synthesized in order to yield a precursor with theragnostic potential. DOTA is an almost universal chelator that is able to complex (with high stability) several metal radionuclides currently applied in clinical practice such as gallium-68 ($t_{1/2}$ = 68 m; $E_{\beta+}$ = 0.89 MeV), indium-111 ($t_{1/2}$ = 2.8 d; E_γ = 173 and 247 keV), lutetium-177 ($t_{1/2}$ = 6.7 d; $E_{\beta(max)}$ = 497, 384, and 176 keV; E_γ = 113 and 208 keV), just to mention a few [18]. Herein, the labelling of the DOTA-functionalized nastorazepide derivative was carried out with indium-111. Indium-111 was preferred among the other diagnostic radionuclides for its established use in pre-clinical and clinical studies, as well as its half-life that is suitable for long-term evaluations. In fact, in our explorative experiment, indium-111 was particularly adapted to shed light on the uptake and pharmacokinetic of the radiotracer for several hours, giving better insights into the structure–metabolism relationship.

2. Results

2.1. Synthesis of the Ligand (IP-001)

The synthesis of the nastorazepide core (**5**) was performed as previously reported [19] with only few modifications, with an overall yield of 23%. The assemblage of the linker started with a condensation reaction between N-Boc-hexanediamine (**1**) and 1,1′-carbonyldiimidazole (CDI)-activated N-Fmoc-N″-succinyl-4,7,10-trioxa-1,13-tridecanediamine (**2**) to yield compound **3**. The selective elimination of the Boc-protecting group was then achieved through a reaction with 30% trifluoroacetic acid (TFA) in dichloromethane (DCM) to obtain the corresponding amino-derivate **4** in a quantitative yield. After the activation of the carboxylic group with CDI, nastorazepide (**5**) was condensed with the linker **4** to yield compound **6**. The fluorenylmethoxycarbonyl (Fmoc)-protecting group was eliminated by a reaction with 50% morpholine in dimethylformamide (DMF) to obtain compound **7**, which was subsequently reacted with the DOTA(tBu)$_3$ ester to give the functionalized nastorazepide-based ligand **8**. The carboxylic functions of the DOTA chelators were finally deprotected with 30% TFA in DCM to yield IP-001 (Scheme 1). The formation of the intermediate compounds was monitored by ^{1}H-NMR (Figures S1–S5 in Supplementary Materials), and the final product of the reaction was also characterized by HRMS (Figures S6 and S7). The overall yield of the process was around 5%. Predictions of lipophilicity and other pharmacokinetic properties for IP-001 and Z-360 are reported in Table 1.

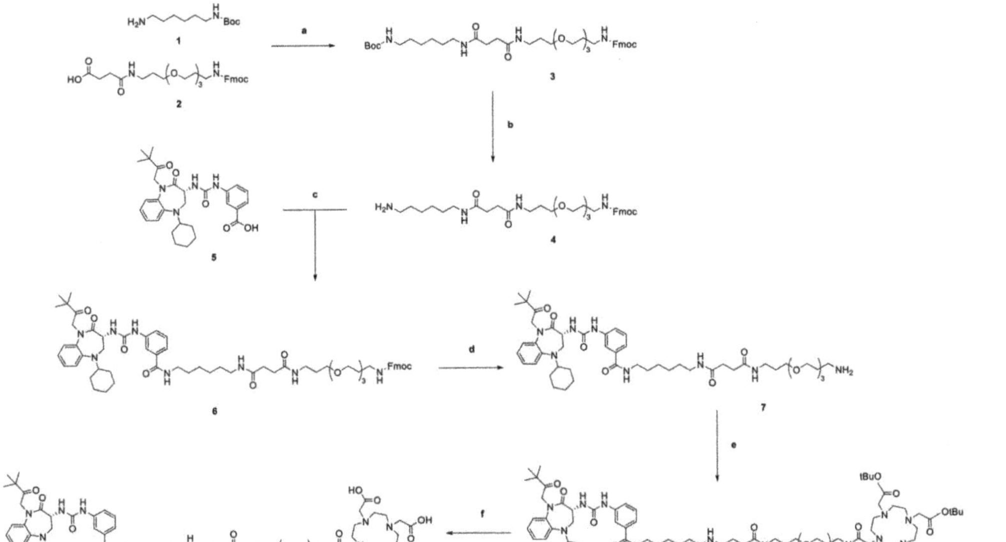

Scheme 1. Reaction conditions: a: **2**, 1,1′-carbonyldiimidazole (CDI), MeCN; triethylamine (TEA), **1**, MeCN. 1 h (reflux); 2 h (RT). b: trifluoroacetic acid (TFA) (30%), dichloromethane (DCM) (0 °C). 1 h (RT). c: **5**, CDI, MeCN; TEA, **4**, MeCN. 1 h (reflux); 1 h (RT) d: Morpholine 50%, **6**, dimethylformamide (DMF) (0 °C). 1.5 h (RT). e: benzotriazol-1-yloxytris(dimethylamino)phosphonium hexafluorophosphate (BOP), TEA, 1,4,7,10-tetraazacyclododecane-1,4,7,10-tetraacetic acid (DOTA)(tBu)$_3$ ester, DMF, **7**. 24 h (RT) f: **8**, TFA (30%), DCM, (0 °C). 16 h (RT).

Table 1. Predicted properties for Z-360 and IP-001. W-LogP: Wildman's logP; TPSA: topological polar surface area; GI: gastrointestinal; BBB: blood brain barrier; ESOL: Estimated aqueous SOLubility; P-gp: P-glycoprotein; CYP: P-cytochrome.

	Z-360	IP-001
W-LogP	2.39	−3.22
TPSA ($\AA^2$; pH = 7.4)	121.88	356.36
logS (ESOL)	−5.49 (moderately soluble)	−1.79 (very soluble)
GI absorption	high	low
BBB permeant	no	no
P-gp substrate	yes	yes
CYP1A2 inhibitor	no	no
CYP2C19 inhibitor	no	no
CYP2C9 inhibitor	yes	no
CYP2D6 inhibitor	yes	no
CYP3A4 inhibitor	yes	no

2.2. Labelling of IP-001 with Indium-111

IP-001 was labelled with a solution of indium-111 chloride in an acetate-buffered environment (pH 4.6). Incorporation was generally higher than 95% when 50 μg of precursor were used, and the total volume of the reaction was maintained at around 1 mL, achieving a molar activity of about 10 MBq/nmol. Impurities were mainly due to unlabeled indium-111 only. Further purification with solid phase extraction (SPE) enabled a radiochemical purity (RCP) higher than 99%, but the procedure was generally avoided for the in vitro and in vivo studies due to the high concentration of EtOH used in the process that may have been detrimental for cells and animals. Quality controls were performed by means of a reverse phase HPLC (R_t: free-$[^{111}In]In^{3+}$ = 1.1 min; $[^{111}In]In$-IP-001 = 6.0 min) and radio-TLC (thin-layer chromatography) (R_f: free-$[^{111}In]In^{3+}$ = 0.0; $[^{111}In]In$-IP-001 = 0.6). Paradigmatic examples of chromatograms are reported in Figures S8 and S9.

2.3. Stability, Serum Proteins Binding, and Lipophilicity Studies

The long-term stability of the radiotracer (up to 120 h) was tested by TLC and RP-HPLC in various media, evaluating the resistance to degradation of the non-peptide-based backbone at different pHs (NaCl 0.9% solution pH 7 and 0.1 M 4-(2-hydroxyethyl)-1-piperazineethanesulfonic acid (HEPES) pH 4) and to enzymatic cleavage (for instance, to proteases in human serum and blood). The kinetic inertness of $[^{111}In]In$–DOTA complexes were also evaluated in presence of a 0.1 M ethylenediaminetetraacetic acid (EDTA) solution or serum proteins as competitors. All these studies revealed a high stability of the radiotracer showing a percentage of intact compound generally higher than 95% after 120 h when incubated with NaCl 0.9%, HEPES and EDTA, and after 48 h when challenged with human serum or blood (HB). Results obtained are summarized in Table 2. Amount of radiotracer bound to serum proteins was computed in the samples incubated with HB after 8, 24 and 48 h. After treatment with a MeCN/H_2O/TFA 50/45/5 $v/v/v$ solution and centrifugation, the radioactivity in the precipitate was found noteworthy being from 12 to 20% of the total. Lipophilicity calculation gave a partition coefficient (LogP) of 0.45.

Table 2. Amount of intact $[^{111}In]In$-IP-001 over time in different media (n = 3; mean ± SD). The percentages were computed by means of radio-TLC (thin-layer chromatography) or RP-HPLC analyses. HEPES: 4-(2-hydroxyethyl)-1-piperazineethanesulfonic acid.

Stability (%)	4 h	8 h	24 h	72 h	120 h
NaCl 0.9%	98 ± 2	98 ± 2	97 ± 2	97 ± 2	97 ± 2
HEPES (0.1 M)	98 ± 2	97 ± 3	97 ± 3	96 ± 1	96 ± 3
EDTA (0.1 M)	98 ± 1	98 ± 1	96 ± 2	96 ± 3	96 ± 3
Human serum	98 ± 2	96 ± 3	95 ± 3	94 ± 2	93 ± 1
Human blood	98 ± 2	95 ± 1	95 ± 2	93 ± 1	93 ± 2

2.4. Selection of CCK-2R-Expressing Human Cancer Cell Lines

To identify the most suitable cell line for the in vitro evaluation of the radiolabeled ligand, CCK-2R protein expression was investigated in human cancer cell lines derived from tumors of different origins (non-small cell lung cancer, skin cancer, prostate cancer, breast cancer, and colon cancer). Western blot analysis revealed that CCK-2R was expressed to a certain extent by all the tested human cancer cell lines, with A549, PC3, and SK-BR-3 exhibiting the highest expression levels (Figure 1, panel A, B). Eventually, the A549 cell line was selected for the following experiments since, upon equal CCK-2R expression, it has the major advantage of ease of cultivation and efficiently forms tumors in nude mice xenograft models. Importantly, to determine whether CCK-2R was also expressed on the A549 cell surface, flow cytometry studies were performed on live cells. Cells were incubated with two different fluorescently-labelled anti-CCK-2R antibodies and, alternatively, with the fluorescent endogenous octapeptide CCK-8 (FAM-CCK-8), which binds CCK-2R with a subnanomolar affinity. Expression was confirmed, to some extent, using both antibodies and the natural ligand (Figure 1, panel C).

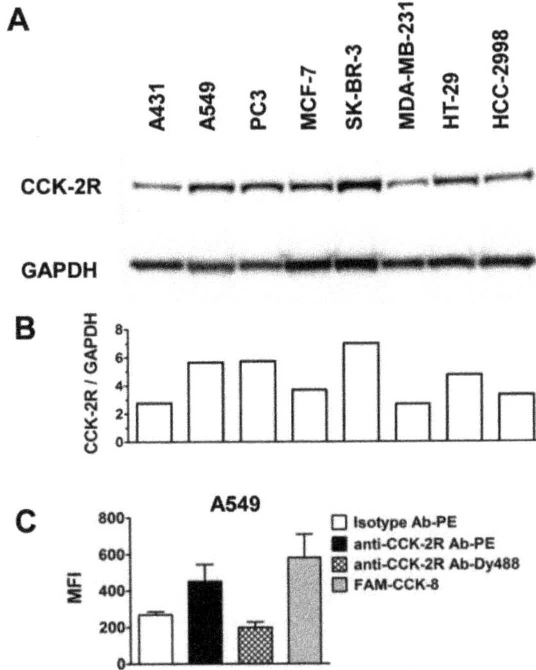

Figure 1. (**A**) Western blot and (**B**) quantification of the total cholecystokinin-2 receptor (CCK-2R) expression on different cell lines. GADPH served as the loading control for total proteins. (**C**) Levels of CCK-2R expression on cell surface by flow cytometry. Dead cells were removed from the analysis using LIVE/DEAD staining. The fluorescence intensity of control cells stained only with the LIVE/DEAD was subtracted. Mean fluorescence intensity plus SEM of three independent experiments is shown.

2.5. [111. In]In-IP-001 Cellular Uptake

To study [^{111}In]In-IP-001 uptake, A549 cells were incubated with 2 MBq of the radio-tracer for different times of up to 24 h. Experiments were performed to gain some insights about the kinetics of the accumulation of the radiolabeled compound in vitro, with the aim to design a further and more exhaustive study in vivo. The maximal accumulation was achieved after 4 h of incubation and was around 5% (expressed as percent of the added radioactivity per million cells). After 8 h, the uptake was only slightly decreased, but it was more than halved after 24 h of incubation. Data elucidated slow kinetics regarding both the uptake and efflux of the radiotracer in vitro. The obtained results are summarized in Figure 2.

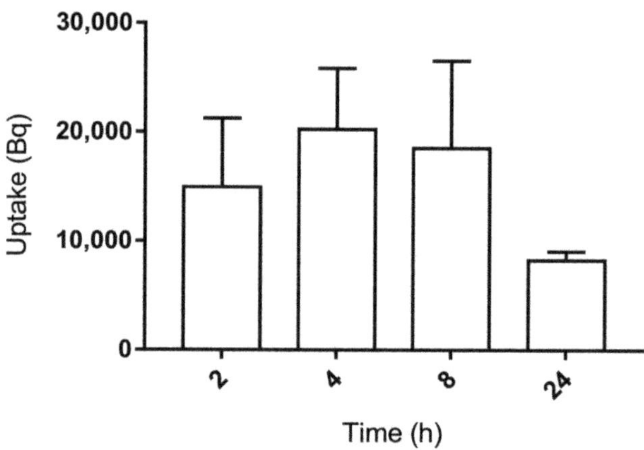

Figure 2. Uptake of [^{111}In]In-IP-001 by A549 cells at different incubation times ($n = 3$; mean $\pm$ SD).

2.6. In Vivo Experiments

The A549 tumor cell line was inoculated, by subcutaneous injection, in the right shoulder of homozygous female BALB/c nude mice. The tumor was allowed to grow under observation for five weeks, and imaging procedures were started when 80% of the animals developed tumors of sufficient size (i.e., >0.5 cm). A first group of mice ($n = 5$) was intravenously injected with [^{111}In]In-IP-001 (2.06 nmol, 7.4 MBq), and tomographic single photon emission tomography (SPECT)/CT images were acquired at 2, 4, 8, and 24 h post injection to monitor the pharmacokinetics of the radiotracer. For blocking experiments, a second group of mice ($n = 5$) was injected with approximately the same amount of [^{111}In]In-IP-001 of the first group with the addition of a 50-fold molar excess of unlabeled IP-001 ligand and imaged at 4 and 24 h. In a parallel independent experiment, three mice belonging to both groups were sacrificed at 4 and 24 h post injection, and their main organs were explanted for measuring the biodistribution of the radiotracer. As per SPECT/CT imaging analysis, the tumor could already be visualized after 4 h, even though [^{111}In]In-IP-001 showed a high accumulation in all the main organs. Blood radioactivity was high, suggesting that a significant amount of the radiotracer was in circulation or bound to blood constituents, such as serum albumin, as found in the in vitro experiments. The kinetics of clearance appeared to be slow with a tumor to background ratio improving over time up to 24 h, when radioactivity accumulation mainly persisted in the tumor, the gastro-intestinal tract, and the kidneys. Paradigmatic SPECT/CT images are reported in Figure 3. In the biodistribution studies, the tumor achieved the maximal uptake (2.36 $\pm$ 0.26% IA/g; $n = 3$) at 4 h post injection despite the greater (blood: 4.21 $\pm$ 0.40% IA/g; liver: 8.25 $\pm$ 2.21%; kidneys: 6.99 $\pm$ 0.97%; spleen: 3.88$\pm$ 0.36%; $n = 3$) or at least comparable (intestines: 2.39 $\pm$ 0.28%; lungs: 2.82 $\pm$ 0.21%; pancreas: 2.03 $\pm$ 0.16%; $n = 3$) accumulation in other organs and tissues, suggesting a low specific biodistribution of the radiotracer at this time. At 24 h post injection, tumor uptake decreased to 1.52 $\pm$ 0.71% IA/g ($n = 3$), but the tumor to organ ratios highly improved with respect to 4 h, as shown in Figure 4. Actually, only the liver and kidneys still exhibited higher uptake as compared to the tumor tissue, with ratios of 0.47 and 0.48, respectively. Conversely, the tumor to muscle (T/M) and tumor to blood (T/B) ratios were highly favorable (15.72 and 10.19, respectively), thus confirming the rapid accumulation and persistence in the tumor visualized by SPECT/CT images, as well as a slow kinetics of clearance that occurred via both renal and hepatobiliary excretion. Residual radioactivity in all the other main organs and, in particular, the intestines and pancreas was low ($0.55 \pm 0.29\%$ IA/g and $0.26 \pm 0.04\%$ IA/g, respectively).

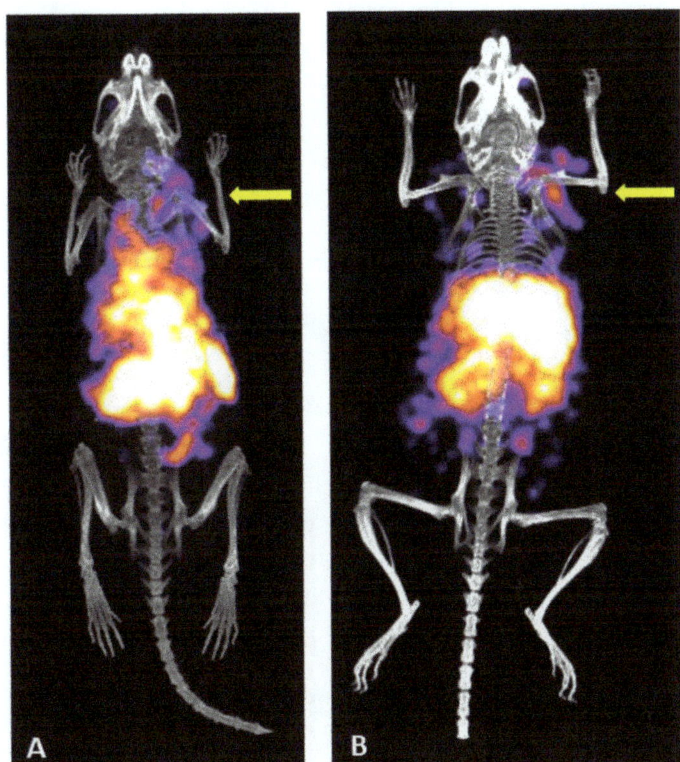

Figure 3. Representative single photon emission tomography (SPECT)/CT scans as maximum intensity projection (MIP) views of an A549 tumor-bearing mouse at 4 (**A**) and 24 h (**B**) post injection of about 7.4 MBq of [^{111}In]In-IP-001. Tumor position in the shoulder is indicated by the arrow.

At 4 h post injection, a blocking experiment reflected a low specific distribution of the radiotracer, as no significant differences were found in the radioactivity measured in the organs and tumors of the normal and blocked groups. After 24 h, a more pronounced gap could be appreciated when, for instance, the tumor uptake was $1.52 \pm 0.71\%$ IA/g ($n = 3$) in the experimental group versus $0.84 \pm 0.33\%$ IA/g ($n = 3$) in the blocked group. However, the differences were still not statistically significant (p = 0.08). Complete comparisons between the two groups at 4 and 24 h are shown in Figure 5. On the other hand, a clear effect of the blocking could be visualized in the SPECT/CT images at 24 h post injection. In Figure 6, for instance, it is shown that a negligible uptake in the tumor was found in the blocked mouse in comparison with the unblocked one.

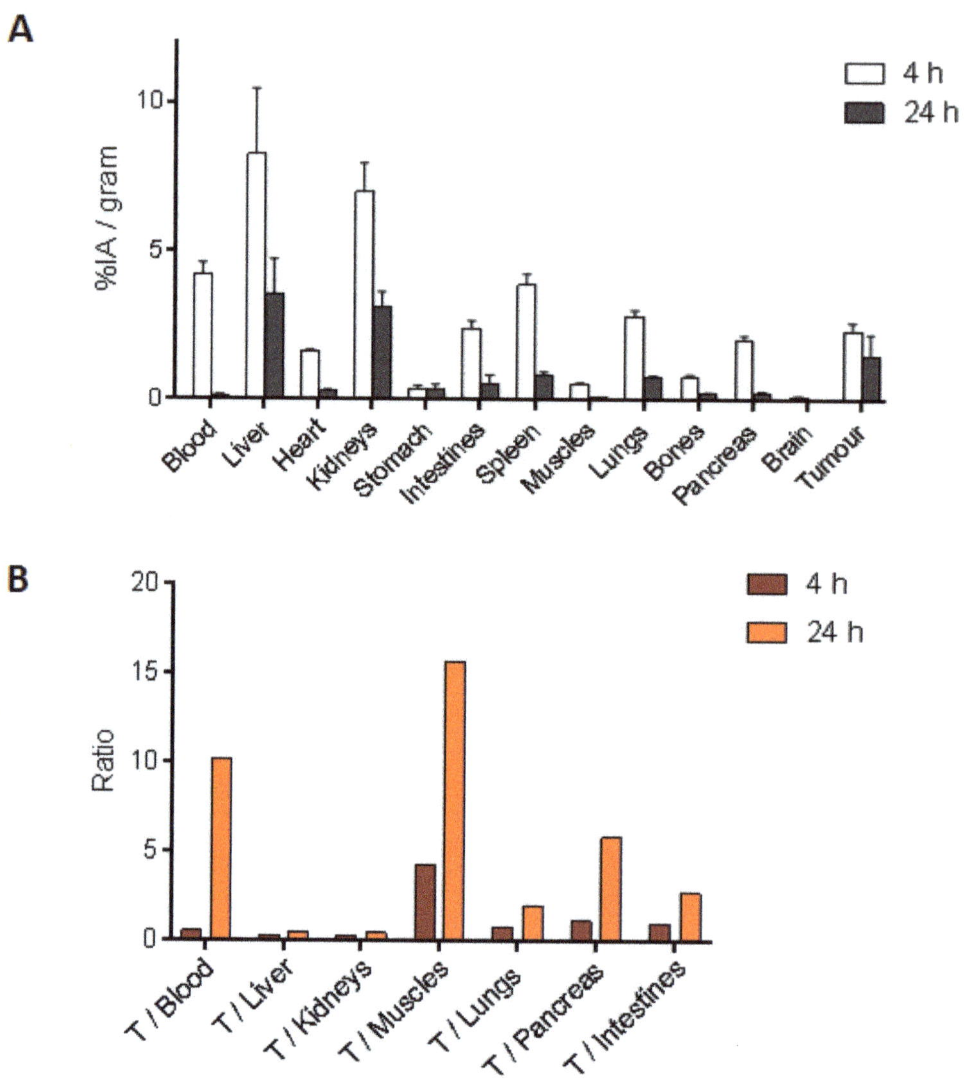

Figure 4. Biodistribution in multiple organs of [^{111}In]In-IP-001 in A549 cancer-bearing nude mice (n = 3; mean ± SD) at 4 and 24 h post injection (**A**); ratios of the tumor uptake to major organs (**B**).

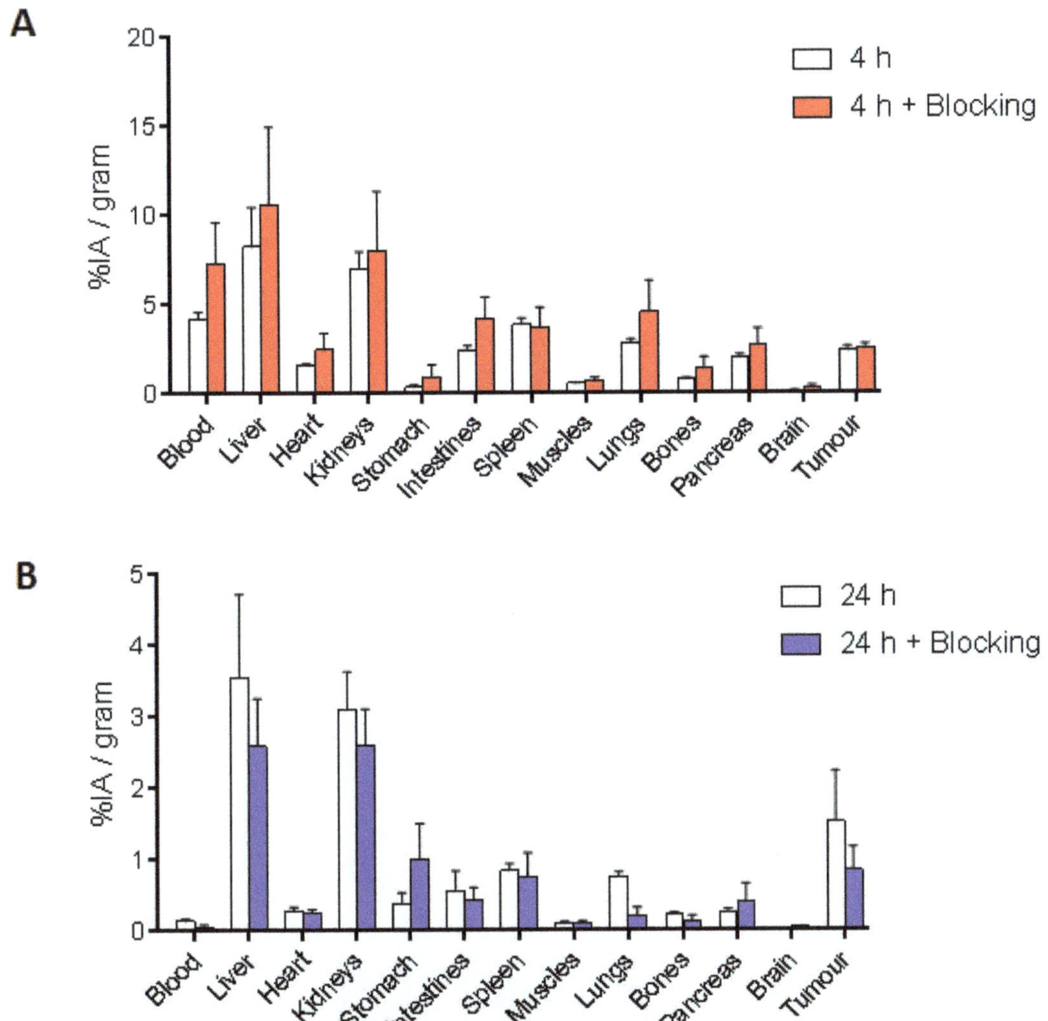

Figure 5. Comparison among the biodistributions in multiple organs of [^{111}In]In-IP-001 and [^{111}In]In-IP-001, as well as unlabeled-IP-001 (blocking) in A549 cancer-bearing nude mice (*n* = 3; mean ± SD) at 4 (**A**) and 24 h (**B**) post injection.

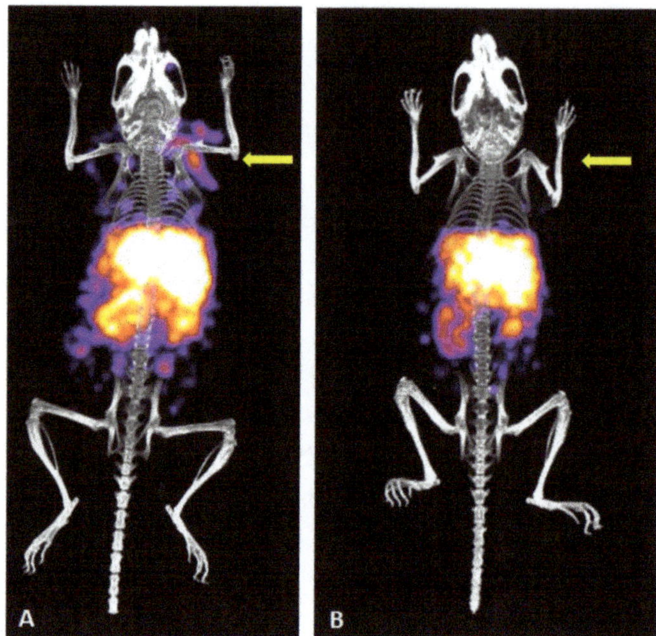

Figure 6. Representative SPECT/CT scans as MIP views of two A549 tumor-bearing mice injected with [^{111}In]-IP-001 (**A**), as well as [^{111}In]-IP-001 and 50-fold molar excess of unlabeled IP-001 (blocking) (**B**) at 24 h post injection. Mice developing tumors approximately of the same size (about 112 mm^3) are shown. Tumor position in the shoulder is indicated by the arrow.

3. Discussion

In the present study, a nastorazepide derivative functionalized with a DOTA chelator was synthesized to provide a potential precursor for the diagnosis and systemic therapy of CCK-2R-expressing tumors. In fact, the presence of a chelator moiety allowed for the coordination of several metal radionuclides, thus triggering the possibility of a radio-theragnostic approach. Nastorazepide is a well-known 1,5-benzodiazepine derivative that acts as an antagonist ligand for CCK-2R, and it is currently under development as a therapeutic drug for pancreatic cancer, gastroesophageal reflux disease, and peptic ulcers. The state of the art in the field of imaging and treatment of CCK-2R-expressing tumors with radiopharmaceuticals is mainly focused on the labelling of the natural ligands of this receptor (i.e., CCK and gastrin) or those derived thereof. However, issues regarding in vivo stability, receptors affinity, and high kidney retention have hindered this approach so far. An alternative and less explored pathway is the labelling of small molecules such as 1,5-benzodiazepine derivatives, although, to the best of our knowledge, this has only been pursued with the diagnostic radionuclide technetium-99 and, despite promising preliminary results, only minor efforts have been devoted to this approach [16,17].

The precursor synthesized here is composed by a nastorazepide moiety connected to a DOTA chelator with a short alkyl chain followed by a PEGylated portion introduced to increase the hydrophilicity of the whole molecule. The linker is an important part of the molecules because it ensures a sufficient separation between the pharmacophore and the radiometal binding moiety, which also influences the affinity. Differently from previously reported works, a linker completely not based on an amino acid sequence was assembled to avoid any possible site of enzymatic cleavage. While a nastorazepide core was synthesized in a six-step reaction pathway, as reported in the literature, the linker was obtained by a condensation reaction between *N*-Boc-1,6-hexanediamine (**1**) and CDI-activated N^{19}-Fmoc-4-oxo-9,12,15-trioxa-5,19-diazanonadecanoic acid (**2**). Consequently, the Boc protecting group was eliminated, and the linker was again condensed to the CDI-activated nastorazepide core. Finally, the Fmoc moiety was cleaved, and the chelator was inserted. The CDI-activation of the succinyl moiety of compound **2** and, more generally, the activation of the carboxylic groups with CDI in the presence of an Fmoc protecting group (such as reaction steps a and c in Scheme 1) were the trickiest steps of the pathways, yielding 46% and 32%, respectively. The moderate yields were supposed to derive from two side-reactions due to the well-known mechanism of action of CDI that leads to the release of an imidazole group [20]. This group, imparting a basic character to the reaction mixture, could have been responsible for a first side-reaction with compounds **2** and **4** that led to the elimination of the Fmoc-protecting group with the consequent formation of a free amine. The suggested mechanism of this reaction occurring for the CDI-activated compound **2** is reported in Figure 7. Moreover, a second imidazole-mediated side-reaction was supposed to involve compound **2** during the first condensation step of the pathway, thus yielding the cyclization of the succinyl group instead of its activation, as reported in Figure 8. The formation of this compound was confirmed by ^{1}H-NMR (Figure S10). To minimize the formation of these by-products, the original work-up or the reactions was modified by adding a 1 M acetic acid solution to the reaction mixtures to fix the pH to around 5 (thus protonating the imidazole and reducing its catalytic behavior) before the evaporation step.

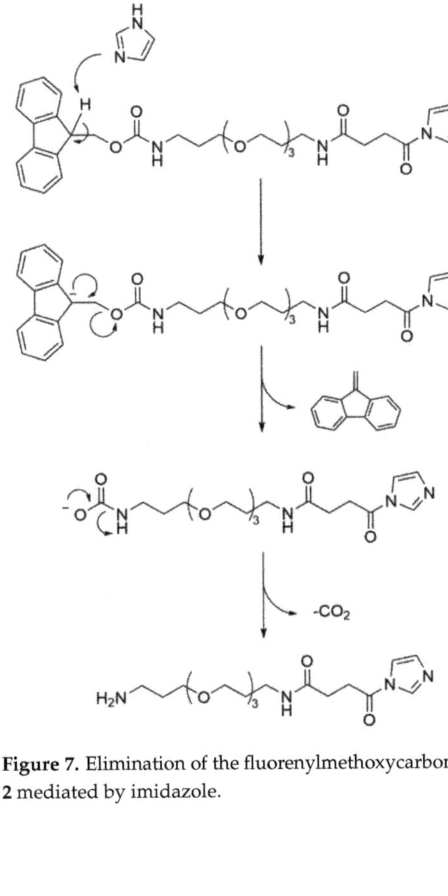

Figure 7. Elimination of the fluorenylmethoxycarbonyl (Fmoc) group from CDI-activated compound **2** mediated by imidazole.

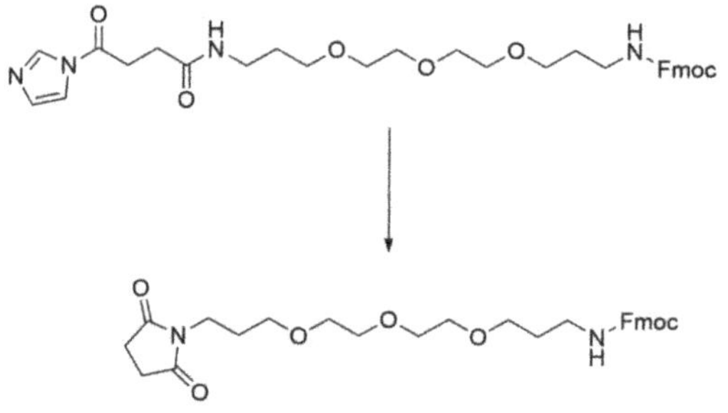

Figure 8. By-product obtained from the cyclization of the CDI-activated succinyl group of compound **2**.

Another step of the synthesis worth attention is the passage from compound **6** to compound **7**, i.e., the elimination of the Fmoc group after the condensation of the linker with Z-360. Conventional elimination conditions suggest the use of a 20% piperidine/DMF solution [21], but, in our case, a very low yield was obtained, probably due to the instability of the linker in a basic environment, as reported for similar chemical structures [22]. For this reason, the use of a milder base, such as a 50% morpholine/DMF solution, was preferred to obtain almost quantitative results [23]. Finally, in an alternative synthetic approach, CDI-activated Z-360 (**5**) was firstly reacted with *N*-Boc-1,6-hexanediamine (**1**), and, after the elimination of the Boc group, the obtained intermediate was further reacted with CDI-activated N^{19}-Fmoc-4-oxo-9,12,15-trioxa-5,19-diaza-onadecanoic acid (**2**). However, this method did not show significant advantages with respect to the former one, exhibiting the same side-reactions described above and, moreover, implying an elevate loss of Z-360 due to the low yield of the first reaction. For these reasons, it was not pursued further. The overall yield of the process to obtain IP-001 was around 5% (22% starting from compound **5**), which was in line with a multi-step synthetic pathway, with the yield-limiting step being the synthesis of compound **5**. In view of further optimization, attention should be paid to the synthesis optimization of the latter.

The radiolabeling of IP-001 with indium-111 was performed in the standard conditions involving the labelling of a DOTA-conjugated small molecule with trivalent radiometals [24], obtaining an incorporation higher than 95% in a quite straightforward way. Stability in all tested conditions, and particularly in human serum and blood, was high and generally comparable at short incubation times (<24 h) to the best peptide-based candidates for the treatment of CCK-2R-expressing tumors, but it was much superior at longer times (intact amount of [^{111}In]In-IP-001 >93% after 120 h in HB). On the other hand, the amount of [^{111}In]In-IP-001 bound to the serum protein was almost comparable to other indium-111-labelled minigastrin analogues [13]. A direct comparison with other non-peptide-based radiotracers was not possible since, to the best of our knowledge, the high stability in HB at a late time found here was the first reported experimental confirmation of previous theoretic assumptions for this kind of molecular structure.

CCK-2R expression has been reported in cancers of different origins by immunohistochemistry on tumor specimens [25]. However, CCK-2R expression in human cancer cell lines has largely not been investigated. CCK-2R expression at the mRNA level has been reported in human PC3 prostate carcinoma, U373 glioma, U2OS osteosarcoma, Colo205 colon carcinoma [26], HepG2 hepatoma [27], gastric and colorectal cell lines, and MCF-7 breast cancer and Molt4 lymphoblastic leukemia cell lines [28]. To date, studies on CCK-2R-targeting probes for cancer imaging have only been performed by exploiting transfected cell lines (the human A431 cancer cell line, HEK293 human embryonic kidney cells, NIH 3T3 mouse embryonic fibroblast cells, and Chinese hamster ovary cells) or rat pancreatic tumor cells [16,24,29–33]. However, the demonstration of CCK-2R overexpression by transfected versus parental cells has generally been lacking.

Herein, we demonstrated that the CCK-2R protein is indeed expressed by various human cancer cell lines, and we decided to perform experiments in vitro and in vivo on A549 cells as representative (Figure 1). We acknowledge the fact that our results are only partially comparable to other studies reported thus far, but we would like to emphasize that non-transfected tumor cells represent a closer model to the real physio-pathological condition than transfected cancer cell lines. With this in mind, we ascertained that [^{111}In]In-IP-001 cell uptake in vitro was two-fold lower than what reported previously in AR4-2J rat pancreatic tumor cells for indium-111-labelled minigastrin derivatives [24] and four-fold lower than the total amount of technectium-99m nastorazepide-based derivatives accumulated by CCK-2R-transfected HEK293 cells [17].

In BALB/c nude mice developing A549 cell-derived tumors, [^{111}In]In-IP-001 exhibited a quite rapid absorption rate in all the main organs (liver: 8.25 ± 2.21%; kidneys: 6.99 ± 0.97%; spleen: 3.88 ± 0.36%) and the tumor (2.36 ± 0.26% IA/g) within 4 h. A high amount of injected radioactivity was also found in the circulation, suggesting a strong interaction with serum albumin (Figures 3 and 4). Clearance was slow and performed by both the renal and hepatobiliary pathways, probably due to the relatively high lipophilicity of the radiotracer. After 24 h, the kidneys and liver (3.11 ± 0.52% IA/g and 3.55 ± 1.17% IA/g, respectively) remained the organs with the highest uptake, followed by the tumor (1.52 ± 0.71% IA/g), which was clearly delineated in the CT/SPECT images (Figure 3). Moreover, at 4 h post-injection, a certain level of radioactivity was also detected in the intestines, lungs, and pancreas (no statistically significant difference with the tumor), with a decrease of the signal of more than 73% over 24 h, indicating a transient accumulation. On the contrary, the tumor tissue still retained 65% of the accumulated [^{111}In]In-IP-001 at 24 h post-injection. The lungs, pancreas, and intestines exhibited a more rapid wash-out with tumor/organ ratios of 2.02, 5.85, and 2.76, respectively (Figure 4). The slow kinetics of radiotracers based on the nastorazepide core was already reported, although a significantly more promising biodistribution was obtained in mice bearing tumors derived from transfected HEK293 cells [16,17], in which kidneys were the only organ with a comparable uptake with respect to the tumor at delayed times. As reported in Table 1, the addition of a PEGylated spacer and a strongly hydrophilic chelator like DOTA to the hydrophobic Z-360 core imparted a quite hydrophilic nature to IP-001. Indeed, at pH = 7.2, the chelator was completely dissociated, and the precursor was in the anionic form (i.e., IP-001^{3-}). However, it is reasonable that the formation of a neutral complex with [^{111}In]In^{3+} increased the lipophilicity of the final radiotracer (experimental value of 0.45), thus giving reason of its unspecific accumulation in the liver and spleen. Hence, further chemical optimizations are required, with a particular focus on the hydrophilicity of the linker, to obtain a novel ligand with a logP remarkably lower than −3.22. In general, [^{111}In]In-IP-001 biodistribution reflected the behavior of a radiotracer with a moderate receptor affinity, as attested by the blocking studies performed with an excess of unlabeled ligand in which, despite the results showing a trend of lower [^{111}In]In-IP-001 accumulation in the blocking setting (45% lower uptake in the tumor tissue), no statistically significant differences between the two groups of injected mice were found. Definitely, although a high stability was achieved by avoiding the use of peptides or peptide-mimetic moieties on the linker, affinity and hydrophilicity could likely be further tailored and enhanced by acting on this portion of the structure.

4. Materials and Methods

4.1. General Procedures and Chemicals

All chemicals (Acros, Aldrich, IrisBiotech suppliers, Marktredwitz, Germany) were reagent grade and used without further purification unless otherwise specified. Solvents (Carlo Erba, Milan, Italy and Lab Scan, Bangkok, Thailand) were obtained as analytical grade and degassed by ultra-sonication for 15–20 min before use. Deuterated solvents (SG Isotec, Pančevo, Republic of Serbia) used for NMR analysis exhibited an isotopic purity of 99.5%. pH measurements were conducted using a calibrated pH-meter (Mettler-Toledo, Biosigma, Venice, Italy). All reaction intermediates were purified as specified in the following procedures and their purity was $\geq$95% by NMR and/or HRMS. Indium-111 chloride solutions (370 MBq in 0.5 mL) were purchased from Curium (Milan, Italy).

4.2. Instrumentations

NMR spectra were recorded by means of a Bruker 400-AMX spectrometers using tetramethylsilane (TMS) as the internal standard. For each sample, 1–2 mg were weighed and diluted up to 0.6 mL with the deuterated solvent described in the specific synthetic paragraph into a 5 mm NMR tube. A 90° pulse was calibrated for each sample, and standard NMR parameters were used.

Mass spectrometry was performed through the direct injection of the sample on an Applied Biosystem Mariner System 5220 equipped with a MALDI TOF/TOF 400 Plus AB Sciex, (Framingham, MA, USA) analyzer.

The purification of the precursor (IP-001) was carried out by an Agilent 1290 Infinity II preparative HPLC equipped with a binary pump and UV–Visible detector. A semi-preparative Zobrax-Eclipse (Agilent) (C18; 5μm; 250 × 21.2 mm) column was used. Eluents were H_2O + 0.1% TFA (A) and ACN (B) at flow = 17mL/min with the following gradient: minutes 0–2 A: 90% B: 10%; minutes 2–23 B: 10% -> 76%; minutes 23–27 B: 76% -> 100%; and minutes 27–28 B: 100% -> 10%. The detection of reaction products was performed, evaluating absorbance at λ = 225 nm. During the purification process, IP-001 showed a retention time of 19 min.

HPLC analyses on indium-111-labelled IP-001 were performed on an S1125 HPLC pump system, and the column used was an Acclaim 120 C18, 3 × 150 mm, 3 μm 120 Å pore size; flow: 0.6 mL/min; gradient: minutes 0–2 A: 70% B: 30%; minutes 2–12 B: 30% -> 95%; and minutes 12–15 B: 95% -> 30%, where A: H_2O (0.1% TFA) and B: MeCN. Detection was carried out with an S3245 UV–Vis and an S3700 gamma detector system.

TLC plates used for the assessment of the preparative reactions were silica gel 60 F245 (0.2 mm, Merck) with the eluents mentioned for the specific product in the following paragraphs. Column chromatography was carried out with silica gel 60 (0.063–0.100 mm, Merck), and elution was performed using the eluent described for the specific preparations. The radioactive incorporation and stability of [^{111}In]In-IP-001 were assessed on ITLC-SG plates, developed in a 1 M ammonium acetate/MeOH 1:1 *v/v* solution, by using a flatbed-imaging scanner (Cyclone, Perkin Elmer).

4.3. Synthesis of N^1-Boc-N^{25}-Fmoc-8,11-dioxo-16,19,22-trioxa-7,12-diazapentacosane-1,25-diamine (3)

CDI (0.2 g, 1.5 mmol) was added to a 32 mL MeCN solution of N^{19}-Fmoc-4-oxo-9,12,15-trioxa-5,19-diazanonadecanoic acid (**2**; 0.5 g, 1.0 mmol), and the mixture was stirred and heated to reflux for 1 h (TLC: $CHCl_3$/MeOH 9:1). After cooling, triethylamine (TEA) (0.2 g, 1.5 mmol) was added to the mixture, followed, after 10 min, by a solution of N-Boc-1,6-hexanediamine (**1**; 0.3 g, 1.2 mmol) in MeCN (20 mL; dropwise addition). The solution was stirred at room temperature for 2 h (TLC: $CHCl_3$/MeOH 9:1). The mixture was than acidified to pH 5 with a 1 M CH_3COOH solution in MeCN and concentrated to dryness. The residue was dissolved in 40 mL of ethyl acetate (EtOAc) and washed with a saturated NH_4Cl solution (20 mL × 2). The organic phase was then concentrated to dryness, and the product was purified by column chromatography (eluent: $CHCl_3$/MeOH 95:5) to yield **3** (0.3 g, 0.46 mmol, 46% yield, chemical structure in Figure 9).

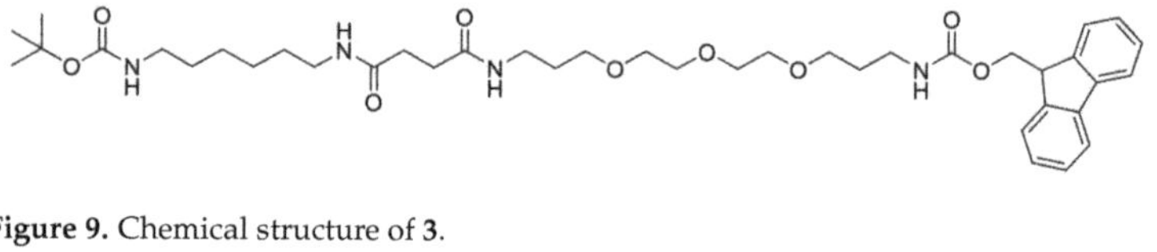

Figure 9. Chemical structure of **3**.

^{1}H-NMR (400 MHz, CDCl$_3$-d)δ ppm 7.75 (d, *J* = 7.6, 2H); 7.60 (d, *J* = 7.6, 2H); 7.39 (t, *J* = 7.6, 2H); 7.30 (t, *J* = 7.6, 2H); 6.66 (broad s, 1H); 6.40 (broad s, 1H); 5.67–5.59 (m, 1H); 4.62 (broad s, 1H); 4.45–4.35 (m, 2H); 4.24–4.17 (m, 1H); 3.65–3.49 (m, 12H); 3.35–3.25 (m, 4H); 3.22–3.14 (m, 2H); 3.12–3.02 (m, 2H); 2.46 (s, 4H); 1.83–1.68 (m, 4H); 1.50–1.44 (m, 4H); 1.42 (s, 9H); 1.34–1.26 (m, 4H).

4.4. Synthesis of N^{25}-Fmoc-8,11-dioxo-16,19,22-trioxa-7,12-diazapentacosane-1,25-diamine (4)

A solution of **3** (0.3 g, 0.5 mmol) in DCM (3.5 mL) was cooled to 0 °C by an ice bath for 20 min, and then TFA (1 mL) was added dropwise. The mixture was stirred for 1 h at room temperature (TLC: CHCl$_3$/MeOH 9:1) and then evaporated to dryness to obtain **4** (0.2 g, 0.5 mmol, quantitative yield, chemical structure in Figure 10).

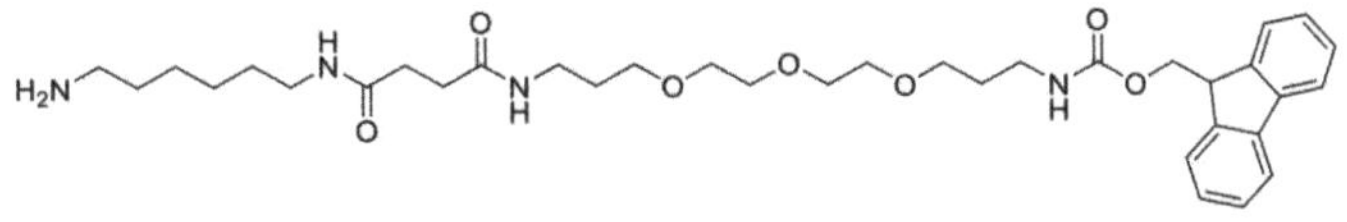

Figure 10. Chemical structure of **4**.

^{1}H-NMR (400 MHz, CDCl$_3$-d)δ ppm 7.76 (d, *J* = 7.6, 2H); 7.58 (d, *J* = 7.6, 2H); 7.40 (t, *J* = 7.6, 2H); 7.31 (t, *J* = 7.6, 2H); 4.56–4.36 (m, 2H); 4.26–4.16 (m, 1H); 3.66–3.44 (m, 12H); 3.36–3.16 (m, 6H); 3.08–2.94 (m, 2H); 2.63 (s, 4H); 1.82–1.60 (m, 6H); 1.54–1.24 (m, 6H).

4.5. Synthesis of N^1-{3-[3-(1-pivaloylmethyl-5-cyclohexyl-2,3,4,5-tetrahydro-1H-benzo[b][1,4]diazepin-2-one-3-yl)]ureido}benzoyl-N^{25}-Fmoc-8,11-dioxo-16,19,22-trioxa-7,12-diazapentacosane-1,25-diamine (6)

CDI (68.0 mg, 0.4 mmol) was added to a 10 mL MeCN suspension of Z-360 (**5**; 0.2 g, 0.3 mmol), and the mixture was stirred and refluxed for 1 h (TLC: CHCl$_3$/MeOH 95:5). After cooling, TEA (42 mg, 0.4 mmol) was added dropwise to the mixture, followed, after 10 min, by a solution of **4** (0.2 g, 0.3 mmol) in MeCN (10 mL; dropwise addition). The solution was stirred at room temperature for 1 h (TLC: CHCl$_3$/MeOH 9:1). The mixture was than acidified to pH 5 with a 1 M CH$_3$COOH solution in MeCN and concentrated to dryness. The residue was dissolved in 40 mL of EtOAc and washed with a saturated NH$_4$Cl solution (20 mL × 2). The organic phase was then concentrated to dryness, and the product was purified by column chromatography (eluent: EtOAc/EtOH 95:5, followed byCHCl$_3$/MeOH 95:5) to yield **6** (0.1 g, 0.1 mmol, 32% yield, chemical structure in Figure 11).

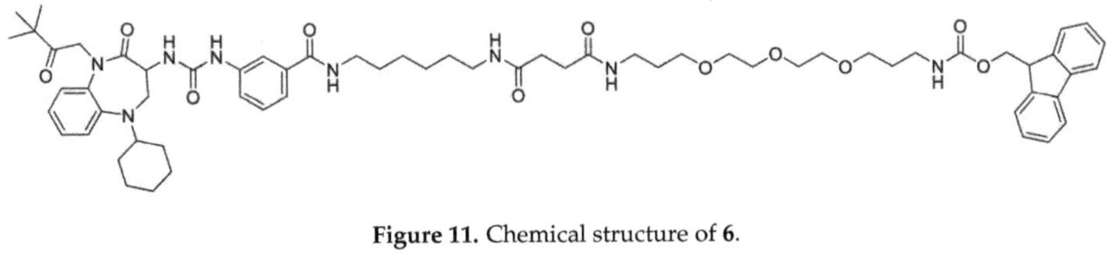

Figure 11. Chemical structure of **6**.

HRMS *m/z* = 1143.6494 [M+H$^+$] (calculated); 1143.6495 [M+H$^+$] (found). ^{1}H-NMR (400 MHz, CDCl$_3$-d)δ ppm 7.75 (d, *J* = 7.6, 2H); 7.67–7.52 (m, 3H); 7.38 (t, *J* = 7.6, 2H); 7.30 (t, *J* = 7.6, 2H); 7.25–7.19 (m, 2H); 7.17–7.01 (m, 4H); 6.98–6.93 (m, 1H); 5.24 (d, *J* = 17.9, 1H); 4.80–4.78 (m, 1H); 4.46–4.32 (m, 2H); 4.24–4.16 (m, 1H); 4.11 (d, *J* = 17.9, 1H); 3.66–3.10 (m, 23H); 2.54–2.40 (m, 4H); 2.10–2.00 (m, 2H); 1.88–1.26 (m, 20H); 1.15 (s, 9H).

4.6. Synthesis of N¹-{3-[3-(1-pivaloylmethyl-5-cyclohexyl-2,3,4,5-tetrahydro-1H-benzo[b][1,4] diazepin-2-one-3-yl)]ureido}benzoyl-8,11-dioxo-16,19,22-trioxa-7,12-diazapentacosane-1,25-diamine (7)

A DMF solution (2 mL) of **6** (0.1 g, 0.1 mmol) was cooled and stirred in an ice bath for 10 min. Then, under constant cooling and stirring, morpholine (2 mL) was added, and the mixture was reacted for 1.5 h at room temperature. The completion of reaction was assessed by TLC (CHCl₃/MeOH 9:1). The mixture was brought to dryness through an azeotropic distillation with toluene, and the residue was purified by column chromatography (eluent: CHCl₃/MeOH 9:1) to obtain **7** (0.1 g, 97 µmol, 97% yield, chemical structure in Figure 12).

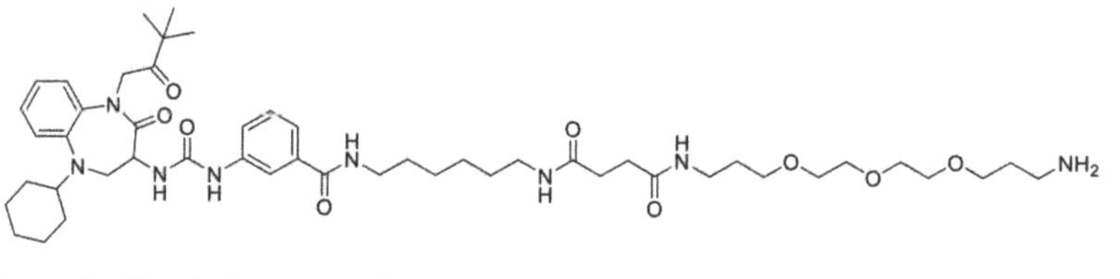

Figure 12. Chemical structure of **7**.

HRMS *m/z* = 921.5814 [M+H⁺] (calculated); 921.5806 [M+H⁺] (found). ¹H-NMR (400 MHz, MeOD-d₄)δ ppm 7.76–7.73 (m, 1H); 7.48–7.43 (m, 1H); 7.40–7.35 (m, 1H); 7.34–7.25 (m, 3H); 7.11–7.03 (m, 2H); 5.19 (d, *J* = 17.9, 1H); 4.61–4.53 (d, *J* = 7.0, 1H); 4.40 (d, *J* = 17.9, 1H); 3.65–3.34 (m, 18H); 3.25–3.18 (m, 3H); 3.10–3.05 (m, 2H); 2.45 (s, 4H); 2.12–2.02 (m, 1H); 1.95–1.26 (m, 21H); 1.23 (s, 9H).

4.7. Synthesis of N¹-{3-[3-(1-pivaloylmethyl-5-cyclohexyl-2,3,4,5-tetrahydro-1H-benzo[b] [1,4]diazepin-2-one-3-yl)]ureido}benzoyl-N²⁵-{2-[4,7,10-tri(tertbutoxycarbonylmethyl)-1,4,7,10-tetrazacyclododecan-1-yl]acetyl}-8,11-dioxo-16,19,22-trioxa-7,12-diazapentacosane-1,25-diamine (8)

BOP (35.0 mg, 80 µmol) and TEA (16.0 mg, 0.2 mmol) were added to a 2 mL DMF solution of DOTA(tBu)₃ ester (40.0 mg, 71.0 µmol). The mixture was stirred for 10 min, and then a 2 mL DMF solution of **7** (74.0 mg, 80.0 µmol) was added dropwise. The solution was reacted under stirring for 24 h at room temperature (TLC: CHCl₃/MeOH 9:1). Then, it was concentrated to dryness, re-dissolved in 10 mL EtOAc, and finally washed with a saturated water solution of NH₄Cl (5 mL × 2). The organic phase was concentrated to dryness, and the obtained residue was purified by column chromatography (eluent: CHCl₃/MeOH 9:1) to yield **8** (30.7 mg, 21.2 mmol, 30% yield, chemical structure in Figure 13).

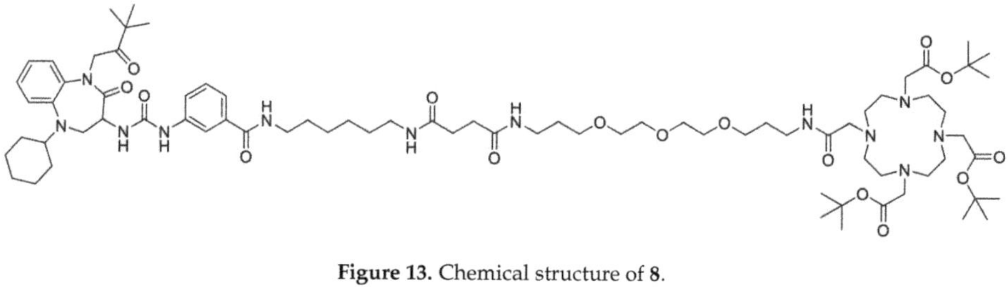

Figure 13. Chemical structure of **8**.

HRMS *m/z* = 1475.9493 [M+H⁺] (calculated); 1475.9584 [M+H⁺] (found). ¹H-NMR (400 MHz, CDCl₃.*d*)δ ppm 8.07 (broad s, 1H); 7.68–7.56 (m, 2H); 7.40–7.28 (m, 2H); 7.24–7.18 (m, 3H); 6.84–6.78 (m, 1H); 6.60–6.52 (m, 1H); 5.25 (d, *J*=17.9, 1H); 4.78–4.66 (m, 1H); 4.13 (d,

J = 17.9, 1H); 3.64–3.12 (m, 35H); 2.68–2.42 (m, 16H); 2.10–2.00 (m, 2H); 1.88–1.50 (m, 16H); 1.46–1.42 (m, 27H); 1.37–1.33 (m, 4H); 1.17 (s, 9H).

4.8. Synthesis of N^1-{3-[3-(1-pivaloylmethyl-5-cyclohexyl-2,3,4,5-tetrahydro-1H-benzo[b] [1,4]diazepin-2-one-3-yl)]ureido}benzoyl-N^{25}-{2-[4,7,10-tri(carboxymethyl)-1,4,7,10-tetrazacyc-lododecan-1-yl]acetyl}-8,11-dioxo-16,19,22-trioxa-7,12-diazapentacosane-1,25-diamine (IP-001)

A solution of **8** (30.7 mg, 21.0 μmol) in 2 mL of DCM was pre-cooled to 0 °C for 20 min in an ice bath. Then, TFA (0.6 mL) was added dropwise, and the mixture was stirred overnight at room temperature. After the assessment of the completion of the reaction by TLC (CHCl$_3$/MeOH 9:1), the solution was evaporated to dryness, and the product was purified by preparative HPLC. IP-001 was obtained in a quantitative yield (27 mg, 21.0 μmol, purity > 95%, chemical structure in Figure 14). Lipophilicity and other pharmacokinetic properties for the IP-001 and Z-360 were predicted by using the SwissADME web tool, as reported in [34].

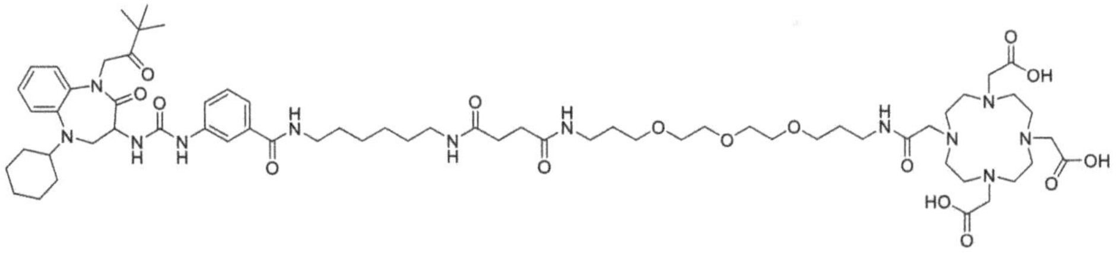

Figure 14. Chemical structure of **IP-001**.

HRMS m/z = 1307.7615 [M+H$^+$] (calculated); 1307.7650 [M+H$^+$] (found). ^{1}H-NMR (400 MHz, MeOD-d_4)δ ppm 7.88 (d, J = 8.1, 1H); 7.74 (d, J = 8.1, 1H); 7.60–7.53 (m, 1H); 7.52–7.46 (m, 1H); 7.44–7.25 (m, 3H); 7.15–7.02 (m, 1H); 5.20 (d, J = 17.9, 1H); 4.60–4.46 (m, 1H); 4.39 (d, J = 17.9, 1H); 3.65–3.45 (m, 14H); 3.40–3.15 (m, 16H); 2.46 (s, 4H); 2.12–2.03 (m, 1H); 1.80–160 (m, 4H); 1.64–1.56 (m, 4H); 1.54–1.44 (m, 4H); 1.40–1.30 (m, 10H); 1.24 (s, 9H).

4.9. Labelling of IP-001 with Indium-111

A 0.8 mL solution of a 0.05 M HCl and a 60 μL 1.5 M sodium acetate solution (pH 4.66) were placed in a 2 mL microtube. To this mixture, 50 μL of a 1 mg/mL IP-001 water solution was added, followed by 0.2 mL of an indium-111 chloride (60–130 MBq) solution. The reaction was heated at 95 °C for 15 min, and aliquots were analyzed by HPLC and radio-TLC to determine the labelling efficiency. When necessary, the labelling mixture was purified by SPE using a light C-18 cartridge. The cartridge was firstly conditioned with 3 mL of EtOH followed by 3 mL of water. Afterwards, sample was loaded and the cartridge was washed with a hydro-alcoholic solution (1.5 mL, water/EtOH 80:20) to remove unlabeled [^{111}In]In^{3+}. The radiotracer ^{111}[In]In-IP-001 was then eluted with EtOH (0.4 mL) and diluted with water (2 mL). All experiments were performed in triplicate.

4.10. Stability, Serum Proteins Binding and Lipophilicity Studies

The stability of [^{111}In]In-IP-001 solutions (0.5 mL, 10 nmol, 18 MBq) was assessed by means of radio-TLC or RP-HPLC at various temperatures (4, 24, and 37 °C) and in the presence of different media up to 120 h after preparation. The studies were performed by incubating the radiotracer with (i) a 0.9% NaCl solution at a 9:1 ratio, (ii) a 0.1 M HEPES solution at a 9:1 ratio, (iii) a 0.1 M EDTA solution at 9:1 and 1:1 ratios, (iv) human serum (HS) at a 9:1 ratio, and (v) HB at a 1:1 ratio. After TLC analysis, samples incubated with HB were centrifuged at 3000 rpm for 10 min to precipitate blood cells, and 200 μL of a MeCN/H$_2$O/TFA 50/45/5 $v/v/v$ solution was added to 400 μL of the supernatant. After another centrifugation under the same conditions, the supernatant was discarded and the residual radioactivity due to the proteins bound fraction was measured in a dose calibrator.

Lipophilicity calculation were performed in octanol–water with the shaking flask method, as already described elsewhere [13]. All experiments were performed in triplicate.

4.11. Selection of the Human Cell Line
4.11.1. Cell Cultures

A series of human cancer cell lines (A549, A431, PC3, MCF-7, SK-BR-3, MDA-MB-23, HT-29 and HCC-2998) were screened regarding the expression of CCK-2R. The A549 non-small cell lung cancer and A431 skin cancer cell lines were purchased from the European Collection of Authenticated Cell Cultures (ECACC 86012804 and ECACC 85090402). The PC3 prostate cancer cell line; the MCF-7, SK-BR-3, and MDA-MB-231 breast cancer cell lines; and the HT-29 and HCC-2998 colon cancer cell lines were kindly provided by Dr. Alessia Ciarrocchi (Laboratory of Translational Research AUSL-IRCCS, Reggio Emilia, Italy). A549, A431, PC3, MCF-7, and HCC-2998 were cultured in RPMI-1640 added of 10% heat-inactivated FBS. SK-BR-3, MDA-MB-231, and HT-29 were cultured in DMEM with high glucose added to 10% heat-inactivated FBS. Media contained 100 U/mL penicillin-streptomycin and glutaMAX. Cell culture reagents were purchased from Thermo Fisher (Milan, Italy). Cell lines were routinely maintained at 37 °C with 5% CO_2.

4.11.2. Screening of the Cell Lines by Western Blot

To avoid the use of trypsin, cells were lysed in situ with a 500 μL RIPA lysis buffer supplemented with protease-inhibitors (RIPA lysis buffer system, sc-24948, Santa Cruz Biotechnologies) at 4 °C for 30 min. Lysates were clarified by centrifugation at 12,000× *g* at 4 °C for 15 min, and then protein concentrations were measured using the DC Protein Assay (Bio-Rad). About 40 μg of proteins were separated using Bolt Bis-Tris Plus precast 8% polyacrylamide gels with an MOPS SDS running buffer (Thermo Fisher) then blotted onto PVDF membranes. After blocking with TBS containing 0.1% Tween 20 and 5% BSA for 2 h at room temperature, and membranes were incubated overnight at 4 °C with (anti-human CCK-2R) mouse monoclonal antibodies conjugated with phycoerythrin (clone E-3, sc-166690, Santa Cruz Biotechnology) diluted 1:400 in TBS containing 0.1% Tween 20 and 5% BSA. Fluorescent signals were detected with the ChemiDoc instrument (Bio-Rad) using a Cy3 filter. To check the protein load, membranes were afterwards incubated overnight at 4 °C with rabbit anti-human GAPDH antibodies (Santa Cruz Biotechnologies) diluted 1:500 in TBS containing 0.1% Tween 20 and 5% BSA. Signals were detected by 1-h incubation at room temperature with HRP-conjugated secondary antibodies (Abcam, AB6013), followed by incubation with an ECL detection reagent (Thermo Fisher) and ChemiDoc imaging (Bio-Rad).

4.11.3. Confirmation of CCK-2R Membrane Expression by Flow Cytometry

Cells of the selected line (A549) were detached with 5 mM EDTA then stained with 100 μL of phosphate buffer saline (PBS) containing 0.1% Live–Dead Fixable Dead Cell Stain near-IR-fluorescent reactive dye (Molecular Probes) on ice for 30 min to exclude dead cells from the analysis. After washing with PBS added to 1% heat-inactivated FBS (PBS/FBS), cells were incubated on ice for 30 min with 50 μL of PBS/FBS containing 20 μg/mL of anti-CCK-2R-phycoerythrin antibodies (clone E-3, sc-166690, Santa Cruz Biotechnology), 40 μg/mL of anti-CCK-2R-Dy488 antibodies (LS-C756504, LSBio), or 40 μg/mL of CCK-8 peptide (Asp-Tyr-Met-Gly-Trp-Met-Asp-Phe) labelled with FAM supplied by GenScript Biotech (Piscataway, New Jersey, USA). Isotype control antibodies labelled with phycoerythrin (20 μg/mL, Santa Cruz Biotechnology) were used to check non-specific binding. After washing with PBS/FBS, cells were analyzed with the FACSCantoII flow cytometer (BD).

4.12. [111. In]In-IP-001 Cellular Uptake

A549 epithelial lung cancer cells were seeded in 6-well plates (20,000/cm^2) and allowed to adhere overnight. The following day, the medium was removed and radiotracer uptake was studied by incubating 2×10^5 cells at 37 °C in 2 mL of culture medium added to 30 μL (ca. 2 MBq, 1.35 μg, 1 nmol) of [^{111}In]In-IP-001. Uptake was monitored at 2, 4, 8, and 24 h. At these time points, the medium was removed and cells were washed twice with 2 mL of ice-cold PBS. Finally, cells were detached with 2 mL of a 0.25% trypsin/EDTA solution at 37 °C and centrifuged to separate the supernatant from the cells pellet. The radioactivity associated with the pellets was measured in a γ-spectrometer and corrected for decay. All experiments were performed in triplicate.

4.13. Animal Hosting, Inoculation and Monitoring

Homozygous female BALB/c nude J:NU (JAX stock number 007850) mice were purchased from Jackson Laboratories (Bar Harbor, ME, USA). Mice were 4 weeks old at arrival, and they were hosted in an IVC system (Smart Flow AHU, Tecniplast) with autoclaved rodent bedding hosting 3–4 mice per cage. Mice were fed with autoclaved rodent food and water ad libitum. The inoculation of A549 tumor cell line was performed by the subcutaneous injection of 5×10^6 cells/mouse in 100 uL of fetal bovine serum-free medium (RPMI added of 10% FBS, 1% penicillin-streptomycin, 1% glutamine) into the right shoulder. The cell line was found to be free of mycoplasma contamination, as visually judged by microscopic inspection and by regular 4′,6-diamidino-2-phenylindole (DAPI) staining of the cell cultures. Inoculation was carried out at 5 weeks of age after one week in the facility for acclimatization purposes. Mice were monitored every 2–3 days by measuring weight and tumor dimensions with an electronic caliper. For this purpose, two perpendicular dimensions were noted and volumes of the tumors were calculated with the formula $0.5 \times L \times W^2$, where L is the measurement of the longest axis and W is the measurement of the axis perpendicular to L, in millimeters [16].

4.14. Animal Injection and Imaging

Tumors were allowed to grow for about five weeks until they reached 0.5–1.0 cm of size in one dimension [35,36] (i.e., around 50–250 mm^3 volume), and then mice were divided in two groups. The experimental group (n = 5) was intravenously injected (retro-orbital venous sinus) [37,38] with 150 μL of a [^{111}]In-IP-001 solution (2.06 nmol, 7.4MBq), while the blocked group (n = 5) was injected with 150 μL of a [^{111}]In-IP-001 solution added to a 50-fold molar excess of cold ligand. All injections were performed with the animals under isoflurane anesthesia (1–3%), using 30G ultra-fine insulin syringes. Real-time, live, and fast dynamic screening studies were performed right after injection and on selected time points post-injection, on a dedicated desktop, mouse-sized, planar scintigraphic system (γ-eye^{TM} by BIOEMTECH, Athens, Greece). The system supports fusion with a digital mouse photograph for anatomical co-registration. The main detector is based on two position-sensitive photomultiplier tubes coupled to a CsI(Na) pixelated scintillator and a medium-energy lead collimator with parallel hexagonal holes that supports a range of SPECT isotopes. The system's field of view is 5×10 cm^2, with a spatial resolution of ~2 mm. For the planar imaging, mice were kept under isoflurane anesthesia and under a constant temperature of 37 °C, and scans had a duration of 10 min. This allowed for a fast test, right after injection, to check whether the injection was successful, and on later time points to quickly evaluate the bio-kinetics of the tracer. Tomographic SPECT/CT imaging was performed with y-CUBE^{TM} and x-CUBE^{TM} (Molecubes, Belgium), respectively, at the same selected time points post-injection. The SPECT system provides a spatial resolution of 0.6 mm for mouse imaging and of 1.5 mm for rat imaging. The CT system performs a spiral scan, it can provide images with 100 um resolution, and images were acquired with 50 kVp. Mouse imaging was performed by keeping the mice anaesthetized under isoflurane and under a constant temperature of 37 °C. SPECT scans were acquired with a 45 min duration, and each SPECT scan was followed by a high-resolution CT scan for co-registration purposes. The SPECT data were reconstructed through a maximum-likelihood expectation-maximization (MLEM) algorithm, with a 500 μm voxel size and 100 μm iterations. CT data were reconstructed through an iterative image space reconstruction algorithm (ISRA) algorithm with a 100 µm voxel size. Tomographic images are presented through maximum intensity projection (MIP) view). The experimental group was imaged at 2, 4, 8, and 24 h post-injection with both imaging systems, while the blocked group was imaged at 4 and 24 h.

4.15. Ex Vivo Biodistribution

For biodistribution studies, the injections were performed as described above, and 3 mice for both groups were sacrificed at 4 and 24 h post injection using 2% isoflurane. Main organs and tissues were removed, weighed, and counted together with blood samples, muscle, and urine by a γ-counter system (Cobra II from Packard, Canberra, ME, USA). Results were expressed as a mean percentage $\pm$ SD of the injected activity per g (%IA/g) per organ or tissue. For total blood radioactivity calculation, blood was assumed to be 7% of the total body weight.

4.16. Ethical Approval

The protocol described and all the animal procedures were approved by the General Directorate of Veterinary Services (Athens, Attica Prefecture, Greece) and by the Bioethical Committee of BIOEMTECH Laboratories (Permit number: EL 25 BIOexp 045) on the basis of the European Directive 2010/63/EU on the protection of animals used for experimental purposes.

4.17. Statistical Analysis.

Student's *t*-test was used to determine whether there were any statistically significant differences between the means of two independent (unrelated) groups. The threshold for statistical significance was set at $p < 0.05$.

5. Conclusions

In this study, a CCK-2R-targeting ligand based on the nastorazepide core was synthetized and functionalized with a DOTA chelator with the aim to provide a suitable platform for a theragnostic approach with radioactive metals. Avoiding the use of peptide-based sequences in the structure, including the linker, allowed us to obtain a molecule with high stability in physiological media and that could be easily labelled with indium-111 as pivotal radionuclide for future studies. The obtained radiotracer was successfully employed in the imaging of CCK-2R-expressing xenograft tumor in mice, but further structural studies are needed for enhancing receptor affinity and biodistribution. Additionally, the presented results are particularly noteworthy since the ability of a targeting probe to image cancers has been demonstrated using human cells expressing physiological levels of CCK-2R instead of transfected cells like the majority of the studies on the topic so far.

Supplementary Materials: The following are available online: Figure S1: ^{1}H NMR spectrum of **3** Figure S2: ^{1}H NMR spectrum of **4**; Figure S3: ^{1}H NMR spectrum of **6**; Figure S4: ^{1}H NMR spectrum of **7**; Figure S5: ^{1}H NMR spectrum of **8**; Figure S6: ^{1}H NMR spectrum of IP-001; Figure S7: HRMS analysis of IP-001; Figure S8: Representative RP-HPLC chromatograms of free [^{111}In]In^{3+} (**A**) and [^{111}In]In-IP-001 (**B**); Figure S9: Representative radio-TLC chromatograms of free-[^{111}In]In^{3+} (**A**) and [^{111}In]In-IP-001 (**B**); Figure S10. ^{1}H NMR spectrum of the compound due to the cyclization of the succinyl-group of compound **2**.

Author Contributions: Conceptualization, M.A., G.M., and F.M.; methodology, M.V., S.R., S.C., S.S., F.B., M.T., and E.V.; writing—original draft preparation, M.A., S.C.; writing—review and editing, M.V., M.A., S.C, G.M., and F.M.; supervision, M.A., G.M., and F.M.; funding acquisition, M.A. All authors have read and agreed to the published version of the manuscript.

Funding: This research was funded by the Italian Ministry of Health as part of the program "5perMille, year 2016" promoted by the AUSL-IRCCS of Reggio Emilia and by the project "ISOL-PHARM_EIRA", an experiment promoted in the framework of the National Scientific Committee 5 (Technological, inter-disciplinary and accelerators research) of INFN.

Institutional Review Board Statement: The study was conducted according to the guidelines of the Declaration of Helsinki, and approved by the General Directorate of Veterinary Services (Athens, Attica Prefecture, Greece) and by the Bioethical Committee of BIOEMTECH Laboratories (protocol code EL 25 BIOexp 045, 12/3/2019).

Informed Consent Statement: Not applicable.

Data Availability Statement: Raw data are available at the following link: https://drive.google.com/drive/u/1/folders/1ohexfcWTNlfLvz5n42_VpdiXBGlh8DlL.

Acknowledgments: Authors thank BIOEMTECH Laboratories (Athens, Greece) for hosting the in vivo and imaging part of this study and Chiara Coruzzi for the bibliographic research.

Conflicts of Interest: S.S. is a BIOEMTECH employees. The funders had no role in the design of the study; in the collection, analyses, or interpretation of data; in the writing of the manuscript, or in the decision to publish the results.

Sample Availability: Samples of the compound IP-001 are available from the authors.

Abbreviations

Cholecystokinin-2 (CCK-2), cholecystokinin-2 receptor (CCK-2R), nastorazepide (Z-360), cholecystokinin-1 receptor (CCK-1R), 1,4,7,10-Tetraazacyclododecane-1,4,7,10-tetraacetic acid (DOTA), 1,1'-carbonyldiimidazole (CDI), *tert*-butoxycarbonyl (Boc-), trifluoroacetic acid (TFA), dichloromethane (DCM), fluorenylmethoxycarbonyl (Fmoc-), dimethylformamide (DMF), nuclear magnetic resonance (NMR), high-resolution mass spectrometry (HRMS), Wildman's logP (W-logP), topological polar surface area (TPSA), predicted solubility (log scale) computed according to the Estimated aqueous SOLubility (ESOL) model (LogS (ESOL)), gastrointestinal (GI), blood brain barrier (BBB), P-glycoprotein (P-gp), P-cytochrome (CYP), solid phase extraction (SPE), radiochemical purity (RCP), high-performance liquid chromatography (HPLC), thin-layer chromatography (TLC), 4-(2-hydroxyethyl)-1-piperazineethanesulfonic acid (HEPES), ethylenediaminetetraacetic acid (EDTA), acetonitrile (MeCN), single photon emission tomography (SPECT), computed tomography (CT), tumor to muscle (T/M), tumor to blood (T/B), maximum intensity projection (MIP), injected dose per cube centimeter (ID/cc), tetramethylsilane (TMS), matrix-assisted laser desorption/ionization time of flight (MALDI TOF/TOF), triethylamine (TEA), ethyl acetate (EtOAc), benzotriazol-1-yloxytris(dimethylamino)phosphonium hexafluorophosphate (BOP), human serum (HS), human blood (HB), 4',6-diamidino-2-phenylindole (DAPI), phosphate buffer saline (PBS), maximum-likelihood expectation-maximization (MLEM), and iterative image space reconstruction algorithm (ISRA).

References

1. Tafreshi, N.K.; Doligalski, M.L.; Tichacek, C.J.; Pandya, D.N.; Budzevich, M.M.; El-Haddad, G.; Khushalani, N.I.; Moros, E.G.; McLaughlin, M.L.; Wadas, T.J.; et al. Development of targeted alpha particle therapy for solid tumours. *Molecules* **2019**, *24*, 4314. [CrossRef]
2. Tashima, T.T. Effective cancer therapy based on selective drug delivery into cells across their membrane using receptor-mediated endocytosis. *Bioorg. Med. Chem. Lett.* **2018**, *28*, 3015–3024. [CrossRef]
3. Opalinska, M.; Hubalewska-Dydejczyk, A.; Sowa-Staszczak, A. Radiolabeled peptides: Current and new perspectives. *Q. J. Nucl. Med. Mol. Imaging* **2017**, *61*, 153–167. [PubMed]
4. Fani, M.; Maecke, H.R. Radiopharmaceutical development of radiolabelled peptides. *Eur. J. Nucl. Med. Mol. Imaging* **2012**, *39*, 11–30. [CrossRef]
5. Reubi, J.C.; Waser, B. Unexpected high incidence of cholecystokinin B/gastrin receptors in human medullary thyroid carcinomas. *Int. J. Cancer* **1996**, *67*, 644–647. [CrossRef]
6. Sethi, T.; Herget, M.; Wu, S.V.; Walsh, J.H.; Rozengurt, E. CCK-A and CCK-B receptors are expressed in small cell lung cancer lines and mediate Ca^{2+} mobilization and clonal growth. *Cancer Res.* **1993**, *53*, 5208–5213.
7. Reubi, J.C.; Waser, B. Cholecystokinin (CCK)-A and CCK-B/gastrin receptors in human tumours. *Cancer Res.* **1997**, *57*, 1377–1386. [PubMed]
8. Smith, J.P.; Verderame, M.F.; McLaughlin, P.; Martenis, M.; Ballard, E.; Zagon, I.S. Characterization of the CCK-C (cancer) receptor in human pancreatic cancer. *Int. J. Mol. Med.* **2002**, *10*, 689–694. [CrossRef]

9. Roosenburg, S.; Laverman, P.; van Delft, F.L.; Boerman, O.C. Radiolabeled CCK/gastrin peptides for imaging and therapy of CCK2 receptor-expressing tumours. *Amino Acids* **2011**, *41*, 1049–1058. [CrossRef]

10. Brom, M.; Joosten, L.; Laverman, P.; Oyen, W.J.; Béhé, M.; Gotthardt, M.; Boerman, O.C. Preclinical evaluation of 68Ga-DOTA-minigastrin for the detection of cholecystokinin-2/gastrin receptor-positive tumours. *Mol. Imaging* **2011**, *10*, 144–152. [CrossRef] [PubMed]

11. Roosenburg, S.; Laverman, P.; Joosten, L.; Eek, A.; Rutjes, F.P.; Van Delft, F.L.; Boerman, O.C. In vitro and in vivo characterization of three 68Ga- and ^{111}In-labeled peptides for cholecystokinin receptor imaging. *Mol. Imaging* **2012**, *11*, 401–407. [CrossRef] [PubMed]

12. Kaloudi, A.; Nock, B.A.; Lymperis, E.; Krenning, E.P.; de Jong, M.; Maina, T. (99m)Tc-labeled gastrins of varying peptide chain length: Distinct impact of NEP/ACE-inhibition on stability and tumour uptake in mice. *Nucl. Med. Biol.* **2016**, *43*, 347–354. [CrossRef]

13. Klingler, M.; Decristoforo, C.; Rangger, C.; Summer, D.; Foster, J.; Sosabowski, J.K.; Von Guggenberg, E. Site-specific stabilization of minigastrin analogs against enzymatic degradation for enhanced cholecystokinin-2 receptor targeting. *Theranostics* **2018**, *8*, 2896–2908. [CrossRef] [PubMed]

14. Sauter, A.W.; Mansi, R.; Hassiepen, U.; Muller, L.; Panigada, T.; Wiehr, S.; Wild, A.-M.; Geistlich, S.; Béhé, M.; Rottenburger, C.; et al. Targeting of the cholecystokinin-2 receptor with the minigastrin analog 177Lu-DOTA-PP-F11N: Does the use of protease inhibitors further improve in vivo distribution? *J. Nucl. Med.* **2019**, *60*, 393–399. [CrossRef]

15. Wayua, C.; Low, P.S. Evaluation of a cholecystokinin 2 receptor-targeted near-infrared dye for fluorescence-guided surgery of cancer. *Mol. Pharm.* **2014**, *11*, 468–476. [CrossRef] [PubMed]

16. Wayua, C.; Low, P.S. Evaluation of a nonpeptidic ligand for imaging of cholecystokinin 2 receptor-expressing cancers. *J. Nucl. Med.* **2015**, *56*, 113–119. [CrossRef]

17. Kaloudi, A.; Kanellopoulos, P.; Radolf, T.; Chepurny, O.G.; Rouchota, M.; Loudos, G.; Andreae, F.; Holz, G.G.; Nock, B.A.; Maina, T. [99mTc]Tc-DGA1, a promising CCK2R-antagonist-based tracer for tumour diagnosis with single-photon emission computed tomography. *Mol. Pharm.* **2020**, *17*, 3116–3128. [CrossRef]

18. De León-Rodríguez, L.M.; Kovacs, Z. The synthesis and chelation chemistry of DOTA-peptide conjugates. *Bioconjug. Chem.* **2008**, *19*, 391–402. [CrossRef]

19. Philip, S. 9 Cholecystokinin b Receptor Targeting for Imaging and Therapy. U.S. Patent WO 2013126797 A1, 29 August 2013.

20. Paul, R.; Anderson, G.W. *N*,*N*′-Carbonyldiimidazole, a new peptide forming reagent. *J. Am. Chem. Soc.* **1960**, *82*, 4596–4600. [CrossRef]

21. Athertohn, E.; Zelf, A.; Harkissc, D.; Logan, C.J.; Sheppard, R.C.; Williams, B.J. A mild procedure for solid phase peptide synthesis: Use of fluorenylmethoxycarbonylamino-acids. *J.C.S. Chem. Comm.* **1978**, *13*, 537–539. [CrossRef]

22. Thomson, S.A.; Josey, J.A.; Cadilla, R.; Gaul, M.D.; Hassman, C.F.; Luzzio, M.J.; Pipe, A.J.; Reed, K.L.; Ricca, D.J.; Wiethe, R.W.; et al. Fmoc mediated synthesis of peptide nucleic acids. *Tetrahedron* **1995**, *51*, 6179–6194. [CrossRef]

23. Fields, G.B. Methods for removing the Fmoc group. *Methods Mol. Biol.* **1994**, *35*, 17–27. [PubMed]

24. Good, S.; Walter, M.A.; Waser, B.; Wang, X.; Müller-Brand, J.; Béhé, M.P.; Reubi, J.-C.; Maecke, H.R. Macrocyclic chelator-coupled gastrin-based radiopharmaceuticals for targeting of gastrin receptor-expressing tumours. *Eur. J. Nucl. Med. Mol. Imaging* **2008**, *35*, 1868–1877. [CrossRef]

25. Roy, J.; Putt, K.S.; Coppola, D.; Leon, M.E.; Khalil, F.K.; Centeno, B.A.; Clark, N.; Stark, V.E.; Morse, D.L.; Low, P.S. Assessment of cholecystokinin 2 receptor (CCK2R) in neoplastic tissue. *Oncotarget* **2016**, *7*, 14605–14615. [CrossRef] [PubMed]

26. Sturzu, A.; Klose, U.; Sheikh, S.; Echner, H.; Kalbacher, H.; Deeg, M.; Enägele, T.; Schwentner, C.; Ernemann, U.; Heckl, S. The gastrin/cholecystokinin-B receptor on prostate cells—A novel target for bifunctional prostate cancer imaging. *Eur. J. Pharm. Sci.* **2014**, *52*, 69–76. [CrossRef] [PubMed]

27. Savage, K.; Waller, H.A.; Stubbs, M.; Khan, K.; Watson, S.A.; Clarke, P.A.; Grimes, S.; Michaeli, D.; Dhillon, A.P.; Caplin, M.E. Targeting of cholecystokinin B/gastrin receptor in colonic, pancreatic and hepatocellular carcinoma cell lines. *Int. J. Oncol.* **2006**, *29*, 1429–1435. [CrossRef] [PubMed]

28. McWilliams, D.F.; Watson, S.A.; Crosbee, D.M.; Michaeli, D.; Seth, R. Coexpression of gastrin and gastrin receptors (CCK-B and delta CCK-B) in gastrointestinal tumour cell lines. *Gut* **1998**, *42*, 795–7958. [CrossRef] [PubMed]

29. Aloj, L.; Aurilio, M.; Rinaldi, V.; D'Ambrosio, L.; Tesauro, D.; Peitl, P.K.; Maina, T.; Mansi, R.; Von Guggenberg, E.; Joosten, L.; et al. Comparison of the binding and internalization properties of 12 DOTA-coupled and ^{111}In-labelled CCK2/gastrin receptor binding peptides: A collaborative project under COST Action BM0607. *Eur. J. Nucl. Med. Mol. Imaging* **2011**, *38*, 1417–1425. [CrossRef]

30. Laverman, P.; Joosten, L.; Eek, A.; Roosenburg, S.; Peitl, P.K.; Maina, T.; Macke, H.R.; Aloj, L.; Von Guggenberg, E.; Sosabowski, J.K.; et al. Comparative biodistribution of 12 ^{111}In-labelled gastrin/CCK2 receptor-targeting peptides. *Eur. J. Nucl. Med. Mol. Imaging* **2011**, *38*, 1410–1416. [CrossRef] [PubMed]

31. Summer, D.; Kroess, A.; Woerndle, R.; Rangger, C.; Klingler, M.; Haas, H.; Kremser, L.; Lindner, H.H.; Von Guggenberg, E.; Decristoforo, C. Multimerization results in formation of re-bindable metabolites: A proof of concept study with FSC-based minigastrin imaging probes targeting CCK2R expression. *PLoS ONE* **2018**, *13*, e0201224. [CrossRef] [PubMed]

32. Brillouet, S.; Dorbes, S.; Courbon, F.; Picard, C.; Delord, J.P.; Benoist, E.; Poirot, M.; Mestre-Voegtlé, B.; Silvente-Poirot, S. Development of a new radioligand for cholecystokinin receptor subtype 2 scintigraphy: From molecular modeling to in vivo evaluation. *Bioorg. Med. Chem.* **2010**, *18*, 5400–5412. [CrossRef]

33. Laverman, P.; Roosenburg, S.; Gotthardt, M.; Park, J.; Oyen, W.J.G.; De Jong, M.; Hellmich, M.R.; Rutjes, F.P.J.T.; Van Delft, F.L.; Boerman, O.C. Targeting of a CCK(2) receptor splice variant with (111)In-labelled cholecystokinin-8 (CCK8) and (111)In-labelled minigastrin. *Eur. J. Nucl. Med. Mol. Imaging* **2008**, *35*, 386–392. [CrossRef] [PubMed]

34. Daina, A.; Michielin, O.; Zoete, V. SwissADME: A free web tool to evaluate pharmacokinetics, drug-likeness and medicinal chemistry friendliness of small molecules. *Sci. Rep.* **2017**, *7*, 42717. [CrossRef]

35. Qin, H.; Zhang, M.R.; Xie, L.; Hou, Y.; Hua, Z.; Hu, M.; Wang, Z.; Wang, F. PET imaging of apoptosis in tumour-bearing mice and rabbits after paclitaxel treatment with (18)F(-)Labeled recombinant human His10-annexin V. *Am. J. Nucl. Med. Mol. Imaging* **2014**, *5*, 27–37.

36. Jiao, H.; Zhao, X.; Liu, J.; Ma, T.; Zhang, Z.; Zhang, J.; Wang, J. In vivo imaging characterization and anticancer efficacy of a novel HER2 affibody and pemetrexed conjugate in lung cancer model. *Nucl. Med. Biol.* **2019**, *68*, 31–39. [CrossRef] [PubMed]

37. Pfister, J.; Summer, D.; Rangger, C.; Petrik, M.; Von Guggenberg, E.; Minazzi, P.; Giovenzana, G.B.; Aloj, L.; Decristoforo, C. Influence of a novel, versatile bifunctional chelator on theranostic properties of a minigastrin analogue. *EJNMMI Res.* **2015**, *5*, 74. [CrossRef] [PubMed]

38. Steel, C.D.; Stephens, A.L.; Hahto, S.M.; Singletary, S.J.; Ciavarra, R.P. Comparison of the lateral tail vein and the retro-orbital venous sinus as routes of intravenous drug delivery in a transgenic mouse model. *Lab. Anim.* **2008**, *37*, 26–32. [CrossRef]

Article

Synthesis of Ionizable Calix[4]arenes for Chelation of Selected Divalent Cations

Markus Blumberg [1,2], Karrar Al-Ameed [3,†], Erik Eiselt [1], Sandra Luber [3] and Constantin Mamat [1,2,*]

[1] Helmholtz-Zentrum Dresden-Rossendorf, Institut für Radiopharmazeutische Krebsforschung, Bautzner Landstraße 400, D-01328 Dresden, Germany; blumberg.markus@yahoo.de (M.B.); erik.eiselt@gmx.de (E.E.)
[2] Fakultät Chemie und Lebensmittelchemie, Technische Universität Dresden, D-01062 Dresden, Germany
[3] Department of Chemistry, University of Zurich, Winterthurerstrasse 190, CH-8057 Zürich, Switzerland; karrar.al-ameed@linacre.ox.ac.uk (K.A.-A.); sandra.luber@chem.uzh.ch (S.L.)
* Correspondence: c.mamat@hzdr.de
† Current address: Department of Chemistry, University of Kufa, Najaf 54001, Iraq.

Abstract: Two sets of functionalised calix[4]arenes, either with a 1,3-crown ether bridge or with an open-chain oligo ether moiety in 1,3-position were prepared and further equipped with additional deprotonisable sulfonamide groups to establish chelating systems for selected cations Sr^{2+}, Ba^{2+}, and Pb^{2+} ions. To improve the complexation behaviour towards these cations, calix[4]arenes with oligo ether groups and modified crowns of different sizes were synthesized. Association constants were determined by UV/Vis titration in acetonitrile using the respective perchlorate salts and logK values between 3.2 and 8.0 were obtained. These findings were supported by the calculation of the binding energies exemplarily for selected complexes with Ba^{2+}.

Keywords: calixarenes; barium; chelation

Citation: Blumberg, M.; Al-Ameed, K.; Eiselt, E.; Luber, S.; Mamat, C. Synthesis of Ionizable Calix[4]arenes for Chelation of Selected Divalent Cations. *Molecules* **2022**, *27*, 1478. https://doi.org/10.3390/molecules27051478

Academic Editors: Petra Martini and Alessandra Boschi

Received: 28 January 2022
Accepted: 17 February 2022
Published: 22 February 2022

Publisher's Note: MDPI stays neutral with regard to jurisdictional claims in published maps and institutional affiliations.

1. Introduction

Molecular baskets based on the calixarene backbone were found to be a promising basis to search for suitable macrocyclic ligands especially for divalent metal ions. Calix[4]arenes are metacyclophanes having a hydrophobic cavity between the lower and upper rim, formed by four phenol units connected with methylene units [1,2]. Due to their numerous possibilities for functionalization on the upper as well as on the lower rim [3], this class of macromolecules is known for a wide range of applications [4,5]. Calix[4]arenes were found to act as biologically active compounds as they are used as antibacterial and even antimalarial agents or in cancer chemotherapy [6–9]. Additionally, they can interact with amino acids or show promising enzyme inhibitory effects [10]. Their ability to form inclusion compounds with neutral molecules [11] or ions [12–16] makes them useful as sensors [17,18], catalysts [19], ligands, or as effective separation agents for ions [20] when resin-bound [21–24]. Since calix[4]arene derivatives interact particularly strongly with heavy group 2 metals [25–31], one major field of application consists in their use as extraction agents for nuclear waste treatment [27,32–35].

In this regard, the calix[4]arene skeleton can be seen as an ideal platform to build an optimized chelator. Two out of the four hydroxy groups on the lower rim are possible to be functionalized with proton-ionizable groups, which leads to the formation of neutral complexes with divalent cations. The remaining two can be furnished with oligo ether groups, either open-chain or bridged, leading to the calixcrown scaffold, which is easily accessible. Using this concept, both the advantages of the electrostatic, macrocyclic, and cryptate effect are unified.

Three divalent cations Sr^{2+}, Ba^{2+}, and Pb^{2+} are in the focus of our interest, because they all possess radioisotopes with useful nuclear properties for various diagnostic or therapeutic applications in nuclear medicine [36], and are therefore suitable to prepare radiopharmaceuticals. The beta-emitter ^{89}Sr is applied as "bone seeker" [37], and ^{90}Sr is used for superficial brachytherapy of some cancers [38,39]. ^{131}Ba ($t_{1/2}$ = 11.5 d) is a γ-emitter for possible diagnostic uses and is discussed as a bone-scanning agent in scintigraphy [40–42]. Furthermore, Ba^{2+} functions as a non-radioactive surrogate [43–45] for both alpha-emitters $^{223/224}$Ra, because of their analogous chemical properties and their radii of similar range [46]. Radium-223 and radium-224, have suitable half-lives (^{223}Ra: 11.4 d, ^{224}Ra: 3.6 d) and nuclear decay properties (decay chain with four alpha and two beta particles) [47] that make them useful tools for alpha-particle therapy [48]. $[^{223}$Ra]RaCl$_2$ is known as Xofigo$^{®}$ and is applied in clinics for the treatment of bone metastases. Furthermore, Pb^{2+} was part of our research. On the one hand, it is the stable end product (^{207}Pb and ^{208}Pb) of both radium decay chains. On the other hand, ^{212}Pb is a promising β^{-}-emitter, and a feasible candidate for radiopharmaceutical applications, since it can also be used as an in vivo generator for ^{212}Bi, which is a strong alpha emitter [49,50]. No indications were found regarding radiopharmaceutical applications of the light alkaline earth metals beryllium, magnesium, and calcium for therapy or diagnosis.

To provide the ideal cavity for heavy group 2 metals for stable complexation, it is essential to choose a suitable oligo ether length or crown size, respectively, combined with an adequate number of donor sites. Recently, the impact of the crown-6-functionalization of two simple calix[4]crown-6 derivatives has been elaborated [29,30]. The group of R. A. Bartsch focused on extraction of alkaline earth metal cations using functionalized calixcrowns [51–53] and found the calix[4]arene-1,3-crown-6 derivatives to be very effective group 2 metal ion extraction agents. The additional impact of two trifluoromethylsulfonylamide groups as proton-ionizing residues was corroborated by them and our research group [26]. Furthermore, these compounds showed high selectivity for Ba^{2+} over the lighter alkaline earth metal or alkali metal ions. However, Ra^{2+} was not investigated. There are only a handful of reports dealing with Ra^{2+} and the efficiency of various ligands including calixarenes as ionophores in nuclear waste management [28,54,55]. Particularly in radiopharmacy, high stability of the complex is urgently important so that a M^{2+}-release and the following accumulation in bone tissues is minimized [42,44].

The objective of this research was to evaluate and compare different open-chain and bridged *p*-tert-butylcalix[4]arene derivatives as possible leading compounds that could, upon further modification, yield viable chelators for the selected divalent metal ions Sr^{2+}, Ba^{2+}, and Pb^{2+} in radiopharmaceutical applications and provide information about comparable stability constants. The existing literature about group 2 metal ligands specifically with radium, is focused mainly on extraction studies. Therefore, the UV titration as reliable and constant method for the calculation of stability constants was used to determine association constants for the respective ions. Additionally, theoretical calculations involving Ba^{2+} as a surrogate for Ra^{2+} were accomplished to underline the results.

2. Results and Discussion

2.1. Preparation of the Functionalized Calix[4]arenes

For a better understanding of the complexation ability, two sets of calix[4]arene derivatives either with open-chain or bridged oligoethers were evaluated. The first set of chelators containing the open-chain oligo ether functions was prepared to start from the basic compound **1**, which contains the same number of oxygen donor atoms as calix[4]crown-6 **14a**. The complete synthesis path is described in Scheme 1. Calix **1** is proposed to form complexes with Na$^+$ and K$^+$ [56]. For the introduction of proton-ionizable groups to improve the complexation behavior, the two remaining free OH groups were modified by alkylation with ethyl bromoacetate to yield calix **2**. In the next step, calix **2** was saponified under basic conditions to yield diacid **3** in quantitative yield without further purification after precipitation with HCl. To introduce the amide functions, compound **3** was then treated with oxalyl chloride to form the dichloride **4**, which was instantly reacted with morpholine or 1,4,7-trioxa-10-azacyclododecane to yield amides **5** and **6**, respectively. Both amines were used for modification to raise the number of donors for complexation and the steric demand. The introduction of proton-ionizable sulfonamides to yield **7a–c** follows the same procedure by using trifluoromethyl sulfonamide, perfluoroisopropyl sulfonamide, and perfluorophenyl sulfonamide, respectively, which were deprotonated prior to the reaction with dichloride **4**.

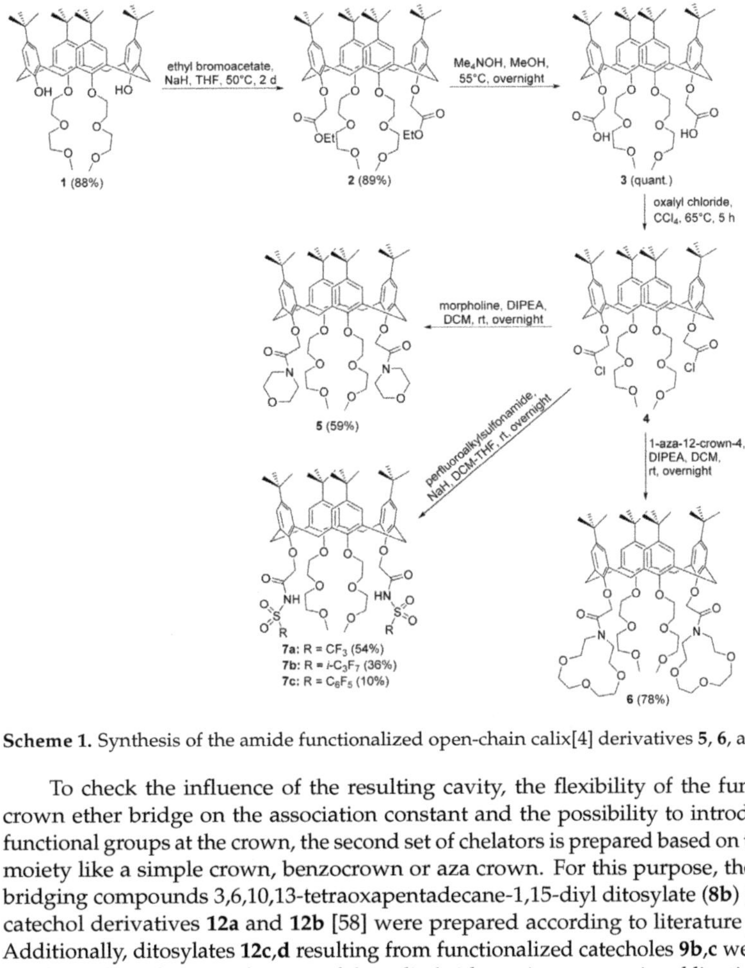

Scheme 1. Synthesis of the amide functionalized open-chain calix[4] derivatives **5**, **6**, and **7a–c**.

To check the influence of the resulting cavity, the flexibility of the functionalized crown ether bridge on the association constant and the possibility to introduce further functional groups at the crown, the second set of chelators is prepared based on the bridging moiety like a simple crown, benzocrown or aza crown. For this purpose, the respective bridging compounds 3,6,10,13-tetraoxapentadecane-1,15-diyl ditosylate (**8b**) [57] and the catechol derivatives **12a** and **12b** [58] were prepared according to literature procedures. Additionally, ditosylates **12c,d** resulting from functionalized catecholes **9b,c** were prepared to allow a later functionalisation of the calix-bridge using conventional ligation reactions. The preparation procedure is outlined in Scheme 2.

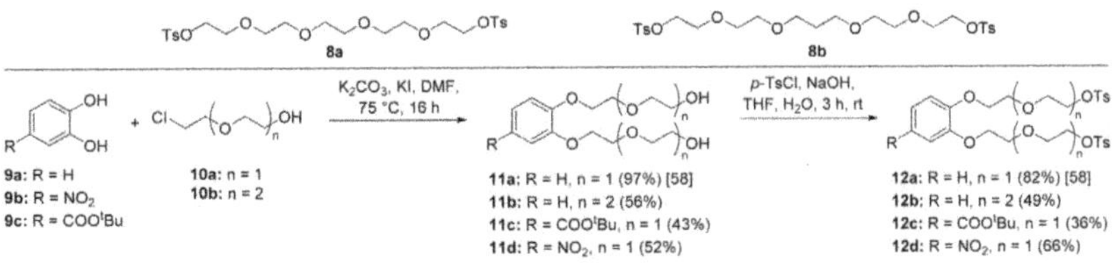

Scheme 2. Structure of alkyl ditosylates **8a,b** and synthesis of the benzo ditosylates **12a–d**.

For the connection of the functionalized crown ethers with the calix[4]arene skeleton, the resulting ditosylates **8a,b** and **12a–d** of the oligo ethers were reacted with ^{t}Bu-calix[4]arene (**13**) under basic conditions using K_2CO_3 in dichloromethane (DCM) to prepare the calix-crown-6 derivatives **14a,b** and the calix-benzocrown-6 derivatives **14c–f** in yields of 36–81%. The complete synthesis path is described in Scheme 3. The two remaining free OH groups of the calix-crowns **14a–f** were alkylated with ethyl bromoacetate to yield **15a–f**. In the next step, they were saponified under basic conditions to yield the respective diacid derivatives **16a–f** in mostly quantitative yields without further purification after precipitation with HCl. To introduce the proton-ionizable fluorinated sulfonamide functions, an amide coupling strategy using EDC and (Cl-)HOBt was applied. Thus, compounds **16a–f** were dissolved in anhydrous acetonitrile and reacted with the respective perfluorinated sulfonamides **17–19** at ambient temperature using the aforementioned coupling agents to form calix-crowns **20–22** and the modified calix-benzocrowns **23–27** in yields of 27–97%.

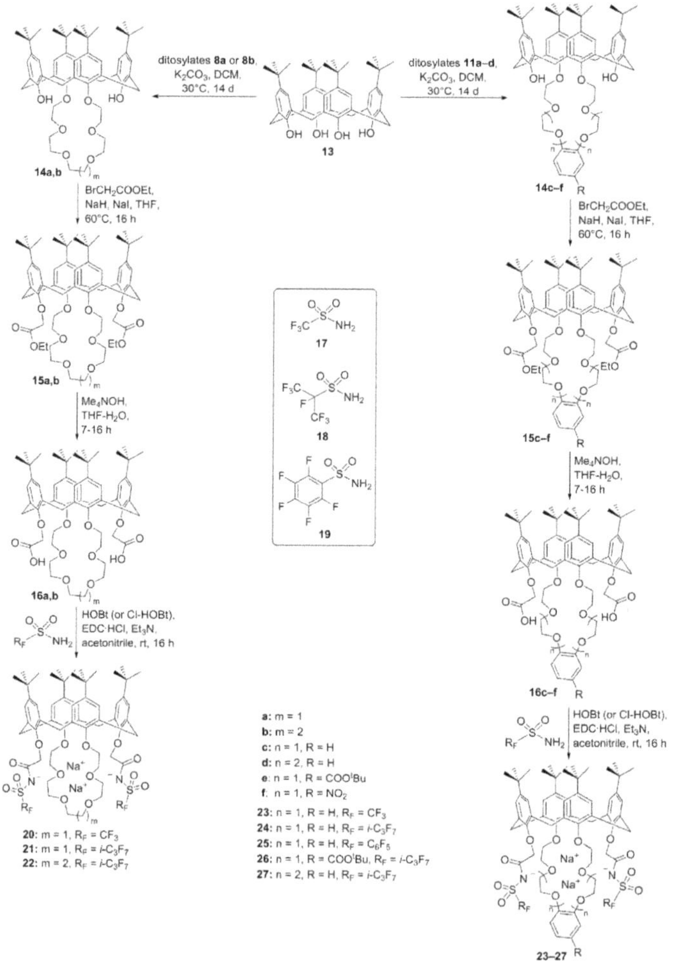

Scheme 3. Preparation of the proton-ionizable calix[4]crown derivatives **20–27** with perfluorosulfonamide functions.

2.2. The Influence of the Crown Type

The influence of lower-rim crown modifications on the metal-calixcrown-coordination was next checked. For this purpose, reaction binding energies of Ba^{2+} with four selected crown-bearing calixarenes were calculated (Figure 1). That involved calix[4]crown-6 **20** and additionally benzo-crown derivative **23** with the modification on adding an aromatic six-membered ring at the bottom of the crown, propylene derivative **22** with an extended crown by a CH_2 group, and calix[4]crown-5 **28** lacking an ethoxy unit (CH_2CH_2O). As the result of the calculated binding energies (corrected for basis set superposition error) shown in Table 1, the energetically most favored environment to host the metal ion is found for calix[4]crown-6 **20**, while in the other three compounds, the metal-crown binding is weakened up to about 10%. Furthermore, the lowest binding energy is found for the barium ion and calix[4]crown-5. At first glance, the crown-5 might appear to feature an optimal size to host ions like Ba^{2+}, as the size of the macrocycle cavity relative to the ionic radius is often used as a common parameter to rationalize and design new ligands in host-guest chemistry [59]. Considering the geometrical parameters given in Table S1, the crown-5 cavity seems to enclose the Ba^{2+} ion better than the crown-6; however, the calculated binding energies show a reversed trend. Our calculations are in accordance with previous studies in the field of host-guest chemistry of crown ethers [60]. Islam et al., for instance, showed that Na^+ binds more tightly to a crown-6 body although the crown-5 hole's size matches the sodium ion radius better compared to the one of crown-6 [61].

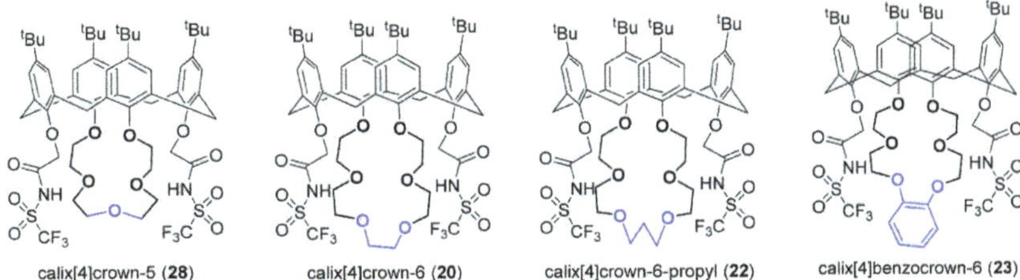

Figure 1. Calix[4]arene derivatives with modified crown ethers were used in the calculations.

Table 1. Computed binding energies of different types of calixcrowns (B3LYP-D3/def2-TZVP).

Calixarene	Crown Type	ΔE (kcal/mol)
20	Crown-6	−100.9
22	Propylene-crown-6	−93.1
23	Benzocrown-6	−90.2
28	Crown-5	−89.9

2.3. NMR Investigations

To determine the stability constants for the complexation of Ba^{2+}, Sr^{2+}, and Pb^{2+}, a reliable method was developed in the past by our group using ^1H NMR spectroscopy. Due to the different ^1H NMR spectra, which were recorded for the respective complexes in comparison with the ligands, a ^1H NMR titration method was established [26,29,30]. For this purpose, the compounds were dissolved in acetonitrile-d_3 and treated in portions with an acetonitrile-d_3 solution containing $Ba(ClO_4)_2$. It was observed, that after addition of 0.5 equivalents of Ba salt, a new set of signals appeared, belonging to the respective Ba-complex exemplarily shown for calix **24** in Figure 2. This leads to two separate species: ligand **24** and complex **Ba-24** in the ^1H NMR spectrum (see, for instance, the difference of the methylene protons of the calix skeleton in the box of Figure 3). After the addition of 1 eq. of $Ba(ClO_4)_2$, no ligand was detectable anymore, which leads to the assumption of a complex with 1:1 stoichiometry. The situation in Figure 2 is a consequence of a slow exchange on the NMR time scale and is therefore not suitable for a logK determination by

NMR titration for our calix compounds [62]. Next, a more reliable titration method based on UV/vis spectroscopy was used instead to determine the association constants.

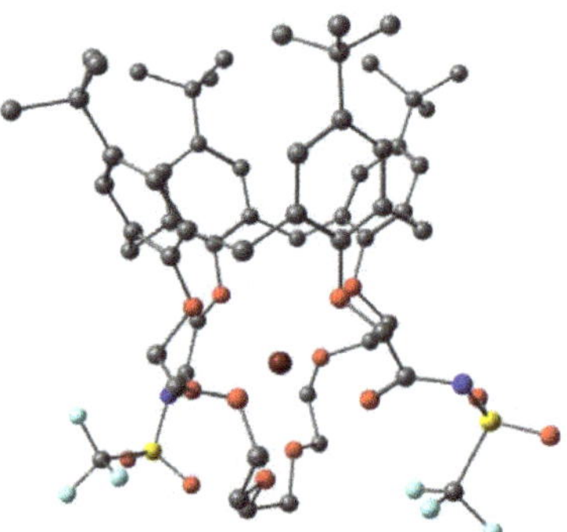

Figure 2. The optimized geometrical structure of the Ba-complex **Ba-20** (hydrogen atoms are hidden for clarity; the colors of the atoms are: carbon; grey, fluorine; green, sulfur; yellow, nitrogen; blue, oxygen; red, barium; brown).

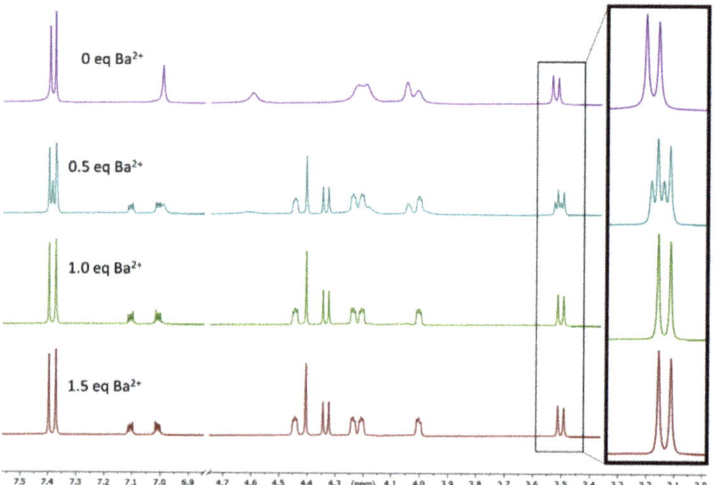

Figure 3. ^{1}H NMR spectra showing the change in chemical shifts by treatment of calix **24** with different equivalents of Ba(ClO$_4$)$_2$ in acetonitrile-d$_3$.

2.4. UV Titration Studies

To determine the association constants by UV titration, the open-chain calix[4]arenes **7a,b** as well as the calix[4]crowns **20–22** and benzocrown compounds **23–25** and **27** were dissolved in acetonitrile and aliquots of M(ClO$_4$)$_2$ with M = Ba, Sr, Pb in acetonitrile were added, which caused a clear change in the UV absorption. This behavior is shown in Figure 4 based on the example of compound **21** and its complexes **Ba-21**, **Sr-21** and **Pb-21**. Considering one wavelength with a high change in absorption, the stoichiometry of the complex was determined from the diagram titration curve (Figure 4, exemplarily for **Ba-21**). All titration experiments showed the presence of a complex with a 1:1 stoichiometry, formed by functionalized calix[4]crowns with M(ClO$_4$)$_2$ except for calix **6** (see Supplementary Materials), as the slope of the graph is changing at the 1:1 molar ratio. The results from the evaluation of the association constants are summarized in Table 2 (all UV titration experiments are expressed in the Supplementary Materials).

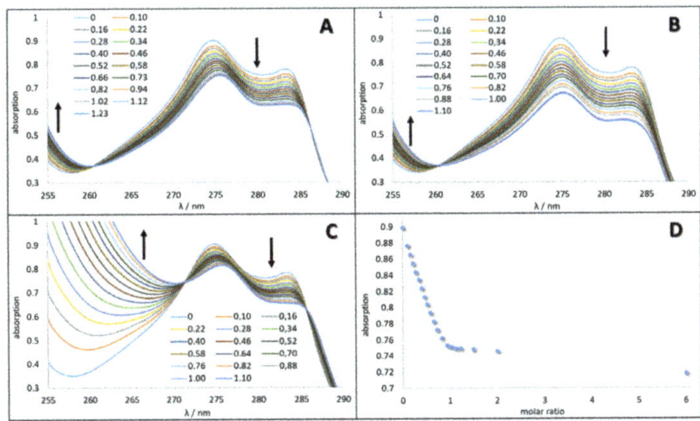

Figure 4. UV/Vis spectra of calix[4]arene **21** at different $M(ClO_4)_2$ concentrations (**A**) M = Ba, (**B**) M = Sr, (**C**) M = Pb for titration measured in acetonitrile. (**D**) Titration curve with change of the absorption using compound **21** at different $Ba(ClO_4)_2$ concentrations, measured in acetonitrile at λ = 275 nm.

Table 2. Determined logK values using the UV titration method.

Calixarene	logK(Ba)	logK(Sr)	logK(Pb)
5	6.1 ± 0.3	3.8	5.7 ± 0.2
6	— [a]	— [a]	— [a]
7a	4.4 ± 0.2	4.9 ± 0.2	5.0 ± 0.6
7b	4.6 ± 0.3	3.2 ± 0.2	5.4 ± 0.1
14a	4.6 ± 0.1 [b]	4.3 ± 0.1 [b]	3.3 ± 1 [b]
20	5.7 ± 0.1	n.d.	n.d.
21	5.0 ± 0.8	5.0 ± 0.8	5.5 ± 0.8
22	4.9 ± 0.7	5.6 ± 0.8	7.4 ± 1.5 [c]
23	6.6 ± 1.0 [c]	7.0 ± 1.4 [c]	4.0 ± 0.2
24	5.8 ± 0.9	6.4 ± 1.0 [c]	6.3 ± 0.9 [c]
25	7.5 ± 1.5 [c]	7.6 ± 1.5 [c]	8.0 ± 1.6 [c]
27	5.2 ± 0.8	7.0 ± 1.4 [c]	3.9 ± 0.1

[a] No value available from UV titration, [b] see ref. [29], [c] Value over the upper method detection limit, n.d.—not determined.

The association constants of the bridged derivatives **20–25** and **27** with Sr^{2+}, Ba^{2+} and Pb^{2+} showing values up to 8.0 and were higher as for the basic complexes **Ba-14a** (logK 4.6), **Sr-14a** (logK: 4.3), and **Pb-14a** (logK 3.3) [26] without additional side functions due to the increasing number of donor atoms and the additional macrocyclic effect. Interestingly, higher association constants were obtained for the benzocrown derivatives **23–25** in comparison to the basic crown derivatives **20–22**. The calix derivatives with benzocrown bridge led to more stable complexes showing higher stability constants of approximately one magnitude (e.g., **Sr-21**: 5.0 vs. **Sr-24**: 6.4 (C_3F_7) or **Ba-20**: 5.7 vs. **Ba-23**: 6.6 (CF_3)) for the group 2 metal ions. Moreover, the propylene bridge in calix **22** has only a marginal influence on the association constant regarding the respective Ba-complexes **Ba-21** and **Ba-22**, but a strong influence when comparing the Pb-complexes ($logK_{(Pb-21)}$ = 5.5 vs. $logK_{(Pb-22)}$ = 7.4).

Additionally, a high value is even found for calix **5** with morpholine modification and its respective complexes with Ba^{2+} and Pb^{2+} possibly due to the higher number of donor atoms. The constants for the open-chain derivatives **Ba-7a,b** are lower or equal to the value found for **14a** in contrast to the values found for **Pb-7a,b**, which were higher. Differences between the M^{2+} complexes from the same ligand arose from the different ion radii as well as from the chemical behavior (HSAB concept), as Sr^{2+} and Ba^{2+} belong to the alkaline earth metals (group 2) and Pb^{2+} is a group 4 metal ion and therefore, show a softer ion character.

3. Materials and Methods

3.1. General

All chemicals were purchased from commercial suppliers and used without further purification unless otherwise specified. Anhydrous THF was purchased from Acros, anhydrous $Ba(ClO_4)_2$ was purchased from Alfa Aesar, and deuterated solvents were purchased from deutero GmbH. Compounds **8b** [57], **9c** [63], **11a**, **12a** [59], **11b**, **12b** [64], **14a**, **15a**, **16a**, and **20** [26] were prepared according to the literature. NMR spectra of all compounds were recorded on an Agilent DD2-400 MHz NMR or an Agilent DD2-600 MHz NMR spectrometer with ProbeOne. Chemical shifts of the ^{1}H, ^{19}F, and ^{13}C spectra were reported in parts per million (ppm) using TMS as an internal standard for ^{1}H/^{13}C and $CFCl_3$ for ^{19}F spectra. Mass spectrometric (MS) data were obtained on a Xevo TQ-S mass spectrometer (Waters) by electron spray ionization (ESI). The melting points were determined on a Galen III melting point apparatus (Cambridge Instruments & Leica) and are uncorrected. TLC detections were performed using Silica Gel 60 F_{254} sheets (Merck, Darmstadt, Germany). TLCs were developed by visualization under UV light (λ = 254 nm). Chromatographic separations were accomplished by using an automated silica gel column chromatography system Biotage Isolera Four and appropriate Biotage KP-SIL SNAP columns. UV/Vis- measurements were realized at a Specord 50 by Analytik Jena. The calculation of the stability constants was accomplished using HypSpec 1.1.18.

3.2. The Computational Methodology

The calculations of calixcrowns were performed using Kohn-Sham DFT. The general gradient approximation functional BP86 [65,66] was used for geometry optimizations and the B3LYP [67,68] to calculate single point energies. With the latter settings, we have used a continuum solvation COSMO model for acetonitrile (dielectric constant ζ = 37.5) in order to approximately include the solvent environment used in the experiment. Ahlrichs' triple zeta valence polarized (def2-TZVP) [69] basis set was employed as well as the resolution of identity (RI) approach and corresponding auxiliary basis sets [70], along with Grimme's D3 dispersion correction [71]. Since the def2-TZVP was not available for radium, we have used the split valence polarized (def-SVP) basis set [72] for both Ra^{2+} and Ba^{2+} in the calculations that involved a comparison between their binding energies. Effective core potentials have been used for the heavy metals in order to account for scalar relativistic effects. The Turbomole 7.3 package [73] was used for all calculations in this study. To prevent the over-stabilization of final energies, basis set superposition error (BSSE) was considered. Binding energies were calculated as differences of electronic energies E via E(complex)-[E(calixcrown)+E(M^{2+})].

3.3. Chemical Syntheses

5,11,17,23-Tetrakis(*tert*-butyl)-25,27-dihydroxy-26,28-bis[2-(2-methoxyethoxy)ethoxy] calix[4]arene 1. Under Ar, *tert*-butylcalix[4]arene (2 g, 3.08 mmol) was dissolved in anhydrous acetonitrile (40 mL) and 2-(2-methoxyethoxy)ethyl tosylate (1.86 g, 6.78 mmol) and K_2CO_3 (1.06 g, 7.7 mmol) added. The reaction mixture was stirred at 72 °C for 4 d. After cooling, the solvent was changed to chloroform (40 mL), the insoluble ingredients were removed by filtration and the organic phase was washed with 10% aqueous HCl (2 × 30 mL) and water (2 × 30 mL). After separation, the organic phase was dried over Na_2SO_4 and the crude product was purified using column chromatography (petroleum ether:acetone = 10:1 → 5:1) to obtain **1** as colorless oil, which tends to crystallize after standing (2.31 g, 88%); mp 100 °C; R_f 0.46 (petroleum ether:acetone = 2:1); ^{1}H NMR (600 MHz, CDCl$_3$): δ 0.95 (s, 18 H, ^{t}Bu), 1.30 (s, 18 H, ^{t}Bu), 3.30 (d, ^2J = 13.1 Hz, 4 H, CH$_2$Ar), 3.39 (s, 6 H, OCH$_3$), 3.61–3.64 (m, 4 H, OCH$_2$), 3.83–3.86 (m, 4 H, OCH$_2$), 3.96–4.00 (m, 4 H, OCH$_2$), 4.16–4.19 (m, 4 H), 4.36 (d, ^2J = 13.1 Hz, CH$_2$Ar), 6.77 (s, 4 H, ArH), 7.06 (s, 4 H, ArH), 7.15 (s, 2 H, OH); ^{13}C NMR (151 MHz, CDCl$_3$): δ 31.2 (^{t}Bu), 31.9 (^{t}Bu), 33.9 (C$_q$), 34.0 (C$_q$), 59.2 (OCH$_3$), 70.1 (OCH$_2$), 71.1 (OCH$_2$), 72.3 (OCH$_2$), 75.4 (OCH$_2$), 125.1, 125.6 (2 × CH$_{Ar}$), 128.0 (C Ar),

132.7, 141.4, 146.9, 150.0, 150.8 (6 × C_{Ar}) ppm; MS (ESI+) m/z = 871 (M^+ + NH_4), 876 (M^+ + Na), 892 (M^+ + K).

5,11,17,23-Tetrakis(*tert*-butyl)-25,27-bis(2-ethoxy-2-oxoethoxy)-26,28-bis[2-(2-methoxyethoxy)ethoxy]calix[4]arene 2. Under Ar, compound **1** (1 g, 1.17 mmol) was dissolved in anhydrous THF (40 mL) and NaH (234 mg, 5.86 mmol, 60% in mineral oil) added and stirred at rt for 30 min. Afterwards, a solution of ethyl bromoacetate (978 mg, 5.86 mmol) in 5 mL of THF was added and the resulting mixture stirred at 50 °C for 2 d. After cooling to rt, the solvent was changed to chloroform (40 mL), the insoluble ingredients were removed by filtration and the organic phase was washed with 10% aqueous HCl (3 × 30 mL) and water (2 × 30 mL). After separation, the organic phase was dried over Na_2SO_4 and the crude product was purified using column chromatography (petroleum ether:acetone = 4:1) to obtain **2** as light-yellow oil (1.07 g, 89%); R_f 0.52 (petroleum ether:acetone = 2:1); ^{1}H NMR (400 MHz, $CDCl_3$): δ 1.03 (s, 18 H, ^{t}Bu), 1.12 (s, 18 H, ^{t}Bu), 1.29 (t, ^3J = 7.1 Hz, 6 H, CH_3), 3.15 (d, ^2J = 12.9 Hz, 4 H, CH_2Ar), 3.37 (s, 6 H, OCH_3), 3.53–3.57 (m, 4 H, OCH_2), 3.66–3.69 (m, 4 H, OCH_2), 3.94 (t, ^3J = 5.6 Hz, 4 H, OCH_2), 4.12 (t, ^3J = 5.6 Hz, 4 H, OCH_2), 4.22 (q, ^3J = 7.2 Hz, 4 H, OCH_2), 4.67 (d, ^2J = 12.9 Hz, 4 H, CH_2Ar), 4.77 (s, 4 H, $CH_2C{=}O$), 6.70 (s, 4 H, ArH), 6.83 (s, 4 H, ArH); ^{13}C NMR (101 MHz, $CDCl_3$): δ 14.4, 31.5, 31.6 (3 × CH_3), 31.7 (CH_2Ar), 33.9 (C_q), 34.0 (C_q), 59.2 (OCH_3), 60.5 (OCH_2), 70.4 (OCH_2), 70.7 (OCH_2), 71.3 ($CH_2C{=}O$), 72.2 (OCH_2), 73.2 (OCH_2), 125.1 (CH_{Ar}), 125.5 (CH_{Ar}), 133.5, 134.0, 144.8, 145.2, 153.2, 153.4 (6 × C_{Ar}), 170.8 (C=O) ppm; MS (ESI+) m/z = 1042 (M^+ + NH_4), 1047 (M^+ + Na).

5,11,17,23-Tetrakis(*tert*-butyl)-25,27-bis(carboxymethoxy)-26,28-bis[2-(2-methoxyethoxy)ethoxy]calix[4]arene 3. Compound **2** (213 mg, 0.21 mmol) was dissolved in THF (20 mL), a solution of $Me_4NOH \cdot 5\,H_2O$ (264 mg, 1.46 mmol) in methanol (1 mL) was added and the reaction mixture was stirred at 55 °C overnight. The major part of the solvent was removed and the product was precipitated with ice-cold aqueous HCl (6 M). The solid residue was filtered and washed with water (20 mL) and dissolved in chloroform (20 mL). The organic phase was washed with brine (3 × 10 mL) and 10% aqueous HCl (2 × 10 mL). After separation, the organic phase was dried over Na_2SO_4, the solvent was removed and the product was obtained as pale-orange solid (200 mg, >99%); mp 104–106 °C; ^{1}H NMR (400 MHz, $CDCl_3$): δ 0.83 (s, 18 H, ^{t}Bu), 1.34 (s, 18 H, ^{t}Bu) 3.24 (d, ^2J = 13 Hz, 4 H, CH_2Ar), 3.38 (s, 6H, OCH_3), 3.23–3.26 (m, 4 H, OCH_2), 3.73–3.75 (m, 4H, OCH_2), 3.83–3.85 (m, 4 H, OCH_2), 3.99–4.01 (m, 4 H, OCH_2), 4.45 (d, ^2J = 13.0 Hz, 4 H, CH_2Ar), 4.75 (s, 4 H, $CH_2C{=}O$), 6.56 (s, 4 H, ArH), 7.17 (s, 4 H, ArH); ^{13}C NMR (101 MHz, $CDCl_3$): δ 31.09 (CH_2Ar), 31.14 (CH_3), 31.8 (CH_3), 33.9 (C_q), 34.4 (C_q), 59.2 (OCH_3), 69.8 (OCH_2), 70.5 (OCH_2), 72.2 (OCH_2), 72.3 ($CH_2C{=}O$), 76.2 (OCH_2), 125.5 (CH_{Ar}), 126.2 (CH_{Ar}), 132.3, 135.0, 146.1, 147.3, 150.2, 153.3 (6 × C_{Ar}), 170.6 (C=O); MS (ESI−) m/z = 968 (M^- − H).

5,11,17,23-Tetrakis(*tert*-butyl)-26,28-bis[2-(2-methoxyethoxy)ethoxy]-25,27-bis(2-morpholino-2-oxoethyl)calix[4]arene 5. Under Ar, compound **3** (100 mg, 0.103 mmol) was dissolved in CCl_4 (5 mL), oxalyl chloride (0.75 g, 5.88 mmol) dropwise added and the reaction mixture stirred at 65 °C for 5 h. After cooling to rt, the remaining oxalyl chloride and the solvent were removed under high vacuum and anhydrous DCM (5 mL) added. After cooling the solution to 0 °C, morpholine (22 mg, 0.26 mmol) and DIPEA (40 mg, 0.31 mmol) were added to resulting mixture was allowed to come to rt and was stirred at rt overnight. Afterwards, DCM (20 mL) was added and the organic phase washed with saturated hydrogen carbonate solution (20 mL), aqueous HCl (10%, 20 mL), and brine (15 mL). After separation, the organic phase was dried over Na_2SO_4 and the crude product was purified using automated column chromatography (chloroform: methanol = 0 → 15%) to obtain **5** as colorless oil (67 mg, 59%); R_f 0.16 (chloroform:methanol = 9:1); ^{1}H NMR (400 MHz, $CDCl_3$): δ 1.04 (s, 18 H, ^{t}Bu), 1.09 (s, 18 H, ^{t}Bu), 3.14 (d, ^2J = 12.8 Hz, 4 H, CH_2Ar), 3.34 (s, 6 H, OCH_3), 3.37–3.44 (m, 4 H, NCH_2), 3.49–3.55 (m, 8 H, OCH_2), 3.59–3.71 (m, 12 H, NCH_2, OCH_2), 3.97 (t, ^3J = 5.7 Hz, 4 H, OCH_2), 4.18 (t, ^3J = 5.7 Hz, 4 H, OCH_2), 4.60 (d, ^2J = 12.8 Hz, 4 H, CH_2Ar), 4.78 (s, 4 H, $CH_2C{=}O$), 6.74 (s, 4 H, ArH), 6.78 (s, 4 H, ArH); ^{13}C NMR (101 MHz, $CDCl_3$): δ 31.5 (CH_3), 31.56 (CH_3), 31.62 (CH_2Ar), 33.9 (C_q), 34.0 (C_q), 42.0 (NCH_2), 45.9 (NCH_2), 59.1 (OCH_3), 66.8 (OCH_2), 67.0 (OCH_2), 70.1 (OCH_2), 70.5 (OCH_2), 71.6 ($CH_2C{=}O$),

72.2 (OCH$_2$), 72.8 (OCH$_2$), 125.2 (CH$_{Ar}$), 125.4 (CH$_{Ar}$), 133.4, 133.9, 144.8, 145.1, 153.1, 153.6 (6 × C$_{Ar}$), 168.3 (C=O) ppm; MS (ESI+): m/z = 1130 (M$^+$ + Na).

5,11,17,23-Tetrakis(tert-butyl)-26,28-bis[2-(2-methoxyethoxy)ethoxy]-25,27-[2-(1,4,7-trioxa-10-azacyclododecan-10-yl)-2-oxoethyl]calix[4]arene 6. Under Ar, compound 3 (100 mg, 0.10 mmol) was dissolved in CCl$_4$ (5 mL), oxalyl chloride (0.75 g, 5.88 mmol) dropwise added and the reaction mixture stirred at 65 °C for 5 h. After cooling to rt, the remaining oxalyl chloride and the solvent were removed under high vacuum and anhydrous DCM (5 mL) added. After cooling the solution to 0 °C, 1-aza-12-crown-4 (45 mg, 0.26 mmol) and DIPEA (40 mg, 0.31 mmol) were added to resulting mixture was allowed to come to rt and was stirred at rt overnight. Afterwards, DCM (20 mL) was added and the organic phase washed with saturated hydrogen carbonate solution (15 mL), aqueous HCl (10%, 15 mL), and brine (15 mL). After separation, the organic phase was dried over Na$_2$SO$_4$ and the crude product was purified using automated column chromatography (chloroform:methanol = 0 → 15%) to obtain **6** as colorless solid (103 mg, 78%); R$_f$ 0.29 (chloroform:methanol = 85:15); ^{1}H NMR (400 MHz, CDCl$_3$): δ = 1.08 (s, 18 H, ^{t}Bu), 1.13 (s, 18 H, ^{t}Bu), 3.26–3.32 (m, 10 H, ArCH$_2$, OCH$_3$), 3.34–3.42 (m, 4 H, CH$_2$), 3.44–3.48 (m, 4 H, CH$_2$), 3.50–3.55 (m, 4 H, CH$_2$), 3.58–3.67 (m, 16 H, CH$_2$), 3.69–3.76 (m, 8 H, CH$_2$), 3.86–3.94 (m, 8 H, CH$_2$), 4.12–4.18 (m, 4 H, CH$_2$), 4.34 (d, 4 H, ^2J = 12.5 Hz, CH$_2$Ar), 4.78 (s, 4 H, CH$_2$C=O), 7.05 (s, 4 H, ArH), 7.07 (s, 4 H, ArH) ppm; ^{13}C NMR (101 MHz, CDCl$_3$): δ = 30.0 (CH$_2$Ar), 31.3 (CH$_3$), 31.4 (CH$_3$), 34.2 (C$_q$), 34.3 (C$_q$), 49.3 (CH$_2$), 50.0 (CH$_2$), 59.0 (OCH$_3$), 68.6, 68.8, 69.1, 69.5, 69.9, 70.0, 70.1, 70.9, 72.0 (9 × CH$_2$), 73.9 (CH$_2$C=O), 75.2 (CH$_2$), 125.7 (CH$_{Ar}$), 125.9 (CH$_{Ar}$), 134.6, 134.8, 147.7, 148.3, 148.5, 150.0 (6 × C$_{Ar}$), 170.1 (C=O) ppm.

5,11,17,23-Tetrakis(*tert*-butyl)-26,28-bis[2-(2-methoxyethoxy)ethoxy]-25,27-[2-(trifluoromethylsulfonamido)-2-oxoethyl]calix[4]arene 7a. Under Ar, compound 3 (100 mg, 0.10 mmol) was dissolved in CCl$_4$ (5 mL), oxalyl chloride (0.75 g, 5.88 mmol) dropwise added and the reaction mixture stirred at 65 °C for 5 h. After cooling to rt, the remaining oxalyl chloride and the solvent were removed under high vacuum, anhydrous DCM (5 mL) added and the solution cooled to 0 °C. In a second flask, trifluoromethanesulfonamide (39 mg, 0.26 mmol) was dissolved in anhydrous THF (2 mL) and NaH (41 mg, 1.03 mmol, 10 eq., 60% in mineral oil) was added. After stirring for 15 min, this solution was added to the DCM solution at 0 °C and the combined mixture was stirred for 1.5 h at rt. The solvent was changed to chloroform (20 mL) and the organic phase washed with saturated hydrogen carbonate solution (15 mL), aqueous HCl (10%, 15 mL), and brine (15 mL). After separation, the organic phase was dried over Na$_2$SO$_4$ and the crude product was purified using automated column chromatography (petroleum ether:ethyl acetate = 0 → 30%) to obtain **7a** as colorless solid (68 mg, 54%); mp 113–115 °C; R$_f$ 0.76 (petroleum ether:ethyl acetate = 1:1); ^{1}H NMR (400 MHz, CDCl$_3$): δ 1.04 (s, 18 H, ^{t}Bu), 1.20 (s, 18 H, ^{t}Bu), 3.34 (d, ^2J = 12.4 Hz, 4 H, CH$_2$Ar), 3.34 (s, 6 H, OCH$_3$), 3.59 (t, ^3J = 4.3 Hz, 4 H, OCH$_2$), 3.69–3.78 (m, 8 H, OCH$_2$), 4.08–4.12 (m, 4 H, OCH$_2$), 4.28 (d, ^2J = 12.4 Hz, 4 H, CH$_2$Ar), 4.46 (s, 4 H, CH$_2$C=O), 6.99 (s, 4 H, ArH), 7.15 (s, 4 H, ArH); ^{13}C NMR (101 MHz, CDCl$_3$): δ 30.0 (CH$_2$Ar), 31.2 (CH$_3$), 31.5 (CH$_3$), 34.2 (C$_q$), 34.3 (C$_q$), 59.2 (OCH$_3$), 68.7 (OCH$_2$), 70.0 (OCH$_2$), 70.6 (OCH$_2$), 76.4 (OCH$_2$), 76.9 (CH$_2$C=O), 120.5 (q, $^1J_{C,F}$ = 323 Hz, CF$_3$), 125.97 (CH$_{Ar}$), 126.02 (CH$_{Ar}$), 134.3, 134.7, 147.6, 148.4, 148.5, 149.6 (6 × C$_{Ar}$), 178.3 (C=O); ^{19}F NMR (376 MHz, CDCl$_3$): δ − 79.0 (CF$_3$) ppm; MS (ESI+) m/z = 1232 (M$^+$ + H), 1254 (M$^+$ + Na).

5,11,17,23-Tetrakis(tert-butyl)-26,28-bis[2-(2-methoxy ethoxy)ethoxy]-25,27-[2-(heptafluoroisopropylsulfonamido)-2-oxoethyl]calix[4]arene 7b. Under Ar, compound 3 (100 mg, 0.10 mmol) was dissolved in CCl$_4$ (5 mL), oxalyl chloride (0.75 g, 5.88 mmol) dropwise added and the reaction mixture stirred at 65 °C for 5 h. After cooling to rt, the remaining oxalyl chloride and the solvent were removed under high vacuum, anhydrous THF (3 mL) added and the solution cooled to 0 °C. In a second flask, heptafluoroisopropylsulfonamide (64 mg, 0.26 mmol) was dissolved in anhydrous THF (2 mL) and NaH (41 mg, 1.03 mmol, 60% in mineral oil) was added. After stirring for 15 min, this solution was added to the other THF solution at 0 °C and the combined mixture was stirred for 1.5 h at rt. The solvent was changed to chloroform (20 mL) and the organic phase washed with saturated hydrogen carbonate solution (15 mL), aqueous HCl (10%, 15 mL), and brine (15 mL). After separation, the organic phase was dried over Na_2SO_4 and the crude product was purified using automated column chromatography (petroleum ether:ethyl acetate = 0 → 20%) to obtain **7b** as colorless solid (53 mg, 36%); mp 113–115 °C; R_f 0.76 (petroleum ether:ethyl acetate = 1:1); ^{1}H NMR (400 MHz, CDCl$_3$): δ 1.04 (s, 18 H, ^{t}Bu), 1.21 (s, 18 H, ^{t}Bu), 3.35 (d, ^2J = 12.5 Hz, 4 H, CH$_2$Ar), 3.41 (s, 6 H, OCH$_3$), 3.58 (t, ^3J = 4.4 Hz, 4 H, OCH$_2$), 3.70–3.77 (m, 8 H, OCH$_2$), 4.06–4.11 (m, 4 H, OCH$_2$), 4.27 (d, ^2J = 12.5 Hz, 4 H, CH$_2$Ar), 4.44 (s, CH$_2$C=O), 6.99 (s, 4 H, ArH), 7.16 (s, 4 H, ArH); ^{13}C NMR (101 MHz, CDCl$_3$): δ 29.9 (CH$_2$Ar), 31.2 (CH$_3$), 31.5 (CH$_3$), 34.2 (C$_q$), 34.4 (C$_q$), 59.2 (OCH$_3$), 68.8 (OCH$_2$), 70.0 (OCH$_2$), 70.6 (OCH$_2$), 76.3 (OCH$_2$), 76.9 (CH$_2$C=O), 96.5 (dsep, $^1J_{C,F}$ = 240 Hz, $^2J_{C,F}$ = 32 Hz, CF), 119.6 (dq, $^1J_{C,F}$ = 288 Hz, $^2J_{C,F}$ = 26 Hz, CF$_3$), 126.0 (CH$_{Ar}$), 126.04 (CH$_{Ar}$), 134.3, 134.7, 147.7, 148.5, 148.52, 149.5 (6 × C$_{Ar}$), 178.0 (C=O); ^{19}F NMR (565 MHz, CDCl$_3$): δ –170.5 (sep, ^3J = 7.0 Hz, 1 F, CF), −71.9 (d, ^3J = 7.0 Hz, 6 F, CF$_3$) ppm.

5,11,17,23-Tetrakis(*tert*-butyl)-26,28-bis[2-(2-methoxyethoxy)ethoxy]-25,27-[2-(pentafluorophenylsulfonamido)-2-oxoethyl]calix[4]arene 7c. Under Ar, compound 3 (100 mg, 0.10 mmol) was dissolved in CCl$_4$ (5 mL), oxalyl chloride (0.75 g, 5.88 mmol) dropwise added and the reaction mixture stirred at 65 °C for 5 h. After cooling to rt, the remaining oxalyl chloride and the solvent were removed under high vacuum, anhydrous THF (3 mL) added and the solution cooled to 0 °C. In a second flask, pentafluorophenylsulfonamide (64 mg, 0.26 mmol) was dissolved in anhydrous THF (2 mL) and NaH (41 mg, 1.03 mmol, 60% in mineral oil) was added. After stirring for 15 min, this solution was added to the other THF-solution at 0 °C and the combined mixture was stirred for 1.5 h at rt. The solvent was changed to chloroform (20 mL) and the organic phase washed with saturated hydrogen carbonate solution (15 mL), aqueous HCl (10%, 15 mL), and brine (15 mL). After separation, the organic phase was dried over Na_2SO_4 and the crude product was purified using automated column chromatography (petroleum ether:ethyl acetate = 0 → 20%) to obtain **7c** as colorless solid (15 mg, 10%); mp 98–100 °C; R_f 0.76 (petroleum ether:ethyl acetate = 1:1); ^{1}H NMR (600 MHz, CDCl$_3$): δ 1.06 (s, 18 H, ^{t}Bu), 1.18 (s, 18 H, ^{t}Bu), 3.31 (d, ^2J = 12.4 Hz, 4 H, ArCH$_2$), 3.39 (s, 6 H, OCH$_3$), 3.51 (t, ^3J = 4.6 Hz, 4 H, OCH$_2$), 3.58 (t, ^3J = 3.8 Hz, 4 H, OCH$_2$), 3.67 (t, ^3J = 4.6 Hz, 4 H, OCH$_2$), 4.02 (t, ^3J = 3.8 Hz, 4 H, OCH$_2$), 4.25 (d, ^2J = 12.4 Hz, 4 H, CH$_2$Ar), 4.41 (s, 4 H, CH$_2$C=O), 7.00 (s, 4 H, ArH), 7.12 (s, 4 H, ArH); ^{13}C NMR (151 MHz, CDCl$_3$): δ 30.0 (CH$_2$Ar), 31.3 (CH$_3$), 31.5 (CH$_3$), 34.2 (C$_q$), 34.3 (C$_q$), 59.3 (OCH$_3$), 68.3 (OCH$_2$), 69.7 (OCH$_2$), 70.4 (OCH$_2$), 75.9 (OCH$_2$), 76.9 (CH$_2$C=O), 125.9 (CH$_{Ar}$), 125.94 (CH$_{Ar}$), 134.5, 134.6, 147.7, 148.3, 148.9, 149.5 (6 × C$_q$), 177.3 (C=O); ^{19}F NMR (564 MHz, CDCl$_3$): δ –160.7 (t, ^3J = 20.0 Hz, 4 F, m-ArF), –150.3 (t, ^3J = 20.5 Hz, 2 F, p-ArF), −130.1 (d, ^3J = 20.0 Hz, 4 F, o-ArF) ppm; MS (ESI+) m/z = 1428 (M$^+$ + H), 1445 (M$^+$ + NH$_4$).

***tert*-Butyl 3,4-bis[2-(2-hydroxyethoxy)ethoxy]benzoate 11c.** Under Ar, compound **9c** (176 mg, 0.84 mmol), K$_2$CO$_3$ (578 mg, 4.19 mmol) and KI (194 mg, 4.2 mmol) were suspended in anhydrous DMF (5 mL) and 2-(2-chloroethoxy)ethanol (354 μL, 417 mg, 3.35 mmol) was added. The mixture was stirred at 75 °C overnight. After cooling to rt, the solvent was changed to DCM (15 mL) and filtered. The organic phase was washed with hydrochloric acid (10 mL, 10%), brine (10 mL) and water (10 mL). After removal of the solvent, the crude product was purified via column chromatography (ethyl acetate:methanol 99:1) to obtain **11c** as colorless oil (140 mg, 43%); R_f = 0.4 (DCM:methanol = 9:1); ^{1}H NMR (400 MHz, CDCl$_3$): δ 1.58 (s, 9H, ^{t}Bu). 2.79 (s, 2H, OH), 3.72–3.65 (m, 4H, OCH$_2$), 3.79–3.71 (m, 4H, OCH$_2$), 3.95–3.88 (m, 4H, OCH$_2$), 4.24–4.17 (m, 4H, OCH$_2$), 6.86 (d, ^3J = 8.4 Hz, 1H, ArH), 7.53 (d, ^4J = 1.9 Hz, 1H, ArH), 7.60 (dd, ^3J = 8.4 Hz, ^4J = 1.9 Hz, 1H, ArH); ^{13}C NMR (101 MHz, CDCl$_3$): δ 28.2 (^{t}Bu), 61.6 (OCH$_2$), 68.3 (OCH$_2$), 68.4 (OCH$_2$), 69.0 (OCH$_2$), 69.2 (OCH$_2$), 72.7 (OCH$_2$), 72.8 (OCH$_2$), 80.8 (C$_q$), 111.7 (CH$_{Ar}$), 113.9 (CH$_{Ar}$), 123.6 (CH$_{Ar}$), 125.0 (C$_{Ar}$), 147.7 (C$_{Ar}$), 151.9 (C$_{Ar}$), 165.5 (C=O) ppm; MS (ESI+): m/z = 409 (M$^+$ + Na).

1,2-Bis[2-(2-hydroxyethoxy)ethoxy]-4-nitrobenzene 11d. Under Ar, 4-nitrocatechol (1.0 g, 6.45 mmol), K$_2$CO$_3$ (8.0 g, 64 mmol) and KI (1.66 g, 9.03 mmol) were suspended in anhydrous DMF (20 mL) and 2-(2-chloroethoxy)ethanol (2.72 g, 29mmol) was added. The mixture was stirred at 75 °C overnight. After cooling to rt, the solvent was changed to DCM (35 mL) and filtered. The organic phase was washed with hydrochloric acid (25 mL, 10%), brine (25 mL) and water (25 mL). After removal of the solvent, the crude product was purified via column chromatography (DCM:methanol 97:3) to obtain **11d** as colorless oil (1.08 g, 52%); R_f 0.4 (DCM:methanol 9:1); ^{1}H NMR (600 MHz, CDCl$_3$): δ 3.41 (s, 2H, OH), 3.63–3.59 (m, 4H, OCH$_2$), 3.69–3.66 (m, 4H, OCH$_2$), 3.90–3.84 (m, 4H, OCH$_2$), 4.20–4.12 (m, 4H, OCH$_2$), 6.85 (d, ^3J = 8.9 Hz, 1H, ArH), 7.67 (d, ^4J = 2.6 Hz, 1H, ArH), 7.80 (dd, ^3J = 8.9 Hz, ^4J = 2.6 Hz, 1H, ArH); ^{13}C NMR (151 MHz, CDCl$_3$): δ 61.4 (OCH$_2$), 61.5 (OCH$_2$), 68.6 (OCH$_2$), 68.7 (OCH$_2$), 68.8 (OCH$_2$), 68.9 (OCH$_2$), 72.7 (OCH$_2$), 72.7 (OCH$_2$), 108.0 (CH$_{Ar}$), 111.3 (CH$_{Ar}$), 118.0 (CH$_{Ar}$), 141.4 (C$_{Ar}$), 148.0 (C$_{Ar}$), 153.9 (C$_{Ar}$) ppm; MS (ESI+): m/z = 354 (M$^+$ + Na).

tert-**Butyl 3,4-bis{2-[2-(tosyloxy)ethoxy]ethoxy}benzoate 12c.** Under Ar, compound **11c** (459 mg, 1.2 mmol) and Et$_3$N (536 µL, 392 mg, 3.87 mmol) were dissolved in anhydrous DCM (15 mL), cooled to 0 °C, *p*-TsCl (1.40 g, 9.9 mmol) added and the reaction mixture stirred at rt overnight. Afterwards, the organic phase was washed with water (2 × 15 mL), dried over Na$_2$SO$_4$. After removal of the solvent, the crude product was purified using automated column chromatography (DCM → DCM:methanol 99:1) to obtain **12c** as colorless oil (298 mg, 36%); R_f 0.6 (DCM:methanol = 99:1); ^{1}H NMR (400 MHz, CDCl$_3$): δ 1.59 (s, 9H, ^{t}Bu), 2.42–2.39 (m, 6H, CH$_3$), 3.85–3.73 (m, 8H, OCH$_2$), 4.15–4.05 (m, 4H, OCH$_2$), 4.21–4.14 (m, 4H, OCH$_2$), 6.84 (d, ^3J = 8.4 Hz, 1H, ArH), 7.32–7.25 (m, 4H, Ar$_{Ts}$), 7.48 (d, ^4J = 2.0 Hz, 1H, ArH), 7.60 (dd, ^3J = 8.4 Hz, ^4J = 2.0 Hz, 1H, ArH), 7.78 (d, ^3J = 8.0 Hz, 4H, Ar$_{Ts}$); ^{13}C NMR (151 MHz, CDCl$_3$): δ 21.6 (CH$_3$), 21.6 (CH$_3$), 28.2 (^{t}Bu), 68.6 (OCH$_2$), 68.8 (OCH$_2$), 68.9 (OCH$_2$), 69.0 (OCH$_2$), 69.3 (OCH$_2$), 69.4 (OCH$_2$), 69.6 (OCH$_2$), 69.8 (OCH$_2$), 80.8 (C$_q$), 112.6 (CH$_{Ar}$), 114.8 (CH$_{Ar}$), 123.8 (CH$_{Ar}$), 125.1 (C$_{Ar}$), 128.0 (CH$_{Ts}$), 129.8 (CH$_{Ts}$), 133.0, 144.8, 148.1, 152.4 (4 × C$_{Ar}$), 165.5 (C=O) ppm; MS (ESI+): m/z 718 (M$^+$ + Na).

1,2-Bis{2-[2-(tosyloxy)hydroxyethoxy]ethoxy}-4-nitrobenzene 12d. Compound **11d** (1.37 g, 4,13 mmol) was dissolved in THF and cooled to 0 °C. NaOH (500 mg, 12.4 mmol) dissolved in water (2.5 mL) and *p*-tosyl chloride (1.97 g, 10.3 mmol) were added and the mixture was stirred 3 h at rt. Afterwards, saturated hydrogen carbonate solution (10 mL) was added and extracted with DCM (2 × 50 mL). The organic phase was washed with water (2 × 50 mL), dried over Na$_2$SO$_4$ and the crude product was purified using automated column chromatography (DCM → DCM:methanol 50:1) to obtain **12d** as yellowish oil (1.74 g, 66%); R_f 0.6 (DCM:methanol = 99:1); ^{1}H NMR (600 MHz, CDCl$_3$): δ 2.42–2.40 (m, 6H, CH$_3$), 3.79–3.76 (m, 4H, OCH$_2$), 3.87–3.81 (m, 4H, OCH$_2$), 4.15–4.11 (m, 2H, OCH$_2$), 4.20–4.15 (m, 6H, OCH$_2$), 6.91 (d, ^3J = 8.9 Hz, 1H, ArH), 7.32–7.28 (m, 4H, Ts), 7.73 (d, ^4J = 2.6 Hz, 1H, ArH), 7.80–7.75 (m, 4H, Ts), 7.88 (dd, ^3J = 8.9 Hz, ^4J = 2.6 Hz, 1H, ArH); ^{13}C NMR (151 MHz, CDCl$_3$): δ 21.7 (CH$_3$), 21.8 (CH$_3$), 69.1 (OCH$_2$), 69.2 (OCH$_2$), 69.2 (OCH$_2$), 69.3 (OCH$_2$), 69.4 (OCH$_2$), 69.5 (OCH$_2$), 69.6 (CH$_2$), 69.7 (OCH$_2$), 109.0 (CH$_{Ar}$), 112.1 (CH$_{Ar}$), 118.3 (CH$_{Ar}$), 128.1 (CH$_{Ts}$), 128.2 (CH$_{Ts}$), 129.9 (CH$_{Ts}$), 130.0 (CH$_{Ts}$), 133.1, 141.7, 145.0, 145.1, 148.5, 154.5 (6 × C$_{Ar}$) ppm; MS (ESI+): m/z = 662 (M$^+$ + Na).

5,11,17,23-Tetrakis(*tert*-butyl)-26,28-dihydroxycalix[4]arene-propylenecrown-6 14b. Under Ar, *tert*-butylcalix[4]arene (**13**, 158 mg, 0.24 mmol) and K$_2$CO$_3$ (39 mg, 0.28 mmol) were suspended in DCM (10 mL) and compound **8b** (273 mg, 0.49 mmol) dissolved in 5 mL of DCM was added dropwise. The resulting mixture was stirred at 30 °C for 11 d. Afterwards, the DCM phase was washed with water (2 × 15 mL) and aqueous HCl (10%, 2 × 15 mL), dried over Na$_2$SO$_4$ and the solvent was removed. The crude product was purified via column chromatography (petroleum ether:ethyl acetate = 2:1 → 1:1) yielding compound **14b** as a colourless solid (85 mg, 40%); R_f 0.2 (petroleum ether:ethyl acetate = 1:1); ^{1}H NMR (600 MHz, CDCl$_3$): δ 0.94 (s, 18H, ^{t}Bu), 1.30 (s, 18H, ^{t}Bu), 1.76 (p, ^3J = 5.8 Hz, 2H, OCH$_2$), 3.30 (d, ^2J = 13.0 Hz, 4H, CH$_2$Ar), 3.67 (t, ^3J = 5.8 Hz, 4H, OCH$_2$), 3.75–3.78 (m, 4H, OCH$_2$), 3.90–3.93 (m, 4H, OCH$_2$), 4.00–4.03 (m, 4H, OCH$_2$), 4.13–4.16 (m, 4H, OCH$_2$), 4.37 (d, ^2J = 13.0 Hz, 4 H, CH$_2$Ar), 6.78 (s, 4H ArH), 7.06 (s, 4H, ArH), 7.18 (s, 2H, OH); ^{13}C NMR (101 MHz, CDCl$_3$): δ 30.4 (OCH$_2$), 31.1 (CH$_3$), 31.6 (CH$_2$Ar), 31.9 (CH$_3$),

34.0 (C_q), 34.1 (C_q), 67.5 (OCH_2), 71.0 (OCH_2), 70.1 (OCH_2), 71.3 (OCH_2), 76.1 (OCH_2), 125.1 (CH_{Ar}), 125.6 (CH_{Ar}), 128.0 (C_{Ar}), 132.7 (C_{Ar}), 141.4 (C_{Ar}), 146.9 (C_{Ar}), 150.0 (C_{Ar}), 150.8 (C_{Ar}) ppm; MS (ESI+): m/z = 883 (M^+ + NH_4), 888 (M^+ + Na).

5,11,17,23-Tetrakis(*tert*-butyl)-26,28-dihydroxycalix[4]arene-benzocrown-6 14c. Under Ar, *tert*-butylcalix[4]arene (**13**, 610 mg, 0.94 mmol) and K_2CO_3 (150 mg, 1.08 mmol) were suspended in DCM (20 mL) and compound **12a** (1.12 g, 1.88 mmol) dissolved in 10 mL of DCM was added dropwise. The resulting mixture was stirred at 30 °C for 11 d. Afterwards, the DCM phase was washed with water (2 × 30 mL) and aqueous HCl (10%, 2 × 30 mL), dried over Na_2SO_4 and the solvent was removed. The crude product was purified via column chromatography (DCM:ethyl acetate = 20:1 → 10:1) yielding compound **14c** as colourless solid (346 mg, 41%); mp 154 °C; R_f 0.2 (DCM:methanol = 96:4); ^{1}H NMR (600 MHz, $CDCl_3$): δ 0.93 (s, 18H, ^{t}Bu), 1.29 (s, 18H, ^{t}Bu), 3.29 (d, ^2J = 13.1 Hz, 4H, CH_2Ar), 4.11 (t, ^3J = 4.7 Hz, 4H, OCH_2), 4.15 (t, ^3J = 4.7 Hz, 4H, OCH_2), 4.23 (t, ^3J = 4.7 Hz, 4H, OCH_2), 4.27–4.33 (m, 8H, OCH_2, CH_2Ar), 6.77 (s, 4H, ArH), 6.91–6.94 (m, 2H, ArH), 6.97 (m, 2H, ArH), 7.05 (s, 4H, ArH), 7.18 (s, 2H, OH); ^{13}C NMR (151 MHz, $CDCl_3$): δ 31.1 (^{t}Bu), 31.6 (CH_2Ar), 31.9 (^{t}Bu), 34.0 (C_q), 34.1 (C_q), 70.3 (OCH_2), 70.5 (OCH_2), 71.1 (OCH_2), 76.5 (OCH_2), 116.5 (CH_{Ar}), 122.2 (CH_{Ar}), 125.1 (CH_{Ar}), 125.7 (CH_{Ar}), 127.9, 132.6, 141.5, 147.1, 149.5, 149.9, 150.8 (7 × C_{Ar}) ppm; MS (ESI+): m/z = 917 (M^+ + NH_4), 922 (M^+ + Na).

5,11,17,23-Tetrakis(*tert*-butyl)-26,28-dihydroxycalix[4]arene-benzocrown-8 14d. Under Ar, *tert*-butylcalix[4]arene (**13**, 265 mg, 0.41 mmol) and K_2CO_3 (65 mg, 0.47 mmol) were suspended in DCM (20 mL) and compound **12b** (557 mg, 0.82 mmol) dissolved in 5 mL of DCM was added dropwise. The resulting mixture was stirred at 30 °C for 11 d. Afterwards, the DCM phase was washed with water (2 × 20 mL) and aqueous HCl (10%, 2 × 20 mL), dried over Na_2SO_4 and the solvent was removed. The crude product was purified via column chromatography (petroleum ether:acetone = 3:1) yielding compound **14d** as colourless solid (325 mg, 81%); R_f 0.2 (petroleum ether:acetone = 3:1); ^{1}H NMR (400 MHz, $CDCl_3$): δ 0.95 (s, 18H, ^{t}Bu), 1.29 (s, 18H, ^{t}Bu), 3.28 (d, ^2J = 13.0 Hz, 4H, CH_2Ar), 3.91–3.96 (m, 8H, OCH_2), 3.97–4.05 (m, 8H, OCH_2), 4.10–4.16 (m, 8H, OCH_2), 4.35 (d, ^2J = 13.0 Hz, 4H, CH_2Ar), 6.78 (s, 4H, ArH), 6.85–6.91 (m, 4H, ArH), 7.05 (s, 4H, ArH), 7.30 (s, 2H, OH) ppm; ^{13}C NMR (101 MHz, $CDCl_3$): δ 31.2 (CH_3), 31.6 (CH_2Ar), 31.9 (CH_3), 34.0 (C_q), 34.1 (C_q), 69.6 (OCH_2), 70.1 (OCH_2), 70.4 (OCH_2), 71.4 (OCH_2), 71.5 (OCH_2), 76.1 (OCH_2), 114.2 (CH_{Ar}), 121.5 (CH_{Ar}), 125.1 (CH_{Ar}), 125.6 (CH_{Ar}), 127.9, 132.8, 141.4, 147.0, 149.2, 149.9, 150.9 (7 × C_{Ar}) ppm; MS (ESI+): m/z = 1009 (M^+ + Na).

5,11,17,23-Tetrakis(*tert*-butyl)-26,28-dihydroxycalix[4]arene-4-*tert*-butoxycarbonyl-b-enzocrown-6 14e. Under Ar, *tert*-butylcalix[4]arene (**13**, 140 mg, 0.22 mmol) and K_2CO_3 (34 mg, 0.25 mmol) were suspended in DCM (10 mL). Compound **12e** (298 mg, 0.43 mmol) dissolved in 5 mL of DCM was added dropwise. The resulting mixture was stirred at 30 °C for 11 d. Afterwards, the DCM phase was washed with water (2 × 20 mL) and aqueous HCl (10%, 2 × 20 mL), dried over Na_2SO_4 and the solvent was removed. The crude product was purified via column chromatography (petroleum ether:acetone = 4:1 → 1:1) yielding compound **14e** as colourless solid (78 mg, 36%); R_f 0.5 (petroleum ether:acetone = 2:1); ^{1}H NMR (400 MHz, $CDCl_3$): δ 0.93 (s, 18H, ^{t}Bu), 1.30 (s, 18H, ^{t}Bu), 1.58 (s, 9H, O ^{t}Bu), 3.30 (d, ^2J = 12.7 Hz, 4H, CH_2Ar), 4.10–4.33 (m, 20H, OCH_2 + CH_2Ar), 6.76 (s, 4H, ArH), 6.90 (d, ^3J = 8.1 Hz, 1H, ArH), 7.05 (s, 4H, ArH), 7.14 (s, 2H, OH), 7.59–7.64 (m, 2H, ArH); ^{13}C NMR (101 MHz, $CDCl_3$): δ 28.4 (O^tBu), 31.6 (CH_2Ar), 31.1 (CH_3), 31.9 (CH_3), 34.0 (C_q), 34.0 (C_q), 69.8 (OCH_2), 70.3 (OCH_2), 70.5 (OCH_2), 70.6 (OCH_2), 70.8 (OCH_2), 76.4 (OCH_2), 76.5 (OCH_2), 80.8 (OC_q), 114.0 (CH_{Ar}), 116.9 (CH_{Ar}), 124.4 (CH_{Ar}), 125.1 (CH_{Ar}), 125.2 (CH_{Ar}), 125.7 (CH_{Ar}), 126.1, 127.8, 127.9, 132.6, 141.8, 141.9, 147.0, 147.1, 148.5, 150.8, 153.8 (11 × C_{Ar}), 165.7 C=O) ppm; MS (ESI+): m/z = 1022 (M^+ + Na).

5,11,17,23-Tetrakis(*tert*-butyl)-26,28-dihydroxycalix[4]arene-4-nitrobenzocrown-6 14f. Under Ar, *tert*-butylcalix[4]arene (**13**, 882 mg, 1.36 mmol) and K_2CO_3 (220 mg, 1.56 mmol) were suspended in DCM (20 mL), and compound **12f** (1.74 g, 2.72 mmol) dissolved in 10 mL of DCM was added dropwise. The resulting mixture was stirred at 30 °C for 11 d. Afterwards, the DCM phase was washed with water (2 × 30 mL) and aqueous HCl (10%,

2 × 30 mL), dried over Na_2SO_4 and the solvent was removed. The crude product was purified via column chromatography (DCM:ethyl acetate = 97:3 → 9:1) yielding compound **14f** as colourless solid (870 mg, 68%); R_f 0.3 (DCM:ethyl acetate = 9:1); ^{1}H NMR (600 MHz, CDCl$_3$): δ 0.91 (s, 18H, CH$_3$), 1.29 (s, 18H, CH$_3$), 3.25–3.30 (m, 4H, ArCH$_2$), 4.08–4.15 (m, 8H, OCH$_2$, ArCH$_2$), 4.18–4.30 (m, 8H, OCH$_2$), 4.34–4.40 (m, 4H, OCH$_2$), 6.71–6.76 (m, 4H, ArH), 6.88–6.92 (m, 1H, ArH), 6.99 (s, 2H, OH), 7.02–7.04 (m, 4H, ArH), 7.77–7.83 (m, 2H, ArH); ^{13}C NMR (151 MHz, CDCl$_3$): δ 31.1 (^{t}Bu), 31.5 (ArCH$_2$), 31.9 (^{t}Bu), 34.0 (C$_q$), 34.1 (C$_q$), 69.9 (OCH$_2$), 70.3 (OCH$_2$), 70.4 (OCH$_2$), 70.5 (OCH$_2$), 70.6 (OCH$_2$), 70.8 (OCH$_2$), 76.2 (OCH$_2$), 76.3 (OCH$_2$), 110.5 (CH$_{Ar}$), 113.3 (CH$_{Ar}$), 118.5 (CH$_{Ar}$), 125.1 (CH$_{Ar}$), 125.2 (CH$_{Ar}$), 125.7 (CH$_{Ar}$), 127.8, 127.9, 132.4, 132.5, 141.6, 141.9, 147.1, 149.8, 149.9, 150.7, 154.9 ppm (11 × C$_{Ar}$); MS (ESI+): m/z = 967 (M$^+$ + Na), 983 (M$^+$ + K).

5,11,17,23-Tetrakis(*tert*-butyl)-25,27-bis(2-ethoxy-2-oxoethoxy)calix[4]arene-propylenecrown-6 15b. Under Ar, compound **14b** (85 mg, 0.12 mmol) was dissolved in anhydrous THF (10 mL) and NaH (23 mg, 0.58 mmol, 60% in mineral oil) was added and the mixture stirred for 30 min at rt. Next, ethyl bromoacetate (65 µL, 0.58 mmol) was added and the mixture stirred at 60 °C overnight. Afterwards, the solvent was changed to DCM (15 mL), the organic phase washed with water (2 × 15 mL) and dried over Na_2SO_4. After removal of the solvent, the crude product was purified via column chromatography (petroleum ether:ethyl acetate = 2:1 → 1:1) yielding compound **15b** as colourless solid (45 mg, 37%); R_f 0.2 (petroleum ether:ethyl acetate = 2:1); ^{1}H NMR (600 MHz, CDCl$_3$): δ 0.88 (s, 18H, ^{t}Bu), 1.26 (s, 18H, ^{t}Bu), 1.30 (t, ^3J = 7.1 Hz, 6H, CH$_3$), 1.83 (p, ^3J = 5.7 Hz, 2H, CH$_2$), 3.16 (d, ^2J = 12.7 Hz, 4H, CH$_2$Ar), 3.67 (t, ^3J = 5.7 Hz, 4H, OCH$_2$), 3.70 (m, 4H, OCH$_2$), 3.75–3.78 (m, 4H, OCH$_2$), 4.11–4.15 (m, 4H, OCH$_2$), 4.21–4.28 (m, 8H, CH$_2$C=O, OCH$_2$), 4.51 (d, ^2J = 12.7 Hz, 4H, CH$_2$Ar), 6.55 (s, 4H, ArH), 4.56 (s, 4H, OCH$_2$), 7.02 (s, 4H, ArH); ^{13}C NMR (151 MHz, CDCl$_3$): δ 14.4 (CH$_3$), 30.5 (CH$_2$), 31.3 (CH$_3$), 31.4 (CH$_2$Ar), 31.8 (CH$_3$), 33.8 (C$_q$), 34.2 (C$_q$), 60.8 (CH 3 CH 2 OCO), 67.1 (OCH$_2$), 70.0 (OCH$_2$), 70.1 (OCH$_2$), 70.4 (OCH$_2$), 72.1 (OCH$_2$C=O), 72.7 (OCH$_2$), 125.0 (CH$_{Ar}$), 125.6 (CH$_{Ar}$), 132.4, 135.1, 145.0, 145.2, 152.3, 154.3 (6 × C$_{Ar}$), 169.9 (C=O) ppm; MS (ESI+): m/z = 1060 (M$^+$ + Na).

5,11,17,23-Tetrakis(*tert*-butyl)-25,27-bis(2-ethoxy-2-oxoethoxy)calix[4]arene-benzocrown-6 15c. Under Ar, compound **14c** (346 mg, 0.39 mmol) was dissolved in anhydrous THF (20 mL) and NaH (77 mg, 1.93 mmol, 60% in mineral oil) was added and the mixture stirred for 30 min at rt. Next, ethyl bromoacetate (214 µL, 1.93 mmol) and NaI in catalytic amounts were added and the mixture stirred at 60 °C overnight. Afterwards, the solvent was changed to DCM (20 mL), the organic phase was washed with water (2 × 20 mL) and dried over Na_2SO_4. After removal of the solvent, the product was recrystallized from hot methanol. After filtration the product was washed with cold methanol yielding compound **15c** as colorless solid (238 mg, 58%); R_f 0.2 (petroleum ether: ethyl acetate = 2:1); ^{1}H NMR (600 MHz, CDCl$_3$): δ 0.92 (s, 18H, CH$_3$), 1.14 (t, ^3J = 7.1 Hz, 6H, CH$_3$), 1.23 (s, 18H, CH$_3$), 3.17 (d, ^2J = 12.7 Hz, 4H, ArCH$_2$), 4.00–4.07 (m, 8H, OCH$_2$, CH$_2$), 4.19–4.28 (m, 12H, OCH$_2$), 4.54 (d, ^2J = 12.7 Hz, 4H, ArCH$_2$), 4.61 (s, 4H, CH$_2$C=O), 6.58 (s, 4H, ArH), 6.91 (s, 4H, ArH), 6.97 (s, 4H, ArH); ^{13}C NMR (151 MHz, CDCl$_3$): δ 28.1 (CH$_3$), 31.2 (ArCH$_2$), 31.3 (^{t}Bu), 31.7 (^{t}Bu), 33.8 (C$_q$), 34.1 (C$_q$), 69.4 (OCH$_2$), 70.2 (OCH$_2$), 70.9 (OCH$_2$), 72.0, (CH$_2$C=O), 77.4 (OCH$_2$), 114.6 (CH$_{Ar}$), 121.6 (CH$_{Ar}$), 123.6 (CH$_{Ar}$), 125.1 (CH$_{Ar}$), 125.6 (CH$_{Ar}$), 128.2, 132.6, 134.7, 145.1, 149.6, 149.6 (6 × C$_{Ar}$), 170.0 (C=O) ppm; MS (ESI+): m/z = 1094 (M$^+$ + Na).

5,11,17,23-Tetrakis(*tert*-butyl)-25,27-bis(2-ethoxy-2-oxoethoxy)calix[4]arene-benzocrown-8 15d. Under Ar, compound **14d** (325 mg, 0.33 mmol) was dissolved in anhydrous THF (20 mL) and NaH (184 mg, 1.7 mmol, 60% in mineral oil) was added and the mixture stirred for 30 min at rt. Next, ethyl bromoacetate (183 µL, 1.7 mmol) and NaI in catalytic amounts were added and the mixture stirred at 60 °C overnight. Afterwards, the solvent was changed to DCM (15 mL), the organic phase was washed with brine (2 × 15 mL) and dried over Na_2SO_4. After removal of the solvent, the crude product was purified via column chromatography (DCM:methanol = 95:5) yielding compound **15d** as colorless solid (170 mg, 45%); R_f 0.2 (petroleum ether:ethyl acetate = 2:1); ^{1}H NMR (600 MHz, CDCl$_3$): δ 0.87 (s, 18H, CH$_3$), 1.23–1.28 (m, 24H, CH$_3$ + OCH$_2$C$\underline{H}_3$), 3.15 (d, ^2J = 12.7 Hz, 4H, ArCH$_2$),

3.79–3.82 (m, 4H, OCH$_2$), 3.87–3.90 (m, 4H, OCH$_2$), 3.94 (t, ^3J = 4.5 Hz, 4H, OC$\underline{H}_2$CH$_3$), 4.10–4.33 (m, 16H, OCH$_2$), 4.48 (d, ^2J = 12.7 Hz, 4H, ArCH$_2$), 4.52 (s, 4H, CH$_2$C=O), 6.53 (s, 4H, ArH), 6.88–6.92 (m, 4H, ArH), 7.02 (m, 4H, ArH); ^{13}C NMR (151 MHz, CDCl$_3$): δ 14.4 (CH$_3$), 31.3 (^{t}Bu), 31.6 (ArCH$_2$), 31.8 (^{t}Bu), 60.9 (CH$_2$), 69.6 (OCH$_2$), 69.9 (OCH$_2$), 70.0 (OCH$_2$), 70.8 (OCH$_2$), 71.2 (OCH$_2$), 72.2 (CH$_2$C=O), 72.5 (OCH$_2$), 77.4 (OCH$_2$), 114.3 (CH$_{Ar}$), 121.6 (CH$_{Ar}$), 125.1 (CH$_{Ar}$), 125.6 (CH$_{Ar}$), 132.3, 135.1, 145.0, 145.2, 149.2, 152.3, 154.4 (7 × C$_{Ar}$), 169.9 (C=O) ppm; MS (ESI+): m/z = 1182 (M$^+$ + Na).

5,11,17,23-Tetrakis(*tert*-butyl)-25,27-bis(2-ethoxy-2-oxoethoxy)calix[4]arene-4-*tert*-butoxycarbonyl-benzocrown-6 15e. Under Ar, compound **14e** (78 mg, 0.078 mmol) was dissolved in anhydrous THF (10 mL) and NaH (16 mg, 0.39 mmol, 60% in mineral oil) was added and the mixture stirred for 30 min at rt. Next, ethyl bromoacetate (44 µL, 0.39 mmol) and NaI in catalytic amounts were added and the mixture stirred at 60 °C overnight. Afterwards, the solvent was changed to DCM (15 mL), the organic phase was washed with brine (2 × 15 mL) and dried over Na$_2$SO$_4$. After removal of the solvent, the crude product was purified via column chromatography (petroleum ether:ethyl acetate = 4:1) yielding compound **15e** as colorless solid (31 mg, 34%); R$_f$ 0.2 (petroleum ether:ethyl acetate = 2:1); ^{1}H NMR (400 MHz, CDCl$_3$): δ 0.93 (s, 18H, CH$_3$), 1.12 (t, ^3J = 7.2 Hz, 6H, CH$_2$C$\underline{H}_3$), 1.21 (s, 18H, CH$_3$), 1.58 (s, 9H, O^{t}Bu), 3.17 (d, ^2J = 12.8 Hz, 4H, ArCH$_2$), 3.97–4.07 (m, 8H, OCH$_2$), 4.20–4.29 (m, 12H, OCH$_2$), 4.50–4.58 (m, 4H, ArCH$_2$), 4.62 (s, 4H, CH$_2$C=O), 6.59 (s, 4H, ArH), 6.86 (d, ^3J = 8.5 Hz, 1H, ArH), 6.95 (s, 4H, ArH), 7.52 (s, 1H, ArH), 7.60 (d, ^3J = 8.5 Hz, 1H, ArH); ^{13}C NMR (101 MHz, CDCl$_3$): δ 14.2 (CH$_2\underline{C}$H$_3$), 28.4 (O^{t}Bu), 31.4 (^{t}Bu), 31.5 (ArCH$_2$), 31.7 (^{t}Bu), 33.9 (C$_q$), 34.1 (C$_q$), 60.7 ($\underline{C}$H$_2$CH$_3$), 69.2 (OCH$_2$), 69.6 (OCH$_2$), 70.0 (OCH$_2$), 70.1 (OCH$_2$), 71.1 (OCH$_2$), 71.2 (OCH$_2$), 71.9 ($\underline{C}$H$_2$C=O), 73.5 (OCH$_2$), 80.8 (COO$\underline{C}$(CH$_3$)$_3$), 112.4 (CH$_{Ar}$), 115.0 (CH$_{Ar}$), 123.8 (CH$_{Ar}$), 124.9 (C$_{Ar}$), 125.1 (CH$_{Ar}$), 125.1 (CH$_{Ar}$), 125.6 (CH$_{Ar}$), 132.7, 134.6, 134.6, 145.1, 148.6, 152.2, 153.1, 154.5, 154.6 (9 × C$_{Ar}$), 165.7 (C=O), 170.0 (C=O) ppm; MS (ESI+): m/z = 1195 (M$^+$ + Na).

5,11,17,23-Tetrakis(*tert*-butyl)-25,27-bis(2-ethoxy-2-oxoethoxy)calix[4]arene-4-nitrobenzocrown-6 15f. Under Ar, compound **14f** (870 mg, 0.92 mmol) was dissolved in anhydrous THF (40 mL) and NaH (184 mg, 4.61 mmol, 60% in mineral oil) was added and the mixture stirred for 30 min at rt. Next, ethyl bromoacetate (513 µL, 1.93 mmol) and NaI in catalytic amounts were added and the mixture stirred at 60 °C overnight. Afterwards, the solvent was changed to DCM (30 mL), the organic phase was washed with brine (2 × 30 mL) and dried over Na$_2$SO$_4$. After removal of the solvent, the crude product was purified via column chromatography (DCM → DCM:methanol = 98:2) yielding compound **15f** as colorless solid (960 mg, 93%); R$_f$ 0.5 (DCM:methanol = 95:5); ^{1}H NMR (600 MHz, CDCl$_3$): δ 0.95 (s, 18H, CH$_3$), 1.11 (t, ^3J = 7.1 Hz, 6H, CH$_2$C$\underline{H}_3$), 1.19 (br. s, 18H, CH$_3$), 3.17 (d, ^2J = 12.8 Hz, 4H, ArCH$_2$), 3.98–4.03 (m, 4H, C$\underline{H}_2$CH$_3$), 4.04–4.09 (m, 4H, OCH$_2$), 4.20–4.30 (m, 12H, OCH$_2$), 4.50–4.58 (m, 4H, ArCH$_2$), 4.65 (s, 4H, OCH$_2$), 6.62 (s, 4H, ArH), 6.91–6.94 (m, 1H, ArH), 6.93 (d, ^3J = 8.9 Hz, 4H, ArH), 7.75 (d, ^4J = 2.7 Hz, 1H, ArH), 7.90 (dd, ^3J = 8.9 Hz, ^4J = 2.7 Hz, 1H, ArH); ^{13}C NMR (151 MHz, CDCl$_3$): δ 14.2 (CH$_2\underline{C}$H$_3$). 31.4 (^{t}Bu), 31.7 (ArCH$_2$), 31.8 (^{t}Bu), 33.9 (C$_q$), 34.2 (C$_q$), 60.7 ($\underline{C}$H$_2$CH$_3$), 69.8 (OCH$_2$), 69.9 (OCH$_2$), 70.0 (OCH$_2$), 71.3 (OCH$_2$), 71.4 (OCH$_2$), 72.0 (OCH$_2$CO), 73.4 (OCH$_2$), 73.5 (OCH$_2$), 109.2 (CH$_{Ar}$), 112.1 (CH$_{Ar}$), 118.1 (CH$_{Ar}$), 125.1 (CH$_{Ar}$), 125.2 (CH$_{Ar}$), 125.7 (CH$_{Ar}$), 132.8, 134.5, 134.6, 141.9, 145.2, 145.3, 152.4, 154.6, 154.7, 155.1 (10 × C$_{Ar}$), 169.9 (C=O) ppm; MS (ESI+): m/z = 1139 (M$^+$ + Na).

5,11,17,23-Tetrakis(*tert*-butyl)-25,27-bis(carboxymethoxy)calix[4]arene-propylenecrown-6 16b. Compound **15b** (45 mg, 43 µmol) was dissolved in THF (5 mL), a solution of Me$_4$NOH · 5 H$_2$O (55 mg, 0.30 mmol) in methanol (2 mL) was added and the reaction mixture was stirred at 55 °C overnight. The major part of the solvent was removed and the product was precipitated with ice-cold aqueous HCl (6 M). The solid residue was filtered and washed with water (20 mL) and dissolved in DCM (10 mL). The organic phase was washed with cold water (5 mL). After separation, the organic phase was dried over Na$_2$SO$_4$, the solvent was removed and **16b** was obtained as grey solid (41 mg, 95%); ^{1}H NMR (600 MHz, CDCl$_3$): δ 0.83 (s, 18H, ^{t}Bu), 1.33 (s, 18H, ^{t}Bu), 1.67 (p, ^3J = 6.7 Hz, 2H,

CH$_2$), 3.21 (d, ^2J = 13.0 Hz, 4H, CH$_2$Ar), 3.58 (t, ^3J = 6.7 Hz, 4H, OCH$_2$), 3.65–3.68 (m, 4H, OCH$_2$), 3.72–3.75 (m, 4H, OCH$_2$), 3.83–3.86 (m, 4H, OCH$_2$), 3.96–4.01 (m, 4H, OCH$_2$), 4.50 (d, ^2J = 13.0 Hz, 4H, CH$_2$), 4.86 (s, 4H, CH$_2$C=O), 6.59 (s, 4H, ArH), 7.14 (s, 4H, ArH); ^{13}C NMR (151 MHz, CDCl$_3$): δ = 29.8 (CH$_2$); 31.0 (CH$_2$Ar), 31.1 (CH$_3$), 31.8 (CH$_3$), 33.9 (C$_q$), 34.3 (C$_q$), 67.3 (OCH$_2$), 70.3 (OCH$_2$), 70.6 (OCH$_2$), 70.7 (OCH$_2$), 72.9 (CH$_2$C=O), 77.1 (OCH$_2$), 125.5 (CH$_{Ar}$), 126.1 (CH$_{Ar}$), 132.6, 135.2, 146.0, 146.9, 150.6, 153.4 (6 × C$_{Ar}$), 171.0 (C=O) ppm; MS (ESI+): m/z = 1048 (M$^+$ − 2 H + 3 Na).

5,11,17,23-Tetrakis(*tert*-butyl)-25,27-bis(carboxymethoxy)calix[4]arene-benzocrown-6 16c. Compound **15c** (446 mg, 0.435 mmol) was dissolved in THF (20 mL), a solution of Me$_4$NOH · 5 H$_2$O (2.45 g, 6.85 mmol) in methanol (5 mL) was added and the reaction mixture was stirred at 55 °C overnight. After filtration, the major part of the solvent was removed and the product was precipitated with ice-cold aqueous HCl (6 M). The solid residue was filtered and washed with water (20 mL) and dissolved in DCM (30 mL). The organic phase was washed with cold water (15 mL). After separation, the organic phase was dried over Na$_2$SO$_4$, the solvent was removed and **16c** was obtained as brownish solid (440 mg, >99%); ^{1}H NMR (400 MHz, CDCl$_3$): δ 0.83 (s, 18H, CH$_3$), 1.31 (s, 18H, CH$_3$), 3.19 (d, ^2J = 12.8 Hz, 4H, ArCH$_2$), 3.80–4.28 (m, 16H, OCH$_2$), 4.45 (d, ^2J = 12.8 Hz, 4H, ArCH$_2$), 4.86 (s, 4H, OCH$_2$CO), 6.60 (s, 4H, ArH), 6.75–6.88 (m, 4H, ArH), 7.12 (s, 4H, ArH); ^{13}C NMR (101 MHz, CDCl$_3$): δ 30.9 (ArCH$_2$), 31.1 (^{t}Bu), 31.8 (^{t}Bu), 33.9 (C$_q$), 34.3 (C$_q$), 67.9 (OCH$_2$), 70.7 (OCH$_2$), 71.0 (OCH$_2$), 72.9 (OCH$_2$C=O), 77.4 (OCH$_2$), 111.5 (CH$_{Ar}$), 120.6 (CH$_{Ar}$), 125.5 (CH$_{Ar}$), 126.0 (CH$_{Ar}$), 132.6, 135.2, 145.9, 146.9, 148.4, 150.5 (6 × C$_{Ar}$), 170.1 (C=O) ppm; MS (ESI+): m/z = 1060 (M$^+$ − H + 2 Na), 1082 (M$^+$ − 2 H + 3 Na).

5,11,17,23-Tetrakis(*tert*-butyl)-25,27-bis(carboxymethoxy)calix[4]arene-benzocrown-8 16d. Compound **15d** (170 mg, 0.15 mmol) was dissolved in THF (10 mL), a solution of Me$_4$NOH · 5 H$_2$O (186 mg, 1.03 mmol) in methanol (2 mL) was added and the reaction mixture was stirred at 55 °C overnight. After filtration, the major part of the solvent was removed and the product was precipitated with ice-cold aqueous HCl (6 M). The solid residue was filtered and washed with water (10 mL) and dissolved in DCM (30 mL). The organic phase was washed with cold water (15 mL). After separation, the organic phase was dried over Na$_2$SO$_4$, the solvent was removed and **16d** was obtained as brownish solid (134 mg, 83%); ^{1}H NMR (400 MHz, CDCl$_3$): δ 0.83 (s, 18H, CH$_3$), 1.33 (s, 18H, CH$_3$), 3.17 (d, ^2J = 13.0 Hz, 4H, ArCH$_2$), 3.72–3.75 (m, 4H, OCH$_2$), 3.78–3.84 (m, 8H, OCH$_2$), 3.86–3.90 (m, 4H, OCH$_2$), 3.95–4.04 (m, 8H, OCH$_2$), 4.46 (d, ^2J = 13.0 Hz, 4H, ArCH$_2$), 4.74 (s, 4H, OCH$_2$CO), 6.56 (s, 4H, ArH), 6.71–6.76 (m, 2H, ArH), 6.81–6.85 (m, 2H, ArH), 7.12 (s, 4H, ArH); ^{13}C NMR (101 MHz, CDCl$_3$): δ 30.9 (ArCH$_2$), 31.1 (^{t}Bu), 31.8 (^{t}Bu), 33.9 (C$_q$), 34.3 (C$_q$), 69.1 (OCH$_2$), 70.1 (OCH$_2$), 70.5 (OCH$_2$), 71.1 (OCH$_2$), 72.4 (OCH$_2$CO), 77.4 (OCH$_2$), 115.2 (CH$_{Ar}$), 121.6 (CH$_{Ar}$), 125.5 (CH$_{Ar}$), 126.1 (CH$_{Ar}$), 132.5, 135.2, 147.1, 149.1, 150.4, 153.2 (6 × C$_{Ar}$), 170.6 (C=O) ppm; MS (ESI+): m/z = 1170 (M$^+$ − 2 H + 3 Na).

5,11,17,23-Tetrakis(*tert*-butyl)-25,27-bis(carboxymethoxy)calix[4]arene-4-*tert*-butyl-carboxy-benzocrown-6 16e. Compound **15e** (31 mg, 27 μmol) was dissolved in THF (4 mL), a solution of Me$_4$NOH · 5 H$_2$O (34 mg, 0.19 mmol) in methanol (2 mL) was added and the reaction mixture was stirred at 55 °C overnight. The solvent was removed and the crude product was dissolved in DCM (30 mL). The organic phase was washed with saturated NH$_4$Cl solution (15 mL) and water (15 mL). After separation, the organic phase was dried over Na$_2$SO$_4$, the solvent was removed and **16e** was obtained as yellowish oil (26 mg, 87%); ^{1}H NMR (400 MHz, CDCl$_3$): δ 0.83 (s, 18H, CH$_3$), 1.31 (s, 18H, CH$_3$), 1.56 (s, 9H, OC(CH$_3$)$_3$), 3.19 (d, ^2J = 12.8 Hz, 4H, ArCH$_2$), 3.83–4.08 (m, 12H, OCH$_2$), 4.26–4.31 (m, 4H, OCH$_2$), 4.44 (d, ^3J = 12.8 Hz, 4H, ArCH$_2$), 4.84 (s, 4H, OCH$_2$CO), 6.60 (s, 4H, ArH), 6.76 (d, ^3J = 8.5 Hz, 1H, ArH), 6.98 (s, 4H, ArH), 7.40 (d, ^4J = 1.9 Hz, 1H, ArH), 7.55 (dd, ^3J = 8.5 Hz, ^4J = 1.9 Hz, 1H, ArH); ^{13}C NMR (101 MHz, CDCl$_3$): δ 28.4 (O^{t}Bu), 31.1 (^{t}Bu), 31.2 (ArCH$_2$), 31.8 (^{t}Bu), 33.9 (C$_q$), 34.3 (C$_q$), 68.2 (OCH$_2$), 68.5 (OCH$_2$), 70.4 (OCH$_2$), 70.5 (OCH$_2$), 70.9 (OCH$_2$), 71.0 (OCH$_2$), 72.0 (OCH$_2$CO), 77.4 (OCH$_2$), 80.6 (OC$_q$), 110.6 (CH$_{Ar}$), 112.2 (CH$_{Ar}$), 123.2 (CH$_{Ar}$), 125.3 (C$_{Ar}$), 125.5 (CH$_{Ar}$), 125.7 (CH$_{Ar}$), 135.2, 136.0, 143.3, 145.8, 145.9, 147.0, 147.9, 150.4,

150.5, 151.7, 152.1 (11 × C$_{Ar}$), 165.9 (COO^{t}Bu), 172.8 (C=O) ppm; MS (ESI+): m/z = 1160 (M$^+$ − H + 2 Na).

Disodium 5,11,17,23-tetrakis(*tert*-butyl)-25,27-bis(*N*-heptafluoropropane-2-sulfonylcarbamoyl-methoxy)calix[4]arene-crown-6 21. Calixarene **16a** (300 mg, 0.31 mmol) was suspended in CCl$_4$ (6 mL), oxalyl chloride (1.5 mL) was added and the mixture was stirred for 5 h at 65 °C. After cooling to rt, the solvent and remaining oxalyl chloride was removed. Under argon, the residue was dissolved in anhydrous DCM (5 mL) and a mixture of perfluoroisopropanesulfonamide (193 mg, 0.77 mmol) and NaH (60% in mineral oil, 124 mg, 3.1 mmol) dissolved in anhydrous DCM was added. After stirring overnight at rt, the mixture was filtered, the organic phase was washed with 10% HCl, the organic phase was separated and dried over Na$_2$SO$_4$. Next, the solvent was removed and the crude product was purified by column chromatography (DCM → DCM/methanol 5:1) to yield calixarene **21** as colorless solid (254 mg, 57%). R$_f$ 0.6 (petroleum ether:ethyl acetate = 2:1); ^{1}H NMR (600 MHz, CDCl$_3$): δ 1.11 (s, 18H, ^{t}Bu), 1.15 (s, 18H, ^{t}Bu), 3.38 (d, ^2J = 12.3 Hz, 4H, ArCH$_2$), 3.71–3.74 (m, 4H, OCH$_2$), 3.82–3.86 (m, 4H, OCH$_2$), 3.86–3.92 (s, 8H, OCH$_2$), 4.10–4.18 (m, 8H, OCH$_2$+ArCH$_2$), 4.63 (s, 4H, CH$_2$C=O), 7.08 (s, 4H, *m*-H$_{Ar}$), 7.11 (s, 4H, *m*-H$_{Ar}$); ^{13}C NMR (151 MHz, CDCl$_3$): δ 30.1 (ArCH$_2$), 31.3, 31.4 (2 x CH$_3$), 34.1, 34.3 (C$_q$), 69.3, 69,4, 72.1, 72.2, (4 × OCH$_2$), 77.0 (CH$_2$C=O), 78.8 (OCH$_2$), 125.8, 126.3 (2 × *m*-C$_{Ar}$), 134.1, 134.4 (2 × *o*-C$_{Ar}$), 147.8, 148.4 (2 × *p*-C$_{Ar}$), 148.3, 151.5 (2 × *i*-C$_{Ar}$), 176.2 (C=O); ^{19}F NMR (565 MHz, CDCl$_3$): δ −169.3 (2F), −71.7 (12F) ppm; MS (MALDI-TOF): m/z = 1429 (M$^+$).

Disodium 5,11,17,23-tetrakis(*tert*-butyl)-25,27-bis(*N*-heptafluoropropane-2-sulfonylcarbamoyl-methoxy)calix[4]arene-propylenecrown-6 22. Under Ar, compound **16b** (41 mg, 42 μmol), 6-chloro-1-hydroxybenzotriazole (16 mg, 92 μmol), EDC · HCl (18 mg, 92 μmol), and Et$_3$N (13 μL, 9.3 mg, 92 μmol) was dissolved in anhydrous acetonitrile (8 mL) and stirred at 0 °C for 30 min. Afterwards, heptafluoroisopropane-2-sulfonamide (42 mg, 167 μmol) and Et$_3$N (12 μL, 8.4 mg, 84 μmol) dissolved in anhydrous acetonitrile (5 mL) were added and the resulting mixture stirred at rt overnight. Next, the solvent was changed to DCM (15 mL) and the organic phase was washed with brine (15 mL), saturated hydrogen carbonate solution (15 mL) and brine (15 mL), dried over Na$_2$SO$_4$ and the solvent was removed. The crude product was purified using column chromatography (petroleum ether:ethyl acetate = 4:1 → 2:1) to obtain **22** as colourless solid (24 mg, 40%); mp >350 °C; R$_f$ = 0.6 (petroleum ether:ethyl acetate = 2:1); ^{1}H NMR (400 MHz, CDCl$_3$): δ 1.07 (s, 18H, ^{t}Bu), 1.18 (s, 18H, ^{t}Bu), 2.09–2.01 (m, 2H, CH$_2$), 3.36 (d, ^2J = 12.3 Hz, 4H, CH$_2$Ar), 3.74–3.69 (m, 4H, OCH$_2$), 3.82–3.74 (m, 8H, 2 × OCH$_2$), 3.89–3.82 (m, 4H, OCH$_2$), 4.07–4.00 (m, 4H, OCH$_2$), 4.16 (d, ^2J = 12.3 Hz, 4H, CH$_2$Ar), 4.58 (s, 4H, CH$_2$C=O), 7.04 (s, 4H, ArH), 7.13 (s, 4H, ArH); ^{13}C NMR (101 MHz, CDCl$_3$): δ 30.0 (CH$_2$Ar), 30.3 (CH$_2$), 31.3 (CH$_3$), 31.5 (CH$_3$), 34.2 (C$_q$), 34.3 (C$_q$), 70.4 (OCH$_2$), 71.1 (OCH$_2$), 71.4 (OCH$_2$), 72.5 (OCH$_2$), 77.4 (CH$_2$C=O), 78.0 (OCH$_2$), 125.9 (CH$_{Ar}$), 126.5 (CH$_{Ar}$), 134.1, 134.5, 147.7, 148.4, 148.8, 151.0 (6 × C$_{Ar}$), 175.9 (C=O); ^{19}F NMR (376 MHz, CDCl$_3$): δ −169.5 (hept, ^3J = 7.5 Hz, 2F), −71.7 (d, ^3J = 7.5 Hz, 12F) ppm; MS (ESI+): m/z = 1466 (M$^+$ + Na).

Disodium 5,11,17,23-tetrakis(*tert*-butyl)-25,27-bis(*N*-trifluormethanesulfonylcarbamoylmethoxy)-calix[4]arene-benzocrown-6 23. Under Ar, compound **16c** (100 mg, 99 μmol), 6-chloro-1-hydroxybenzotriazole (37 mg, 0.22 mmol), EDC · HCl (42 mg, 0.22 mmol), and Et$_3$N (30 μL, 22 mg, 0.22 mmol) were dissolved in anhydrous acetonitrile (10 mL) and stirred at 0 °C for 30 min. Afterwards, trifluoromethanesulfonamide (59 mg, 0.39 mmol) and Et$_3$N (27 μL, 20 mg, 0.20 mmol) dissolved in anhydrous acetonitrile (5 mL) were added and the resulting mixture stirred at rt overnight. Next, the solvent was changed to DCM (15 mL) and the organic phase was washed with water (15 mL), saturated hydrogen carbonate solution (15 mL) and water (15 mL), dried over Na$_2$SO$_4$ and the solvent was removed. The crude product was purified using column chromatography (petroleum ether:ethyl acetate = 2:1 → 3:2) to obtain **23** as yellowish solid (122 mg, 97%); mp 320 °C (decomp.); R$_f$ 0.6 (petroleum ether:ethyl acetate = 1:1); ^{1}H NMR (400 MHz, CDCl$_3$): δ 1.13 (s, 18H, ^{t}Bu), 1.15 (s, 18H, ^{t}Bu), 1.10–1.16 (m, 36H, CH$_3$), 3.40 (d, ^2J = 12.3 Hz, 4H, CH$_2$Ar), 3.96–4.02 (m, 4H, OCH$_2$), 4.04–4.21 (m, 12H, OCH$_2$, CH$_2$Ar), 4.27–4.33 (m, 4H, OCH$_2$), 4.64 (s, 4H,

CH$_2$C=O), 6.88–6.98 (m, 4H, ArH), 7.07–7.14 (m, 8H, ArH) ppm; ^{13}C NMR (101 MHz, CDCl$_3$): δ 30.3 (CH$_2$Ar), 31.3 (CH$_3$), 31.4 (CH$_3$), 34.2 (C$_q$), 34.3 (C$_q$), 67.0 (OCH$_2$), 70.6 (OCH$_2$), 72.6 (OCH$_2$), 77.4 (CH$_2$C=O), 78.3 (OCH$_2$), 111.7 (CH$_{Ar}$), 121.6 (CH$_{Ar}$), 125.8 (CH$_{Ar}$), 126.3 (CH$_{Ar}$), 134.2, 134.3, 147.0, 148.0, 148.1, 148.4, 151.6 (7 × C$_{Ar}$), 176.3 (C=O); ^{19}F NMR (376 MHz, CDCl$_3$): δ − 76.6 (CF$_3$) ppm; MS (ESI+): m/z = 1278 (M$^+$ + H), 1295 (M$^+$ + NH$_4$), 1295 (M$^+$ + Na).

Disodium 5,11,17,23-tetrakis(*tert*-butyl)-25,27-bis(*N*-heptafluoropropane-2-sulfonylcarbamoyl-methoxy)calix[4]arene-benzocrown-6 24. Under Ar, compound **16c** (67 mg, 66 µmol), 6-chloro-1-hydroxybenzotriazole (20 mg, 0.15 mmol), EDC · HCl (29 mg, 0.15 mmol), and Et$_3$N (21 µL, 15 mg, 0.15 mmol) were dissolved in anhydrous acetonitrile (10 mL) and stirred at 0 °C for 30 min. Afterwards, heptafluoroisopropane-2-sulfonamide (66 mg, 0.26 mmol) and Et$_3$N (18 µL, 13 mg, 0.13 mmol) dissolved in anhydrous acetonitrile (5 mL) were added and the resulting mixture stirred at rt overnight. Next, the solvent was changed to DCM (15 mL) and the organic phase was washed with water (15 mL), saturated hydrogen carbonate solution (15 mL) and water (15 mL), dried over Na$_2$SO$_4$ and the solvent was removed. The crude product was purified using column chromatography (DCM → DCM:methanol = 74:1) to obtain **24** as colorless solid (50 mg, 50%); mp 340 °C (decomp.); R$_f$ 0.7 (DCM:methanol = 98:2); ^{1}H NMR (400 MHz, CDCl$_3$): δ 1.10 (s, 18H, ^{t}Bu), 1.17 (s, 18H, ^{t}Bu), 3.41 (d, ^2J = 12.4 Hz, 4H, CH$_2$Ar), 3.98–4.06 (m, 8H, OCH$_2$), 4.17 (d, ^2J = 12.4 Hz, 4H, CH$_2$Ar), 4.20–4.25 (m, 8H, OCH$_2$), 4.65 (s, 4H, CH$_2$C=O), 6.84–6.88 (m, 2H, ArH), 6.93–6.97 (m, 2H, ArH), 7.06 (s, 4H, ArH$_{Calix}$), 7.14 (s, 4H, ArH$_{Calix}$); ^{13}C NMR (101 MHz, CDCl$_3$): δ 30.3 (CH$_2$Ar), 31.4 (CH$_3$), 31.4 (CH$_3$), 34.2 (C$_q$), 34.3 (C$_q$), 66.5 (OCH$_2$), 70.2 (OCH$_2$), 72.8 (CH$_2$C=O), 76.8 (OCH$_2$), 78.1 (OCH$_2$), 110.9 (CH$_{Ar}$), 119.0 (dd, $^1J_{C\text{-}F}$ = 288 Hz, $^3J_{C\text{-}F}$ = 26 Hz, CF), 119.5 (qd, $^1J_{C\text{-}F}$ = 288 Hz, $^3J_{C\text{-}F}$ = 26 Hz, CF$_3$), 121.4 (CH$_{Ar}$), 125.8 (CH$_{Ar}$), 126.3 (CH$_{Ar}$), 134.1, 134.2, 146.8, 147.9, 148.0, 148.5, 151.9 (7 × C$_{Ar}$), 176.1 (C=O); ^{19}F NMR (564 MHz, CDCl$_3$): δ − 72.0 (d, ^3J = 7.0 Hz), −169.9 (hept, ^3J = 7.0 Hz) ppm; MS (MALDI-TOF): m/z = 1499 (M$^+$ + Na).

Disodium 5,11,17,23-tetrakis(*tert*-butyl)-25,27-bis(*N*-pentafluorobenzenesulfonylcarbamoylmethoxy)-calix[4]arene-benzocrown-6 25. Under Ar, compound **16c** (100 mg, 99 µmol), 6-chloro-1-hydroxybenzotriazole (37 mg, 0.22 mmol), EDC · HCl (42 mg, 0.22 mmol), and Et$_3$N (30 µL, 22 mg, 0.22 mmol) were dissolved in anhydrous acetonitrile (10 mL) and stirred at 0 °C for 30 min. Afterwards, pentafluorobenzenesulfonamide (97 mg, 0.39 mmol) and Et$_3$N (27 µL, 20 mg, 0.20 mmol) dissolved in anhydrous acetonitrile (5 mL) were added and the resulting mixture stirred at rt overnight. Next, the solvent was changed to DCM (15 mL) and the organic phase was washed with water (15 mL), saturated hydrogen carbonate solution (15 mL) and water (15 mL), dried over Na$_2$SO$_4$ and the solvent was removed. The crude product was purified using column chromatography (petroleum ether:ethyl acetate = 2:1 → 3:2), dissolved again in DCM and washed with 10% HCl to obtain **25** as yellowish solid (122 mg, 97%); mp 320 °C (decomp.); R$_f$ 0.6 (petroleum ether:ethyl acetate = 1:1); ^{1}H NMR (400 MHz, CDCl$_3$): δ 0.83 (s, 18H, CH$_3$), 1.24 (s, 18H, CH$_3$), 3.14 (d, ^2J = 12.8 Hz, 4H, ArCH$_2$), 4.00–4.10 (m, 8H, OCH$_2$), 4.17–4.23 (m, 4H, OCH$_2$), 4.32 (d, ^2J = 12.8 Hz, 4H, ArCH$_2$), 4.41–4.47 (m, 4H, OCH$_2$), 5.17 (s, 4H, OCH$_2$CO), 6.47 (s, 4H, ArH), 6.93 (s, 4H, ArH), 6.96–7.02 (m, 4H, ArH), 11.12 (s, 2H, C=O); ^{13}C NMR (101 MHz, CDCl$_3$): δ 31.1 (CH$_3$), 31.6 (CH$_3$), 32.1 (ArCH$_2$), 33.8 (C$_q$), 34.2 (C$_q$), 68.1 (OCH$_2$), 70.0 (OCH$_2$), 70.3 (OCH$_2$C=O), 71.7 (OCH$_2$), 73.4 (OCH$_2$), 114.0 (CH$_{Ar}$), 122.1 (CH$_{Ar}$), 125.0 (CH$_{Ar}$), 125.4 (CH$_{Ar}$), 131.9, 135.1, 145.4, 146.1, 148.0, 151.4, 152.5 (7 × C$_{Ar}$), 168.8 (C=O); ^{19}F NMR (376 MHz, CDCl$_3$): δ −138.6 (d, ^3J = 20.4 Hz), −148.6 (d, ^3J = 20.4 Hz), −163.5 (t, ^3J = 20.4 Hz) ppm; MS (ESI +): m/z = 1474 (M$^+$ + H), 1491 (M$^+$ + NH$_4$), 1518 (M$^+$ − H + 2 Na), 1534 (M$^+$ − H + Na + K).

Disodium 5,11,17,23-tetrakis(*tert*-butyl)-25,27-bis(*N*-heptafluoro-2-propanesulfonylcarbamoyl-methoxy)calix[4]arene-4-*tert*-butylcarboxybenzocrown-6 26. Under Ar, compound **16e** (26 mg, 42 µmol), 6-chloro-1-hydroxybenzotriazole (9 mg, 51 µmol), EDC · HCl (10 mg, 51 µmol), and Et$_3$N (7,1 µL, 5 mg, 51 µmol) were dissolved in anhydrous acetonitrile (8 mL) and stirred at 0 °C for 30 min. Afterwards, heptafluoropropane-2-sulfonamide

(23 mg, 0.093 mmol) and Et_3N (6.5 µL, 5 mg, 0.084 mmol) dissolved in anhydrous acetonitrile (5 mL) were added and the resulting mixture stirred at rt overnight. Next, the solvent was changed to DCM (15 mL) and the organic phase was washed with water (15 mL), saturated hydrogen carbonate solution (15 mL) and water (15 mL), dried over Na_2SO_4 and the solvent removed. The crude product was purified using column chromatography (petroleum ether:ethyl acetate 3:1 → 2:1) to obtain **26** as colourless solid (13 mg, 35%); mp 340 °C (decomp.); R_f 0.3 (petroleum ether:ethyl acetate 2:1); ^{1}H NMR (400 MHz, $CDCl_3$): δ 1.10 (s, 18H, CH_3), 1.17 (s, 18H, CH_3), 1.61 (s, 9H, $OC(CH_3)_3$), 3.41 (d, ^2J = 12.3 Hz, 4H, $ArCH_2$), 3.98–4.10 (m, 8H, OCH_2), 4.16 (d, ^2J = 12.3 Hz, 4H, $ArCH_2$), 4.21–4.31 (m, 8H, OCH_2), 4.64 (s, 4H, OCH_2CO), 6.86 (d, ^3J = 8.5 Hz, 1H, ArH), 7.06 (s, 4H, ArH), 7.14 (s, 4H, ArH), 7.50 (d, ^4J = 1.9 Hz, 1H, ArH), 7.67 ppm (dd, ^3J = 8.5 Hz, ^4J = 1.9 Hz, 1H, ArH); ^{13}C NMR (101 MHz, $CDCl_3$): δ 28.4 (O^tBu), 30.3 ($ArCH_2$), 30.4 ($ArCH_2$), 31.3 (^tBu), 31.4 (^tBu), 34.2 (C_q), 34.3 (C_q), 66.9 (OCH_2), 67.2 (OCH_2), 69.9 (OCH_2), 70.0 (OCH_2), 72.8 (OCH_2), 72.9 (OCH_2), 76.7 (OCH_2CO), 78.0 (OCH_2), 78.1 (OCH_2), 80.9 (OC_q), 110.2 (CH_{Ar}), 111.8 (CH_{Ar}), 124.0 (CH_{Ar}), 125.4 (C_{Ar}), 125.8 (CH_{Ar}), 126.3 (CH_{Ar}), 134.1, 134.2, 146.4, 147.9, 148.0, 148.1, 148.5, 150.3, 151.8, 151.9 (10 × C_{Ar}), 165.6 (C=OOtBu), 176.1 (C=ONH); ^{19}F NMR (376 MHz, $CDCl_3$): δ −72.0 (m, 6F), −169.9 (hept, ^3J = 7.3 Hz, 1F) ppm; MS (ESI +): m/z = 1595 ($M^+ + NH_4$), 1600 ($M^+ + Na$), 1616 ($M^+ + K$), 1622 ($M^+ - H + 2\,Na$).

5,11,17,23-Tetrakis(*tert*-butyl)-25,27-bis(*N*-heptafluoro-2-propanesulfonylcarbamoyl-methoxy)calix[4]arene-benzocrown-8 27. Under Ar, compound **16d** (135 mg, 0.12 mmol), 1-hydroxybenzotriazole (36 mg, 0.27 mmol), EDC · HCl (52 mg, 0.27 mmol), and Et_3N (37 µL, 27 mg, 0.27 mmol) were dissolved in anhydrous acetonitrile (20 mL) and stirred at 0 °C for 30 min. Afterwards, heptafluoropropane-2-sulfonamide (122 mg, 0.49 mmol) and Et_3N (34 µL, 25 mg, 0.24 mmol) dissolved in anhydrous acetonitrile (5 mL) were added and the resulting mixture stirred at rt overnight. Next, the solvent was changed to DCM (20 mL) and the organic phase was washed with water (20 mL), saturated hydrogen carbonate solution (20 mL) and water (20 mL), dried over Na_2SO_4 and the solvent removed. The crude product was purified using column chromatography (DCM:petroleum ether:ethyl acetate 3:8:2 → DCM:ethyl acetate 1:1) and washed with 10% aqueous HCl to obtain **27** as colourless solid (51 mg, 27%); mp 162 °C; R_f 0.25 (DCM:petroleum ether:ethyl acetate 3:8:2); ^{1}H NMR (400 MHz, $CDCl_3$): δ 0.82 (s, 18H, CH_3), 1.31 (s, 18H, CH_3), 3.22 (d, ^2J = 13.0 Hz, 4H, $ArCH_2$), 3.77–3.79 (m, 4H, OCH_2), 3.79–3.82 (m, 8H, OCH_2), 3.98–4.00 (m, 4H, OCH_2), 4.07–4.10 (m, 4H, OCH_2), 4.20–4.23 (m, 4H, OCH_2), 4.55 (d, ^2J = 13.0 Hz, 4H, $ArCH_2$), 5.28 (s, 4H, OCH_2CO), 6.50 (s, 4H, ArH), 6.84–6.89 (m, 2H, ArH), 6.90–6.95 (m, 2H, ArH), 7.08 (s, 4H, ArH); ^{13}C NMR (151 MHz, $CDCl_3$): δ 33.7 (^tBu), 34.2 (^tBu), 34.7 ($ArCH_2$), 36.2 (C_q), 36.7 (C_q), 69.4 (OCH_2), 71.5 (OCH_2), 71.9 (OCH_2), 72.9 (OCH_2), 73.3 (OCH_2CO), 73.4 (OCH_2), 114.5 (CH_{Ar}), 123.7 (CH_{Ar}), 127.5 (CH_{Ar}), 128.4 (CH_{Ar}), 134.7, 137.8, 147.7, 148.1, 150.4, 154.0 (6 × C_{Ar}); ^{19}F NMR (564 MHz, $CDCl_3$): δ −71.7 (d, ^3J = 8.0 Hz, 6F), −168.5 (m, 1F) ppm; MS (ESI +): m/z = 1394 ($M^+ - H - NHSO_2CF(CF_3)_2 + K + Na$), 1625 ($M^+ - H + K + Na$).

4. Conclusions

Two sets of new functionalized calixarene ligands were prepared and analysed via UV/vis titration regarding their potential to chelate divalent cations like Ba, Sr and Pb. These ions are of high importance, e.g., as diagnostic or therapeutic radionuclides in radiopharmacy, or they are found in environmental chemistry. Thus, there exists a keen necessity for a strong complexation of these ions. In summary, the additional proton-ionizable side functions connected together with a benzocrown ether structure to the calixarene skeleton led to a considerable improvement of the complex stability resulting in high association constants. There is still room to improve the stability of these complexes for future application as chelators for radiometal ions in radiopharmacy.

Supplementary Materials: The following supporting information can be downloaded online. Table S1: Calculated bond lengths for Ba-O bonds of calix[4]crown-6 (20) and calix[4]crown-5 (BP86-D3/def2-TZVP), Figure S1: Partly labelled schemes of calix[4]crown-6 (20) and calix[4]crown-5, Table S2: the optimized coordinates (in Angstrom) of Ba-calix[4]crown complex calculated using BP86/TZVP, Table S3: The optimized coordinates (in Angstrom) of Ra-calix[4]crown complex calculated using BP86/TZVP, Figures S2–S34: ^{1}H and ^{13}C NMR spectra of compounds, Figures S35–S64: UV spectra and UV titration plot of compounds with Ba^{2+}, Sr^{2+} and Pb^{2+}.

Author Contributions: Conceptualization, C.M. and S.L.; methodology, C.M. and S.L.; software, S.L. and K.A.-A.; validation, C.M., S.L. and K.A.-A.; formal analysis, M.B., E.E. and K.A.-A.; investigation, C.M., K.A.-A. and S.L.; resources, C.M. and S.L.; writing—original draft preparation, C.M. and S.L.; writing—review and editing, all authors; visualization, M.B., E.E., K.A.-A.; supervision, C.M. and S.L.; project administration, C.M.; All authors have read and agreed to the published version of the manuscript.

Funding: This research was funded by the University of Zurich, and the Swiss national science foundation, grant no. PP00P2_170667. K.A.-A. received a Swiss Government Excellence Scholarship.

Institutional Review Board Statement: Not applicable.

Informed Consent Statement: Not applicable.

Data Availability Statement: All data are included in this paper.

Acknowledgments: The authors thank Linda Belke for their extensive experimental support.

Conflicts of Interest: The authors declare no conflict of interest.

Sample Availability: Samples of the compounds are not available from the authors.

References

1. Gutsche, C.D.; Dhawan, B.; No, H.K.; Muthukrishnan, R. Calixarenes. 4. The synthesis, characterization, and properties of the calixarenes from *p-tert*-butylphenol. *J. Am. Chem. Soc.* **1981**, *103*, 3782–3792. [CrossRef]
2. Gutsche, C.D. *Calixarenes—An Introduction*; RSC Publishing: Cambridge, UK, 2008; Volume 2.
3. Sliwa, W.; Kozlowski, C. *Calixarenes and Resorcinarenes*; Wiley-VCH: Weinheim, Germany, 2009.
4. McMahon, G.; O'Malley, S.; Nolan, K.; Diamond, D. Important Calixarene Derivatives—Their Synthesis and Applications. *Arkivoc* **2003**, 23–31. [CrossRef]
5. Deska, M.; Dondela, B.; Sliwa, W. Selected applications of calixarene derivatives. *Arkivoc* **2015**, 93–416. [CrossRef]
6. Mourer, M.; Duval, R.E.; Finance, C.; Regnouf-de-Vains, J.B. Functional organisation and gain of activity: The case of the antibacterial tetra-*para*-guanidinoethyl-calix[4]arene. *Bioorg. Med. Chem. Lett.* **2006**, *16*, 2960–2963. [CrossRef]
7. Yousaf, A.; Hamid, S.A.; Bunnori, N.M.; Ishola, A.A. *Drug Des. Dev. Ther.* **2018**, *9*, 2831–2838.
8. Fahmy, S.A.; Ponte, F.; Fawzy, I.M.; Sicilia, E.; El-Said Azzazy, H.M. Betaine host–guest complexation with a calixarene receptor: Enhanced In Vitro anticancer effect. *RSC Adv.* **2021**, *11*, 24673–24680. [CrossRef]
9. Da Silva, E.; Lazar, A.N.; Coleman, A.W. Biopharmaceutical applications of calixarenes. *J. Drug Deliv. Sci. Technol.* **2004**, *14*, 3–20. [CrossRef]
10. Vovk, A.I.; Kononets, L.A.; Tanchuk, V.Y.; Cherenok, S.O.; Drapailo, A.B.; Kalchenko, V.I.; Kukhar, V.P. Inhibition of Yersinia protein tyrosine phosphatase by phosphonate derivatives of calixarenes. *Bioorg. Med. Chem. Lett.* **2010**, *20*, 483–487. [CrossRef]
11. Arena, G.; Casnati, A.; Contino, A.; Magri, A.; Sansone, F.; Sciotto, D.; Ungaro, R. Inclusion of naturally occurring amino acids in water soluble calix[4]arenes: A microcalorimetric and ^{1}H NMR investigation supported by molecular modeling. *Org. Biomol. Chem.* **2006**, *4*, 243–249. [CrossRef]
12. Harrowfield, J. Calixarenes and cations. *Chem. Commun.* **2013**, *49*, 1578–1580. [CrossRef]
13. Arora, V.; Chawla, H.M.; Singh, S.P. Calixarenes as sensor materials for recognition and separation of metal ions. *Arkivoc* **2007**, 72–200. [CrossRef]
14. Sliwa, W.; Deska, M. Calixarene complexes with soft metal ions. *Arkivoc* **2008**, 7–127. [CrossRef]
15. Sliwa, W.; Girek, T. Calixarene complexes with metal ions. *J. Incl. Phenom. Macrocycl. Chem.* **2010**, *66*, 15–41. [CrossRef]
16. Gramage-Doria, R.; Armspach, D.; Matt, D. Metallated cavitands (calixarenes, resorcinarenes, cyclodextrins) with internal coordination sites. *Coord. Chem. Rev.* **2013**, *257*, 776–816. [CrossRef]
17. Depauw, A.; Kumar, N.; Ha-Thi, M.H.; Leray, I. Calixarene-Based Fluorescent Sensors for Cesium Cations Containing BODIPY Fluorophore. *J. Phys. Chem. A* **2015**, *119*, 6065–6073. [CrossRef]
18. Chen, Y.-J.; Chen, M.-Y.; Lee, K.-T.; Shen, L.-C.; Hung, H.-C.; Niu, H.-C.; Chung, W.-S. 1,3-Alternate Calix[4]arene Functionalized With Pyrazole and Triazole Ligands as a Highly Selective Fluorescent Sensor for Hg2+ and Ag+ Ions. *Front. Chem.* **2020**, *8*, 593261. [CrossRef]

19. Maksimov, A.L.; Buchneva, T.S.; Karakhanov, E.A. Supramolecular calixarene-based catalytic systems in the Wacker-oxidation of higher alkenes. *J. Mol. Catal. A Chem.* **2004**, *217*, 59–67. [CrossRef]
20. Ferreira, A.S.D.; Ascenso, J.R.; Marcos, P.M.; Schurhammer, R.; Hickey, N.; Geremia, S. Calixarene-Based lead receptors: An NMR, DFT and X-Ray synergetic approach. *Supramol. Chem.* **2021**. *accepted*. [CrossRef]
21. Adhikari, B.B.; Kanemitsu, M.; Kawakita, H.; Ohto Jumina, K. Synthesis and application of a highly efficient polyvinylcalix[4]arene tetraacetic acid resin for adsorptive removal of lead from aqueous solutions. *Chem. Eng. J.* **2011**, *172*, 341–353. [CrossRef]
22. Kunsagi-Mate, S.; Szabo, K.; Desbat, B.; Bruneel, J.L.; Bitter, I.; Kollar, L. Complexation of Phenols by Calix[4]arene Diethers in a Low-Permittivity Solvent. Self-Switched Complexation by 25,27-Dibenzyloxycalix[4]arene. *J. Phys. Chem. B* **2007**, *111*, 7218–7223. [CrossRef]
23. Zadmard, R.; Hokmabadi, F.; Jalali, M.R.; Akbarzadeh, A. Recent progress to construct calixarene-based polymers using covalent bonds: Synthesis and applications. *RSC Adv.* **2020**, *10*, 32690–32722. [CrossRef]
24. Lumetta, G.J.; Rogers, R.D.; Gopalan, A.S. *Calixarenes for Separations*; American Chemical Society: Washington, DC, USA, 2000.
25. Chen, X.Y.; Ji, M.; Fisher, D.R.; Wai, C.M. Ionizable Calixarene-Crown Ethers with High Selectivity for Radium over Light Alkaline Earth Metal Ions. *Inorg. Chem.* **1999**, *38*, 5449–5452. [CrossRef] [PubMed]
26. Steinberg, J.; Bauer, D.; Reissig, F.; Köckerling, M.; Pietzsch, H.-J.; Mamat, C. Modified Calix[4]crowns as Molecular Receptors for Barium. *ChemistryOpen* **2018**, *7*, 432–438. [CrossRef] [PubMed]
27. Henriksen, G.; Hoff, P.; Larsen, R.H. Evaluation of potential chelating agents for radium. *Appl. Radiat. Isot.* **2002**, *56*, 667–671. [CrossRef]
28. Van Leeuwen, F.W.B.; Beijleveld, H.; Miermans, C.J.H.; Huskens, J.; Verboom, W.; Reinhoudt, D.N. Ionizable (Thia)calix[4]crowns as Highly Selective 226Ra2+ Ionophores. *Anal. Chem.* **2005**, *77*, 4611–4617. [CrossRef]
29. Bauer, D.; Gott, M.; Steinbach, J.; Mamat, C. Chelation of heavy group 2 (radio)metals by p-tert-butylcalix[4]arene-1,3-crown-6 and logK determination via NMR. *Spectrochim. Acta A* **2018**, *199*, 50–56. [CrossRef]
30. Bauer, D.; Blumberg, M.; Köckerling, M.; Mamat, C. A comparative evaluation of calix[4]arene-1,3-crown-6 as a ligand for selected divalent cations of radiopharmaceutical interest. *RSC Adv.* **2019**, *9*, 32357–32366. [CrossRef]
31. Bauer, D.; Stipurin, S.; Köckerling, M.; Mamat, C. Formation of calix[4]arenes with acyloxycarboxylate functions. *Tetrahedron* **2020**, *76*, 131395. [CrossRef]
32. Arnaud-Neu, F.; Schwing-Weill, M.-J.; Dozol, J.-F. Calixarenes for Nuclear Waste Treatment. In *Calixarenes 2001*; Springer: Berlin, Germany, 2001; pp. 642–662.
33. Malone, J.F.; Marrs, D.J.; McKervey, M.A.; O'Hagan, P.; Thompson, N.; Walker, A.; Arnaud-Neu, F.; Mauprivez, O.; Schwing-Weill, M.-J.; Dozol, J.-F.; et al. Calix[n]arene phosphine oxides. A new series of cation receptors for extraction of europium, thorium, plutonium and americium in nuclear waste treatment. *J. Chem. Soc. Chem. Commun.* **1995**, 2151–2153. [CrossRef]
34. Mokhtari, B.; Pourabdollah, K.; Dallali, N. A review of calixarene applications in nuclear industries. *Radioanal. Nucl. Chem.* **2011**, *287*, 921–934. [CrossRef]
35. Mohapatra, P.K.; Raut, D.R.; Iqbal, M.; Huskens, J.; Verboom, W. Separation of carrier-free ^{90}Y from ^{90}Sr using flat sheet supported liquid membranes containing multiple diglycolamide-functionalized calix [4] arenes. *Supramol. Chem.* **2016**, *28*, 360–366. [CrossRef]
36. Mjos, K.D.; Orvig, C. Metallodrugs in Medicinal Inorganic Chemistry. *Chem. Rev.* **2014**, *114*, 4540–4563. [CrossRef]
37. Kuroda, I. Effective use of strontium-89 in osseous metastases. *Ann. Nucl. Med.* **2012**, *26*, 197–206. [CrossRef]
38. Kearsley, J.H.; Fitchew, R.S.; Taylor, R.G. Adjunctive radiotherapy with strontium-90 in the treatment of conjunctival squamous cell carcinoma. *Int. J. Radiat. Oncol. Biol. Phys.* **1988**, *14*, 435–443. [CrossRef]
39. Laskar, S.; Gurram, L.; Laskar, S.G.; Chaudhari, S.; Khanna, N.; Upreti, R. Superficial ocular malignancies treated with strontium-90 brachytherapy: Long term outcomes. *J. Contemp. Brachyther.* **2015**, *7*, 369–373. [CrossRef]
40. Reissig, F.; Kopka, K.; Mamat, C. The impact of barium isotopes in radiopharmacy and nuclear medicine—From past to presence. *Nucl. Med. Biol.* **2021**, *98*, 59–68. [CrossRef]
41. Ram, P.C.; Fordham, E.W. An historical survey of bone scanning. *Sem. Nucl. Med.* **1979**, *9*, 190–196. [CrossRef]
42. Reissig, F.; Bauer, D.; Ullrich, M.; Kreller, M.; Pietzsch, J.; Mamat, C.; Kopka, K.; Pietzsch, H.-J.; Walther, M. Recent Insights in Barium-131 as a Diagnostic Match for Radium-223: Cyclotron Production, Separation, Radiolabeling, and Imaging. *Pharmaceuticals* **2020**, *13*, 272. [CrossRef]
43. Kirby, H.W.; Salutsky, M.L. *Radiochemistry of Radium*; Clearinghouse for Federal Scientific and Technical Information (U.S.): Springfield, VA, USA, 1964.
44. Harrison, G.E.; Carr, T.E.F.; Sutton, A. Distribution of Radioactive Calcium, Strontium, Barium and Radium Following Intravenous Injection into a Healthy Man. *Int. J. Radiat. Biol.* **1968**, *13*, 235–242. [CrossRef]
45. Gmelin, L. *Gmelins Handbuch der Anorganischen Chemie, Radium*; Springer: Berlin, Germany, 1977.
46. Wiberg, N.; Holleman, A.F.; Wiberg, E. *Holleman-Wiberg's Inorganic Chemistry*, 1st ed.; Academic Press: San Diego, CA, USA; London, UK, 2001; pp. 1063–1067.
47. Gott, M.; Steinbach, J.; Mamat, C. The Radiochemical and Radiopharmaceutical Applications of Radium. *Open Chem.* **2016**, *14*, 118–129. [CrossRef]
48. Bruland, Ø.S.; Jonasdottir, T.J.; Fisher, D.R.; Larsen, R.H. Radium-223: From Radiochemical Development to Clinical Applications in Targeted Cancer Therapy. *Curr. Radiopharm.* **2008**, *1*, 203–208. [CrossRef]

49. Yong, K.; Brechbiel, M. Application of ^{212}Pb for Targeted α-particle Therapy (TAT): Pre-clinical and Mechanistic Understanding through to Clinical Translation. *AIMS Med. Sci.* **2015**, *2*, 228–245. [CrossRef]

50. Yong, K.; Brechbiel, M.W. Towards translation of ^{212}Pb as a clinical therapeutic; getting the lead in! *Dalton Trans.* **2011**, *40*, 6068–6076. [CrossRef] [PubMed]

51. Zhou, H.; Surowiec, K.; Purkiss, D.W.; Bartsch, R.A. Proton di-ionizable p-tert-butylcalix[4]arene-crown-6 compounds in cone, partial-cone and 1,3-alternate conformations: Synthesis and alkaline earth metalcation extraction. *Org. Biomol. Chem.* **2005**, *3*, 1676–1684. [CrossRef]

52. Zhou, H.; Liu, D.; Gega, J.; Surowiec, K.; Purkiss, D.W.; Bartsch, R.A. Effect of para-substituents on alkaline earth metal ion extraction by proton di-ionizable calix[4]arene-crown-6 ligands in cone, partial-cone and 1,3-alternate conformations. *Org. Biomol. Chem.* **2007**, *5*, 324–332. [CrossRef] [PubMed]

53. Boston, A.L.; Lee, E.K.; Surowiec, K.; Gega, J.; Bartsch, R.A. Comparison of upper and lower rim-functionalized, di-ionizable calix[4]arene-crown-6 structural isomers in divalent metal ion extraction. *Tetrahedron* **2012**, *68*, 8789–8794. [CrossRef]

54. Van Leeuwen, F.W.; Verboom, W.; Reinhoudt, D.N. Selective extraction of naturally occurring radioactive Ra^{2+}. *Chem. Soc. Rev.* **2005**, *34*, 753–761. [CrossRef]

55. Van Leeuwen, F.W.; Beijleveld, H.; Velders, A.H.; Huskens, J.; Verboom, W.; Reinhoudt, D.N. Thiacalix[4]arene derivatives as radium ionophores: A study on the requirements for Ra^{2+} extraction. *Org. Biomol. Chem.* **2005**, *3*, 1993–2001. [CrossRef]

56. Rojanathanes, R.; Pipoosananakaton, B.; Tuntulani, T.; Bhanthumnavin, W.; Orton, J.B.; Cole, S.J.; Hursthouse, M.B.; Grossel, M.C.; Sukwattanasinitt, M. Comparative study of azobenzene and stilbene bridged crown ether p-tert-butylcalix[4]arene. *Tetrahedron* **2005**, *61*, 1317–1324. [CrossRef]

57. Casnati, A.; Ca', N.D.; Sansone, F.; Ugozzoli, F.; Ungaro, R. Enlarging the size of calix[4]arene-crowns-6 to improve Cs+/K+ selectivity: A theoretical and experimental study. *Tetrahedron* **2004**, *60*, 7869–7876. [CrossRef]

58. Surowiec, M.; Custelcean, R.; Surowiec, K.; Bartsch, R.A. Mono-ionizable calix[4]arene-benzocrown-6 ligands in 1,3-alternate conformations: Synthesis, structure and silver(I) extraction. *Tetrahedron* **2009**, *65*, 7777–7783. [CrossRef]

59. Dalley, N.K. *Synthetic Multidentate Macrocyclic Compounds*; Izatt, R.M., Christensen, J.J., Eds.; Academic Press: New York, NY, USA, 1978; pp. 207–243.

60. Varadwaj, P.R.; Varadwaj, A.; Marques, H.M. DFT-B3LYP, NPA-, and QTAIM-Based Study of the Physical Properties of $[M(II)(H_2O)_2(15-crown-5)]$ (M = Mn, Fe, Co, Ni, Cu, Zn) Complexes. *J. Phys. Chem. A* **2011**, *115*, 5592–5601. [CrossRef] [PubMed]

61. Islam, M.S.; Pethrick, R.A.; Pugh, D.; Wilson, M.J. Theoretical study of the substituent effects of 4-substituted monobenzo crown ethers and the effects of ring size of 3n-crown-n (n = 4–7) ethers on the cation selectivity. *J. Chem. Soc. Faraday Trans.* **1997**, *93*, 387–392. [CrossRef]

62. Hirose, K. A Practical Guide for the Determination of Binding Constants. *J. Incl. Phenom. Macrocycl. Chem.* **2001**, *39*, 193–209. [CrossRef]

63. Kovalenko, S.V.; Alabugin, I.V. Lysine–enediyne conjugates as photochemically triggered DNA double-strand cleavage agents. *Chem. Commun.* **2005**, 1444–1446. [CrossRef]

64. Lazarides, T.; Miller, T.A.; Jeffery, J.C.; Ronson, T.K.; Adams, H.; Ward, M.D. Luminescent complexes of Re(I) and Ru(II) with appended macrocycle groups derived from 5,6-dihydroxyphenanthroline: Cation and anion binding. *Dalton Trans.* **2005**, 528–536. [CrossRef]

65. Becke, A.D. Density-functional exchange-energy approximation with correct asymptotic behavior. *Phys. Rev. A* **1988**, *38*, 3098–3100. [CrossRef]

66. Perdew, J.P. Density-functional approximation for the correlation energy of the inhomogeneous electron gas. *Phys. Rev. B* **1986**, *33*, 8822–8825. [CrossRef]

67. Becke, A.D. Density-functional thermochemistry. III. The role of exact exchange. *J. Chem. Phys.* **1993**, *98*, 5648–5652. [CrossRef]

68. Lee, C.; Yang, W.; Parr, R.G. Development of the Colle-Salvetti correlation-energy formula into a functional of the electron density. *Phys. Rev. B* **1988**, *37*, 785–789. [CrossRef]

69. Weigend, F.; Ahlrichs, R. Balanced basis sets of split valence, triple zeta valence and quadruple zeta valence quality for H to Rn: Design and assessment of accuracy. *Chem. Phys. Phys. Chem.* **2005**, *7*, 3297–3305. [CrossRef]

70. Weigend, F. A fully direct RI-HF algorithm: Implementation, optimised auxiliary basis sets, demonstration of accuracy and efficiency. *Phys. Chem. Chem. Phys.* **2002**, *4*, 4285–4291. [CrossRef]

71. Grimme, S.; Anthony, J.; Ehrlich, S.; Krieg, H. A consistent and accurate ab initio parametrization of density functional dispersion correction (DFT-D) for the 94 elements H-Pu. *J. Chem. Phys.* **2010**, *132*, 154104. [CrossRef]

72. Eichkorn, K.; Weigend, F.; Treutler, O.; Ahlrichs, R. Auxiliary basis sets for main row atoms and transition metals and their use to approximate Coulomb potentials. *Theor. Chem. Acc.* **1997**, *97*, 119–124. [CrossRef]

73. Ahlrichs, R.; Bär, M.; Häser, M.; Horn, H.; Kölmel, C. Electronic structure calculations on workstation computers: The program system turbomole. *Chem. Phys. Lett.* **1989**, *162*, 165–169. [CrossRef]

Review

^{67}Cu Production Capabilities: A Mini Review

Liliana Mou [1], Petra Martini [2], Gaia Pupillo [1], Izabela Cieszykowska [3], Cathy S. Cutler [4] and Renata Mikołajczak [3,*]

1   Legnaro National Laboratories, National Institute for Nuclear Physics, Legnaro, 35020 Padova, Italy; liliana.mou@lnl.infn.it (L.M.); gaia.pupillo@lnl.infn.it (G.P.)

2   Department of Environmental and Prevention Sciences, University of Ferrara, 44121 Ferrara, Italy; petra.martini@unife.it or mrtptr1@unife.it

3   National Centre for Nuclear Research, Radioisotope Centre POLATOM, 05-400 Otwock, Poland; izabela.cieszykowska@polatom.pl

4   Brookhaven National Laboratory, Collider Accelerator Department, Upton, NY 11973, USA; ccutler@bnl.gov

*   Correspondence: renata.mikolajczak@polatom.pl

Abstract: Is the ^{67}Cu production worldwide feasible for expanding preclinical and clinical studies? How can we face the ingrowing demands of this emerging and promising theranostic radionuclide for personalized therapies? This review looks at the different production routes, including the accelerator- and reactor-based ones, providing a comprehensive overview of the actual ^{67}Cu supply, with brief insight into its use in non-clinical and clinical studies. In addition to the most often explored nuclear reactions, this work focuses on the ^{67}Cu separation and purification techniques, as well as the target material recovery procedures that are mandatory for the economic sustainability of the production cycle. The quality aspects, such as radiochemical, chemical, and radionuclidic purity, with particular attention to the coproduction of the counterpart ^{64}Cu, are also taken into account, with detailed comparisons among the different production routes. Future possibilities related to new infrastructures are included in this work, as well as new developments on the radiopharmaceuticals aspects.

Keywords: Cu-67; copper radionuclides production; radiopharmaceuticals; theranostics

Citation: Mou, L.; Martini, P.; Pupillo, G.; Cieszykowska, I.; Cutler, C.S.; Mikołajczak, R. ^{67}Cu Production Capabilities: A Mini Review. *Molecules* **2022**, *27*, 1501. https://doi.org/10.3390/molecules27051501

Academic Editor: Bohumír Grűner

Received: 31 January 2022
Accepted: 18 February 2022
Published: 23 February 2022

Publisher's Note: MDPI stays neutral with regard to jurisdictional claims in published maps and institutional affiliations.

1. Introduction

Copper-67 (^{67}Cu) ($t_{1/2}$ = 2.58 d), the longest-living radioisotope of Cu, is of paramount importance because of its simultaneous emissions of β^- radiation (mean β^- energy: 141 keV; E_β-max: 562 keV), useful for therapeutic treatments and γ-rays (93 and 185 keV), suitable for single-photon emission computed tomography (SPECT) imaging. In fact, the ^{67}Cu mean β^--emission energy of 141 keV (Eβ−max: 562 keV) is slightly higher than that of Lutetium-177 (^{177}Lu, β^--emission energy of 133.6 keV, Eβ−max: 497 keV). ^{67}Cu decay characteristics make it one of the most promising theranostic radionuclides and its long half-life makes it suitable for imaging in vivo slow pharmacokinetics, such as monoclonal antibodies (MoAbs) or large molecules [1]. ^{67}Cu, studied for decades for radioimmunotherapy [2–4], is currently under the spotlight in the international community, as highlighted by the recent IAEA Coordinated Research Project (CRP) on "Therapeutic Radiopharmaceuticals Labelled with New Emerging Radionuclides (^{67}Cu, ^{186}Re, ^{47}Sc)" (IAEA CRP no. F22053) [5,6]. ^{67}Cu can also be paired with the β^+ emitters ^{64}Cu, ^{61}Cu, and ^{60}Cu to perform pretherapy biodistribution determinations and dosimetry using positron emission tomography (PET) systems. Table 1 presents the decay characteristics of ^{67}Cu and $^{64/61/60}$Cu-radionuclides, as extracted from the NuDat 3.0 database [7].

Among copper radionuclides, only ^{64}Cu has been widely used for preclinical and clinical PET studies due to its moderate half-life ($t_{1/2}$ = 12.7 h), low positron energy, and availability [8]. Given that copper radioisotopes are chemically identical, the same bifunctional chelators that have been developed for ^{64}Cu radiopharmaceuticals can be used directly for ^{67}Cu (and $^{61/60}$Cu) labeling. While the production techniques for ^{64}Cu are well known and are usually based on the $^{64}Ni(p,n)^{64}Cu$, and $^{64}Ni(d,2n)^{64}Cu$ reactions [9], the use of ^{67}Cu has been prevented by a lack of regular availability of sufficient quantities for preclinical and clinical studies. Only recently has ^{67}Cu become available in the U.S. through the Department of Energy Isotope Program (DOE-IP), in quantities and purities that are sufficient for medical research applications [10]. The investigation of ^{67}Cu supply worldwide is, therefore, a crucial point, and this review presents the state-of-art in ^{67}Cu production and medical applications.

Table 1. Main decay characteristics of ^{67}Cu and $^{64/61/60}Cu$-radionuclides [7].

	Half-Life	Main γ-ray Energy, Intensity (keV) (%)	Mean β⁺ Energy, Intensity (keV) (%)	Mean β⁻ Energy, Intensity (keV) (%)	Auger and IC Electrons
^{67}Cu	61.83 h	184.577 (48.7)	-	141 (100)	Yes
^{64}Cu	12.701 h	1345.77 (0.475)	278 (17.6)	191 (38.5)	Yes
^{61}Cu	3.336 h	282.956 (12.7) 656.008 (10.4)	500 (61)	-	Yes
^{60}Cu	23.7 m	826.4 (21.7) 1332.5 (88.0) 1791.6 (45.4)	970 (93)	-	Yes

2. Production Methods of ^{67}Cu

2.1. Accelerator-Based Production

2.1.1. Charged-Particle Induced Reactions

While $^{64/61/60}Cu$ radionuclides can be produced via low energy medical cyclotrons [11], for ^{67}Cu production, intermediate proton energies are needed, as shown by the well-known cross sections on zinc targets. The main route is the $^{68}Zn(p,2p)^{67}Cu$ reaction, studied for decades, whose excitation function has been recommended by the International Atomic Energy Agency (IAEA) [12]. This production method is feasible at intermediate proton-beams (E < 100 MeV), but it can be exploited also at higher energies [13]. The use of enriched ^{68}Zn targets is mandatory to reduce the coproduction of other Cu-radionuclides affecting the radionuclidic purity (RNP) of the final product [14]. However, the coproduction of ^{64}Cu cannot be avoided, as also indicated by the $^{68}Zn(p,x)^{64}Cu$ cross section reported by the IAEA up to 100 MeV.

At low-energy proton beams, the $^{70}Zn(p,\alpha)^{67}Cu$ reaction is feasible up to 30 MeV, without the coproduction of ^{64}Cu [15]. Due to the low value of the cross section (the maximum at 15 MeV is ca. 15 mb), this production route provides quite a low yield, i.e., 5.76 MBq/μAh for the 30–10 MeV energy range (corresponding to a 1.84 mm thick target). On the other hand, when using ^{70}Zn targets and higher proton energies, it is possible to reach a higher ^{67}Cu yield, though with the coproduction of ^{64}Cu [16,17].

Considering gallium as target material, it is possible to investigate the use of ^{71}Ga, whose natural abundance is 39.892% [7]. The cross section of the $^{71}Ga(p,x)^{67}Cu$ nuclear reaction is very low, about 2.6 mb at 33 MeV, and so far few literature data are covering the 20–60 MeV energy range [18]; however, the low cross section makes this route of ^{67}Cu production not feasible.

In order to estimate the influence of various parameters in the proton-based production of ^{67}Cu, we used the IAEA tool ISOTOPIA [19] and the recommended nuclear cross sections for calculations; when the IAEA cross sections values were not available, a fit of the literature data were used [20]. The ^{67}Cu activity was calculated for 62 h irradiation, equivalent to a ^{67}Cu Saturation Factor SF = 50% and a ^{64}Cu SF = 97%. Results are shown in Table 2, calculated at the end of bombardment (EOB) and 24 h after the EOB [13]. Table 2 also presents the ^{67}Cu yield when 70 MeV protons are exploited on a multi-layer target composed of enriched ^{68}Zn and ^{70}Zn layers, a configuration that optimizes ^{67}Cu production and minimizes ^{64}Cu coproduction [21].

Table 2. ^{67}Cu and ^{64}Cu production yields obtained using proton-beams on ^{70}Zn and ^{68}Zn enriched materials for each target material individually and the multi-layer target configuration (I = 1 µA; T_{IRR} = 62 h).

Target	Energy Range (MeV)	^{67}Cu @ EOB (MBq/µA)	^{64}Cu @ EOB (MBq/µA)	^{67}Cu/(^{64}Cu + ^{67}Cu) @ EOB	^{67}Cu/(^{64}Cu + ^{67}Cu) @ 24 h Post EOB
^{70}Zn	25–10 [9]	2.13×10^2	-	100%	100%
^{68}Zn	70–35 [9]	1.24×10^3	6.51×10^3	16%	35%
^{70}Zn + ^{68}Zn	70–55 + 55–35 [16]	1.86×10^3	5.71×10^3	25%	48%

Deuteron beams can also be exploited for ^{67}Cu production. As in the case of proton beams, the use of enriched targets is mandatory to limit the coproduction of contaminant Cu-radionuclides. In addition to this, when using ^{nat}Zn targets, the low production cross section [20] also leads to a low ^{67}Cu yield. The reported data on the ^{70}Zn(d,x)^{67}Cu cross section include a measurement from 2012 [22] and a recent work of 2021, describing an experimental campaign up to 29 MeV [23]. The use of ^{70}Zn targets seems promising in the energy window 26–16 MeV, giving a ^{67}Cu yield of 6.4 MBq/µAh [23]. In order to use higher deuteron beam energies it is important to measure the ^{70}Zn(d,x)^{64}Cu cross section (^{64}Cu threshold energy is E_{THR} = 26.5 MeV [7]), whose values are not reported in the literature [20].

Among the charged-particle induced reactions, the α-beams have also been explored. The main ^{67}Cu production route relies on the use of enriched ^{64}Ni targets (natural abundance 0.9255%). The enrichment level of the target material is again a key parameter, because of the coproduction of ^{64}Cu through the ^{nat}Ni(α,x)^{64}Cu reaction. Recently published new data of the ^{64}Ni(α,p)^{67}Cu reaction up to 50 MeV [24] are in agreement with previous results [25–28]. This latest work also reports the ^{67}Cu yield, after 24 h irradiation and 30 µA of beam current.

To compare the α-based production route with proton- and deuteron-induced nuclear reactions, Table 3 presents the calculated ^{67}Cu (and ^{64}Cu) production yields, assuming 100% enriched targets and the same irradiation conditions (^{67}Cu SF = 24%, ^{64}Cu SF = 73%). The ^{61}Cu and ^{60}Cu contaminants are not included in the calculations, since their half-lives are significantly shorter than ^{67}Cu half-life; thus, their impact on the RNP is relevant only soon after the EOB.

Table 3. ^{67}Cu and ^{64}Cu production yields obtained by using proton-, deuteron-, and α-beams on ^{70}Zn, ^{68}Zn, and ^{64}Ni enriched target materials assuming I = 30 µA and T_{IRR} = 24 h.

Beam	Target	Energy Range (MeV)	Thickness (mm)	^{67}Cu @ EOB (MBq)	^{64}Cu @ EOB (MBq)
	^{70}Zn	25–10	1.22	3.01×10^3	-
Protons	^{68}Zn	70–35	6.43	1.75×10^4	1.48×10^5
	^{70}Zn + ^{68}Zn	70–55 + 55–35	3.26 + 3.27	2.62×10^4	1.30×10^5
Deuterons	^{70}Zn	26–16	0.58	4.01×10^3	-
Alpha	^{64}Ni	30–10	0.16	1.00×10^3	-

The calculations reported in Table 3 show that the most convenient route to obtain pure ^{67}Cu (without ^{64}Cu coproduction) is by using deuteron beams and ^{70}Zn targets. Moreover, the use of α-beams and ^{64}Ni targets provides pure ^{67}Cu. Intense linear accelerators (ca. mA current) for α-particles, requiring specific targets able to withstand such high currents that are still to be designed, are soon foreseen but not yet available. The proton-induced reactions with a ^{68}Zn target and a $^{68/70}$Zn multi-layer target configuration seem to be a promising option if some ^{64}Cu coproduction is acceptable, since these routes provide a larger ^{67}Cu yield in comparison with the ^{70}Zn(d,x)^{67}Cu route. In all cases, by changing the projectile type and/or its energy and/or the target material, it is possible to adapt the ^{67}Cu production yield and the profile of contaminants.

2.1.2. Photonuclear Production

The photonuclear production for ^{67}Cu using bremsstrahlung photons with an e-LINAC accelerator has been studied for decades [29–36]. Recently, a large enriched ^{68}Zn target (55.5 g) was irradiated with 40 MeV e-LINAC for 53.5 h, obtaining 62.9 GBq (1.7 Ci) without detecting ^{64}Cu [10]. The threshold energy of the ^{68}Zn(γ,x)^{64}Cu reaction is $E_{THR} = 27.6$ MeV [37]. For the photonuclear reaction, the use of enriched target material is mandatory to avoid the coproduction of the following radionuclide impurities: ^{63}Zn, ^{65}Zn, and ^{64}Cu. According to the DOE Isotope Program, ^{67}Cu is being routinely produced by the Argonne National Laboratory via a photonuclear reaction at its low-energy accelerator facility (LEAF). Approximately 1 curie per batch can be provided, with radionuclidic purity of >99% and a specific activity >1850 GBq/mg (>50 Ci/mg) ^{67}Cu/total Cu at EOB [38]. Commercial entities have indicated they are pursuing this route of production, although no commercial suppliers are online as of the date of this publication.

2.2. Reactor-Based Production

^{67}Cu can be produced in a reactor via the ^{67}Zn(n,p)^{67}Cu nuclear reaction with fast neutrons and, after the separation process, it is possible to obtain ^{67}Cu in a n.c.a. form. The enrichment of the ^{67}Zn target material is crucial to obtain quantities of ^{67}Cu suitable for medical application. In natural zinc, the abundance of ^{67}Zn is only 4.04%, while ^{64}Zn contributes to 49.17% [7] and the neutron cross-section for the ^{64}Zn(n,p)^{64}Cu nuclear reaction is much higher, hence the amount of ^{64}Cu produced is an order of magnitude higher than that of ^{67}Cu. The use of enriched ^{67}Zn not only limits the contribution of ^{64}Cu but also increases the irradiation yield of ^{67}Cu. However, due to the low cross section value, the ^{67}Zn(n,p)^{67}Cu nuclear reaction requires a high flux of fast neutrons exceeding 10^{14} n cm^{-2}s^{-1} [3,39–41]. In the fast neutron flux of 4.4×10^{14} n cm^{-2}s^{-1} (En > 1 MeV) the saturation yield of ^{67}Cu at the EOB was 4.14 ± 0.37 GBq/mg of ^{67}Zn, these values were dependent on the neutron flux and the position of the target in the reactor [37,40]. When the thermal and the fast neutron fluxes were 1.3×10^{12} n cm^{-2}s^{-1} and 1.5×10^{12} n cm^{-2}s^{-1}, respectively, only 630 kBq/mg Zn of ^{67}Cu was produced after a 5 h irradiation of target containing around 50 mg of 93.4% enriched ^{67}ZnO. In the obtained mixture of ^{64}Cu and ^{67}Cu, the latter contributed to less than 30% of the total radioactivity. In order to increase the amount of ^{67}Cu produced by this reaction, irradiation in a nuclear reactor with higher fast neutron flux, and for longer periods of irradiation are required [42]. More detailed summation of reported neutron irradiation yields for the ^{67}Zn(n,p)^{67}Cu nuclear reaction has been previously reported [43,44].

In nuclear reactors with a high ratio of thermal to fast neutrons, the coproduction of ^{65}Zn ($t_{1/2} = 243.93$ d) is unavoidable because of the presence of ^{64}Zn, particularly in the natural zinc target material, and the relatively high cross section of ^{64}Zn(n,γ)^{65}Zn nuclear reaction with thermal neutrons. Although ^{65}Zn can be separated from ^{67}Cu during the target processing, due to its long half-life, ^{65}Zn contaminates the recycled target material. Thermal neutron shielding made of materials with high neutron capture cross sections, such as boron, cadmium or hafnium, may reduce this contamination [45–47]. A boron nitride shield of 3.48 g/cm^3 density and around 4 mm thickness reduced the thermal flux in the sample holder from about 10^{13} n/cm^2/s to approximately 10^{10} n/cm^2/s, resulting in a production of about 11 times higher for ^{67}Cu than ^{65}Zn and in reducing the ^{65}Zn production in ^{67}Zn targets by a factor of 66 [45]. Other reported by-products included ^{58}Co produced via the ^{58}Ni(n,p)^{58}Co reaction, ^{67}Ga, and interestingly, ^{182}Ta, which can be produced via thermal neutron capture on ^{181}Ta present in the target material [45]. Potentially, all these contaminants can be removed in the chemical separation of ^{67}Cu.

The measurements of the ^{71}Ga(n,n + α)^{67}Cu nuclear reaction show an increasing trend from 13 MeV to 20 MeV, reaching a maximum value of ca. 20 mb [20]. Because of these quite low cross section values, ^{71}Ga targets are impracticable for ^{67}Cu production.

2.3. Targetry

In the charged-particle induced nuclear reactions, enriched Zn is the most commonly used target material, although Ni has also been used with α-beams. The accelerator-based production requires the use of highly enriched target material to achieve high radionuclidic and chemical purity of the product thus making the target very expensive and necessitating target recovery and recycling to minimize the production costs. Targets composed by a set of thin foils in the well-know "stacked-foils" configuration have been used in preliminary studies for nuclear cross section measurements, with enriched Zn foils produced by electrodeposition [48] or by lamination from Zn metal powder [16]. In the accelerator production of ^{67}Cu, mainly thick solid targets have been used, either in the form of metallic foil/coin or in the oxide form [13], despite the low melting point of Zn.

A multilayer target configuration, based on the use of both ^{68}Zn and ^{70}Zn enriched materials, has been recently patented for ^{67}Cu proton induced production [21]. This target configuration, shown in Figure 1, is beneficial with an increase of 48% in the ^{67}Cu production and a 12% decrease of ^{64}Cu coproduction, with respect to a thick monolayer of ^{68}Zn in the energy range 70–35 MeV and 24 h irradiation (Table 3). In addition to this, in the low energy range (E < 30 MeV), it is possible to add another ^{70}Zn layer to exploit the (p,α) reaction to increase the ^{67}Cu yield. It is important to underline that, in this final ^{70}Zn layer (covering the low-energy region), there is no coproduction of ^{64}Cu. In the patent, it is suggested to apply a radiochemical process to each target layer individually, in order to recover each enriched material separately (^{68}Zn and ^{70}Zn). Moreover, the user can decide to combine the final solutions (containing the ^{67}Cu/^{64}Cu mix of radionuclides and the pure ^{67}Cu) or to use them separately to label the desired radiopharmaceuticals.

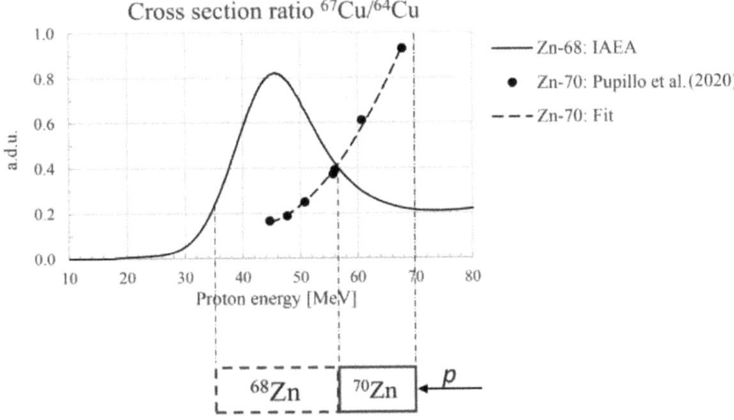

Figure 1. Plot of the nuclear cross section ratio for the production of ^{67}Cu and ^{64}Cu radionuclides: the continuous line is the IAEA recommended value for ^{68}Zn targets; the dashed line refers to the measured values for ^{70}Zn targets. The vertical dashed lines refer to the favorable energy range for ^{67}Cu production. A scheme of the multi-layer target configuration described in the international INFN patent is shown at the bottom [21].

Electroplating of enriched Zn or Ni on a gold, titanium, aluminum, or gold-plated copper backing is the most popular target fabrication technique. With this method, the target thickness can be adapted to the optimal specific beam energy range and thicknesses up to 80 mm have been reported [6,13,49,50]. For the production of ^{67}Cu via the ^{64}Ni(α,p)^{67}Cu nuclear reaction, Ohya et al., used enriched ^{64}Ni target material (^{64}Ni 99.07%) [50]. Sublimation and the casting process have also been used to produce massive zinc ingot targets (50–100 g) for photonuclear production [10,31].

Zinc oxide targets have been used for both photonuclear- and nuclear reactor-based production of ^{67}Cu. Recently, ZnO was pressed and wrapped in aluminum foil and used as a target for the photonuclear production [24]. In the targets for neutron irradiation in nuclear reactors, zinc oxide powder was encapsulated in a quartz ampule and aluminum cans or sealed in a polyethylene bag and sandwiched between two Ni foils [6,51–54].

^{nat}ZnO nanoparticles were compared to ^{nat}ZnO powder in the target irradiation at a fast neutron flux. Both targets of the same mass, 1.0 g each, were irradiated for 30 min in the fast neutron flux of 1.4×10^{13} n·cm^{-2}s^{-1}, showing an increase in the ^{67}Cu activity produced that almost doubled (0.0168 MBq vs. 0.0326 MBq) when ^{nat}ZnO nanoparticles target was used [51].

2.4. Radiochemistry

Strictly related to the target configuration is radiochemical target processing, aimed at transforming the produced ^{67}Cu in the solution suitable for radiolabeling. Since the commercially available target materials, even with the highest achievable enrichment of the desired isotope, still contain elemental and isotopic contaminants, their irradiation leads to collateral production of several zinc, cobalt, and gallium radionuclides. The method for the separation of Cu from the dissolved target material must therefore ensure not only the removal of the bulk target material but also of these specific side products. The process should also aim to recover the enriched target material, which, in general, is very expensive. The separation process can be accomplished in one step or in a combination of more steps using conventional separation methods, such as ion chromatography, solvent extraction, precipitation, sublimation, etc., to isolate the desired radionuclide and eliminate contaminants.

Solvent extraction with dithizone (diphenylthiocarbazone) dissolved in a water immiscible medium, e.g., CCl$_4$, CHCl$_3$, was proposed because of its selectivity for ions of various metals depending on the pH. Dithizone is selective for Cu in the pH range 2–5, for Zn in the pH range 6.5–9.5, and for Ni in the pH range 6–9. Using this method ^{67}Cu could be separated with dithizone even from large amounts of Zn, up to 5 g, and from other co-produced impurities [36,53,55,56]. The Cu is then back extracted into an aqueous phase by shaking the organic solution with 7 M HCl mixed with H$_2$O$_2$, resulting in dissociation of Cu-chelate [36,53,56]. This technique is proposed as a first step for Cu separation from bulk target material, which can be then followed by further purification using chromatographic methods [57,58]. Recently, the dithizone-based solvent extraction and back extraction of Cu from large excess of Zn have been implemented in a microfluidic system [59].

A new method combining the solvent extraction and anion exchange separation techniques into a single separation was proposed by Dolley et al., by using an Amber-Chrom CG-71 dithizone-impregnated resin. At first, the coproduced ^{66,67}Ga were removed with an untreated AmberChrom CG-71 resin. Next, ^{67}Cu was separated from ^{65}Zn and ^{56,57,58}Co on the dithizone-based solid phase extraction chromatographic column, which retained Cu radioisotopes [60]. Improvements of this method were lately proposed for the separation of Cu from large amounts of zinc by using a single modified dithizone (diphenylthiocarbazone) Amberlite$^{®}$ XAD-8 (20–60 mesh) chelating resin [61]. Dithizone is sensitive to oxidation, forming diphenylthiocarbodiazone when exposed to light and heat. Thus, dithizone impregnated resin must be relatively freshly prepared before use. A semi-automated separation module adjusted to operate in a shielded facility employing liquid–liquid extraction of Cu from Zn target was developed [55].

Another extractant proposed for the separation of ^{67}Cu from ^{67}ZnO irradiated in nuclear reactor was thenoyltrifluoroacetone (TTA) in benzene [44]. Despite the high separation yield, the content of organic extractant residue in the final product is a drawback of this extraction methods. In the past, other chelating agents for Cu were proposed, such as cupferron and diethyldithiocarbamate [57].

Electrolysis has also been used for the separation of Cu from Zn and Ni target material [39,56,62]. Although the purity of the final Cu solution isolated by conventional electrolysis was adequate for antibody labeling, the process was very time consuming and

^{67}Cu losses occurred at each electrolytic step. Thus, this approach was suggested to be unsuitable for routine production.

While most researchers have reported on the application of electrolytic separation under an external electromotive force (EMF), spontaneous electrochemical separation of Cu from proton- or neutron- irradiated zinc targets has been investigated [39]. The process is simple and can be easily automated and adopted to work in hot cells, which is advantageous over conventional electrolysis. Moreover, without an externally applied voltage, the process is more selective in separating Cu from interfering metal ions, such as Fe, Co, or Ni, due to the elimination of hydrogen overvoltage and the preservation of the Pt electrode. Time of the process was only 30 min. As result of the spontaneous electrodeposition in the processing of the proton-irradiated ZnO target, the separation factors of ^{67}Cu from isotopes of Co, Cr, Fe, Ga, Mn, Ni, and V has been reported [39]. Separation factors of $>1 \times 10^7$ from grams of Zn can be achieved using this method, obtaining highly pure n.c.a. ^{67}Cu. The overall separation factors ranged from 7×10^3 for ^{57}Ni to 9×10^4 for ^{58}Co, and the separation factor from ^{67}Ga was $>1.3 \times 10^4$.

Ion exchange is frequently used in the separation of radionuclides for medical application due to its high efficiency, reproducibility, and ease in automation. Though, the separation of Cu from Zn and contaminants could be efficiently accomplished (recovery yield 92–95%) using three ion exchange columns (cation exchange resin AG50W-X8, anion exchange resin AG1-X8 and chelating resin Chelex 100). The method is laborious and includes several time consuming evaporation steps [13,63]. An alternative co-precipitation of ^{67}Cu from bulk Zn using H_2S gas with an excess of silver nitrate and consecutive separation of precipitate by filtration was proposed [63]. Compared with ion exchange, this new process of ^{67}Cu separation was completed in less than 3 h with similar recovery.

In the separation of ^{67}Cu produced by irradiation of a Ni target, the same procedures that have been used in the routine production of ^{64}Cu by the ^{64}Ni(p,n)^{64}Cu nuclear reaction can be adopted. The Cu/Ni separation and purification is typically a one-step procedure, in which an ion exchange resin (in anionic or cationic form, AG1-X8 and AG50W-x8 respectively) is used for eluting Ni, Co, and Cu with various acid concentration or acid/organic solvent ratio [50,64–66].

A summary of the procedures that have been investigated for the separation of ^{67}Cu from an irradiated zinc target, either in metallic or oxide form, is presented in Table 4.

Table 4. Cu/Zn separation and purification procedures (SE = solvent extraction, IE = ion exchange; TTA = thenoyltrifluoroacetone), processing time and process yield are included, if available.

Ref.	Target	Dissolution	Radiochemical Separation Method	Processing Time	Yield
[36,53]	^{nat}Zn foil or ZnO	conc. HCl	SE with dithizone	-	>90%
[56]	^{nat}Zn plates	30% HCl (400 K)	SE with dithizone +IE with AG 50 W +IE with AG1-X8	5 h	85 ± 20% for SE
[57]	ZnO (28–30 g)	conc. HCl	SE with dithizone +IE with AG1-X8	5–7 h	>90%
[44]	^{67}ZnO	1 N HCl +30% H_2O_2	SE with TTA	-	-
[51,54]	^{68}ZnO (100 mg)	conc. HCl	IE with Dowex 1×8	4 h	94%
[52]	^{nat}Zn (1–2 g)	8 M HCl	IE with AG1-X8	2 h	95%
[42]	^{67}ZnO, 93.4% (50 mg)	4 M HCl	IE with Diaion SA-100	-	95%

Table 4. *Cont.*

Ref.	Target	Dissolution	Radiochemical Separation Method	Processing Time	Yield
[13]	^{68}Zn, 99.7% (0.7–4 g) electroplated on Ti or Al	12 M HCl	IE with AG50-X4 +Chelex-100 +AG1-X8	-	92–95%
[56]	^{nat}Zn plate	37% HCl	IE with AG50W +Chelex-100 +AG1-X8	4.5 h	90%
[16]	^{70}Zn, >95% metal foils	10 M HCl	IE with AG50W-X4 +AG1-X8	4 h	$95 \pm 2\%$
[63]	^{nat}ZnO powder (3.5 g)	10 M HCl (100 °C)	Double coprecipitation with AgNO$_3$	<3 h	$81 \pm 6\%$
[31]	^{68}Zn metal ingot target (100 g)	-	Sublimation	Rate of Zn separation from Cu: >50 g/h	Removing of >99% of Cu (and other metals) from Zn in each sublimation cycle
[56]	^{nat}Zn plate	30% HCl (400 K)	Electrolysis	12 h	60%
[2]	^{nat}Zn foil	conc. HCl +HNO$_3$	Electrolysis + IE with MP-1	-	80%
[39]	^{67}ZnO, $\geq$94% (50–100 mg)	1 M H$_2$SO$_4$	Spontaneous electrochemical separation	1.5 h	95%

2.5. Recovery

When highly enriched target materials are needed, their cost makes the target recovery and recycling mandatory [6]. The choice of the recovery technique depends on the method used for ^{67}Cu separation and the expected chemical form of the recovered material.

When electrodeposition is the procedure selected for target production, the recovery of either Zn or Ni target materials is rather simple and requires only a few chemical steps for re-establishing the electrodeposition conditions, as described by Medvedev et al. [13]. A pre-cleaning of the target material before its electrodeposition can be performed to minimize the impurities in the recycled target and to improve the specific activity of the final product. Before Zn target material recovery, for example, the radiogallium can be removed by using a cation exchange column AG50W-X8 as described by Ohya et al. [63].

In the procedure reported by Shikata in 1964, zinc was recovered in the oxide form by evaporating to dryness the 0.01 M HCl zinc-rich solution eluted from the anion exchange resin, dissolving the obtained zinc chloride in water, heating up the solution to about 70 °C and adding ammonium oxalate to induce precipitation of zinc oxalate. The precipitate was then filtered, washed and ignited to constant weight [42].

Similarly, after the isolation of ^{67}Cu using solvent extraction with TTA/benzene, zinc oxide target material was recovered after evaporation of the aqueous solution under a stream of argon, followed by dissolution of the obtained solids in 2 N HCl, and finally loading the solution on a Bio-Rad resin column [44]. Elution of Zn^{2+} with 6 N HNO$_3$ was then evaporated under argon stream. The resulting dry ZnO was transferred to a furnace and heated at 350 °C for 2 h.

Another recovery technique was reported by Ehst et al., through the sublimation Cu/Zn separation process by recovering sublimed Zn with negligible losses [31].

2.6. Quality of ^{67}Cu as Radiopharmaceutical Precursor

According to Ph. Eur. monograph (0125) [67], a radionuclide precursor is any radionuclide produced for radiolabeling of another substance prior to administration.

In the absence of a pharmacopoeia monograph specific for ^{67}Cu, the proposed quality control protocol for produced ^{67}Cu includes identity, radioactivity, specific activity, radionuclidic purity, radiochemical purity, and chemical purity [5,6,50,63].

2.6.1. Identity

Radionuclide identity is determined by assessing the physical characteristics of the radionuclide emissions. The γ-line at the energy Eγ = 184.6 keV (intensity Iγ = 48.7%) has been used to evaluate the activity of ^{67}Cu [31]. In order to check that there is no contamination of ^{67}Ga in the product, we suggest verifying that the same activity of ^{67}Cu is obtained from both 184.6 keV (48.7%) and the 300.2 keV (0.797%) γ-lines. This additional control is due to the γ-lines emitted by ^{67}Ga (Eγ = 184.6 keV, Iγ = 21.41% and the Eγ = 300.2 keV, Iγ = 16.64%) having the same energies of ^{67}Cu but with different intensities [7].

2.6.2. Specific Activity

Specific activity (SA) is defined as radioactivity per unit mass of the product [68]. Several factors that could affect the specific activity of the final product include cross-sections, target impurities, secondary nuclear reactions, target burn-up, and post-irradiation processing periods.

The chosen production route may be optimized to limit the production of stable copper isotopes. However, the contamination with stable copper (^{63,65}Cu) coming from the materials and chemicals involved in the production and separation is difficult to control. Special care has to be taken employing metal free chemicals and tools. It is therefore fundamental to determine the quantity of stable copper in the final product through inductively coupled plasma with optical emission spectrometry (ICP-OES) or mass spectrometry (ICP-MS) detection techniques.

2.6.3. Radionuclidic Purity

Radionuclidic purity (RNP) is determined by γ-spectrometry using a HPGe detector calibrated using standard sources, coupled with a multichannel analyzer [50]. The coproduced short-lived ^{60,61,62,66,68,69}Cu radioisotopes do not significantly affect the RNP of ^{67}Cu [5]. However, the coproduction of the longer-lived ^{64}Cu in most production routes is unavoidable and the ^{64}Cu/^{67}Cu ratio is an important factor defining the quality of ^{67}Cu. Ohya et al. [50] determined the activities of ^{61}Cu, ^{64}Cu, ^{67}Cu, and ^{65}Zn considering the ^{64}Ni(α,p)^{67}Cu reaction through the measurement of the following γ-lines, respectively: Eγ = 656.0 (10.8%), 1345.8 (0.473%), 184.6 (48.7%), and 1115.5 (50.60%) keV. Based on the analytical method proposed by Van So et al. [69], the activities of ^{67}Cu and ^{67}Ga can be evaluated from the measurements of their mixed signal in the γ-lines at 184.6 keV (48.7% for ^{67}Cu and 21.2% for ^{67}Ga), 209.0 keV (0.115% ^{67}Cu and 2.4% ^{67}Ga), 300.2 keV (0.797% ^{67}Cu and 16.8% ^{67}Ga), and 393.5 keV (0.22% ^{67}Cu and 4.68% ^{67}Ga) [63]. With this method, it is possible to infer the ^{67}Cu and ^{67}Ga activities without a radiochemical separation process. Similarly, Pupillo et al., proposed the method to correct for the residual Ga-activity in copper-solution [16] and Nigron et al., performed the measurement of the ^{70}Zn(d,x)^{67}Cu cross section [23]. The presence of the following impurities can be determined in the ^{67}Cu-solution considering the main radionuclides γ-lines: ^{66}Ga (1039 keV, 37%) ^{58}Co (810.8 keV, 99.45%), ^{65}Zn (1116.0 keV, 50.04%), ^{57}Ni (1377.6 keV, 81.7%), and ^{105}Ag (344.5 keV, 41.4%). It is important to remember that the impurity profile depends on the nuclear reaction route and on the specific irradiation conditions, in addition to the radiochemical processing of the target.

2.6.4. Chemical Purity

The chemical purity of radionuclides for medical applications refers to the determination of metal cations, which may compete with the radiometal in the chelator for complex formation. Impurities, such as Zn^{2+}, Fe^{3+}, Co^{2+}, Ni^{2+} can dramatically influence the radiolabeling yield of common chelators such as DOTA (1,4,7,10-tetraazacyclododecane-1,4,7,10-tetraacetic acid). They can arise from solvents used in target processing or can be eluted from resins used for isolation of radiometal from bulk target solution. Another possible contaminant is the residual organic solvent or other organics remaining in the product after liquid–liquid or solid phase extraction.

Limits for individual metal impurities for radionuclides intended for radiopharmaceutical preparation are described in Ph. Eur. monographs. For example, in the case of ^{177}Lu solution for labeling the limit of Cu is 1.0 μg/GBq, Fe: 0.5 μg/GBq, Pb: 0.5 μg/GBq, and Zn: 1.0 μg/GBq [70]. Such requirements have not been yet established for ^{67}Cu. Metallic impurities present in radionuclide solutions are typically determined using ICP-OES or ICP-MS techniques. For evaluation of the chemical purity of ^{67}Cu the ICP-OES, ICP-MS, and anodic stripping voltammetry were used for the determination of zinc and copper impurities [5,6]. Commercially available colorimetric test kits for the determination of Zn, Fe, or Ni in the final product were also mentioned [5]. Although this method is less sensitive, it enables sub-ppm analysis and can be used as a preliminary analysis [5]. The macrocyclic copper chelator of 1,4,8,1-tetraazacyclotetradecane-N,N',N'',N''-tetraacetic acid (TETA) has been used for assessment of ^{64}Cu specific activity as it binds copper 1:1; the same method can be adopted for ^{67}Cu analysis [5,65]. The effective specific activity of ^{64}Cu was assessed based on the percent complexation of ^{64}Cu-TETA as a function of TETA concentration, monitored by radio-TLC. Other metallic impurities can't be determined in this way; therefore, Ohya et al. [50] developed the so-called post-column analysis for the evaluation of ^{67}Cu chemical purity, based on ion chromatography in combination with an ion-pair reagent and UV detector.

3. The Use of ^{67}Cu for Medical Applications

3.1. Chelators for Copper

The coordination chemistry of the transition metal copper in an aqueous solution is limited to three accessible oxidation states (I–III). The lowest oxidation state, Cu(I) allows formation of complexes with various ligands, which are however labile and lack sufficient stability for radiopharmaceutical applications. Cu(II) is less labile toward ligand exchange and is optimal for incorporation into radiopharmaceuticals. The Cu(III) oxidation state is relatively rare and difficult to attain and thus not of importance for radiopharmaceuticals [71,72]. To evaluate ^{67}Cu for radioimmunotherapy (RIT), the macrocyclic chelating agent 1,4,7,11-tetraazacyclotetradecane-N,N',N'',N'''-tetraacetic acid (TETA) was designed specifically to bind copper rapidly and selectively for conjugation to MoAbs. The synthesis of this "bifunctional" metal chelator, 6(p-bromoacetamidobenzyl)-1,4,8,11-tetraazacyclotetradecane-N,N',N'',N'''-tetraacetic acid, which can be covalently attached to proteins and which binds copper stably in human serum under physiological conditions, was reported by Moi et al., in 1985 [73]. In contrast, other chelators based on ethylenediaminetetraacetic acid or diethylenetriaminepentaacetic acid rapidly lose copper to serum albumin under the same conditions.

Despite their use in several pre-clinical studies, the stability of Cu(II) metal complexes with common macrocyclic ligands, such as DOTA (1,4,7,10-tetraazacyclododecane-1,4,7,10-tetraacetic acid) and TETA was evaluated and considered not satisfactory for biomedical applications, demonstrating that nearly 70% of the ^{64}Cu trans-chelated in the liver 20 hours post injection [74–76].

The addition of a bond (or Cross-Bridge, CB) between two opposite nitrogen atoms in the tetraazamacrocyclic skeleton of the macrocycle gives greater stability. The Cu(II) complex with the ligand CB-TE2A (4,11-bis(carboxymethyl)-1,4,8,11-tetraazabicyclo [6.6.2] hexadecane) has been demonstrated to be particularly stable in vivo and to have exceptional inertness kinetics in aqueous solution thanks to the stabilizing effects given by the CB-macrocycle and by the two hanging carboxymethyl arms [77–80]. Unfortunately, the extreme labeling conditions at 95 °C do not make it a good candidate for the development of radiopharmaceuticals labile at high temperatures, e.g., antibodies [81]. The replacement

of one or both of the hanging carboxymethyl arms with groups derived from phosphonic acid ($-CH_2-PO_3H_2$), CB-TE1A1P (4,8,11-tetraazacyclotetradecane-1-(methanophosphonic acid)-8-(methanocarboxylic acid)), and CB-TE2P (1,4,8,11-tetraazacyclotetradecano-1,8-bis(methanephosphonic acid)), has been shown to confer greater thermodynamic and kinetic stability, as well as greater selectivity and faster complexation kinetics [82].

Another class of potential sarcophagus-like cage structure chelators for copper are hexaazamacrobicyclic ligands [83,84]. Labeling conditions of copper complexes with sarcophagine-like ligands are particularly favorable in a wide pH range (4–9) with yields of 100% in a few minutes at room temperature [75]. The complexes obtained possess excellent in vitro and in vivo stability [85–87].

Some N2S2-type acyclic chelators have also been studied, such as PTSM (pyruvaldehyde-bis(N4-methylthiosemicarbazone)) and ATSM (diacetyl-bis(N4-methylthiosemicarbazone)). This class of chelators, bis (thiosemicarbazones), act as tetradentate ligands and coordinate Cu^{2+} forming stable and neutral complexes, with a square planar geometry, which show a high permeability of the cell membrane. The preparation of these complexes with radioactive copper isotopes can be carried out under mild conditions in high yield [74,87–89].

3.2. Pre-Clinical Studies

In the early studies on RIT, the ^{67}Cu labeled radioimmunoconjugate, ^{67}Cu-2IT-BAT-Lym-1, was prepared by conjugating the bifunctional TETA derivative BAT to antibody Lym-1, monoclonal antibody against human B cell lymphoma, via 2-iminothiolane (2IT). This modification did not significantly alter immunoreactivity of the antibody [90]. Nude mice bearing human Burkitt's lymphoma (Raji) xenografts treated with ^{67}Cu-2IT-BAT-Lym-1 achieved high rates of response and cure with modest toxicity [91]. In the pre-clinical radioimmunotherapy comparison with another antibody conjugate, BAT-2IT-IA3, the therapeutic effect of ^{67}Cu was similar to that of ^{64}Cu [92].

Searching for optimal radiocopper labeling of anti-L1-CAM antibody chCE7, five bifunctional copper chelators were synthesized and characterized (CPTA-*N*-hydroxysuccinimide, DO3A-L-p-isothiocyanato-phenylalanine, DOTA-PA-L-p-isocyanato-phenylalanine, DOTA-glycyl-L-p-isocyanato-phenylalanine and DOTA-triglycyl-L-p-isocyanato-phenylalanine). CPTA-labeled antibody achieved the best tumor to normal tissue ratios when biodistributions of the different ^{67}Cu-chCE7 conjugates were assessed in tumor-bearing mice. High resolution PET imaging with ^{64}Cu-CPTA-labeled MAb chCE7 showed uptake in lymph nodes and heterogeneous distribution in tumor xenografts [93]. The therapeutic value of ^{67}Cu was also demonstrated pre-clinically in the treatment of bladder cancer [94]. It has been demonstrated that the chelator makes the difference in the behavior of conjugates [95]; therefore, new chelator/biological vector compositions are under investigation.

In recent years, new inputs came from the use of the cage ligand, sarcophagine-like (Sar), as a chelator for ^{67}Cu. The studies on ^{67}Cu-CuSarTATE, a somatostatin receptor targeting ligand and sarcophagine-containing PSMA ligand, labeled either with ^{64}Cu- or ^{67}Cu, supported translation of these radiopharmaceuticals to the clinics [96–99].

3.3. Clinical Studies

As early as in 1972, the absorption of copper was determined by the simultaneous administration of ^{64}Cu orally and ^{67}Cu intravenously to patients with Wilson's disease. The disease results in excess accumulation of copper in tissues, such as the liver and brain. However, in this study, copper radionuclides were administered in a simple cationic form [100]. Two decades later, following the promising pre-clinical results, the radioimmunoconjugate ^{67}Cu-2IT-BAT-Lym-1 was introduced to the clinics [101]. Then the phase I/II clinical trial of ^{67}Cu-2IT-BAT-Lym-1 was conducted, in an effort to further improve the therapeutic index of Lym-1-based radioimmunotherapy patients with B-cell non-Hodgkin's lymphoma (NHL). ^{67}Cu-2IT-BAT-Lym-1 provided good imaging of NHL and favorable radiation dosimetry. The mean radiation ratios of tumor to body and tumor to marrow were 28:1 and 15:1, respectively. Tumor-to-lung, -kidney, and -liver radiation dose ratios

were 7.4:1, 5.3:1, and 2.6:1, respectively. This ^{67}Cu-2IT-BAT-Lym-1 trial for patients with chemotherapy-resistant NHL had a response rate of 58% (7/12). No significant nonhematologic toxicity was observed. Hematologic toxicity, especially thrombocytopenia, was dose limiting [102]. Recently, there have been several new completed and ongoing clinical trials with ^{67}Cu radiolabeled compounds. These included the theranostic trial for pre-treatment dosimetry of somatostatin analog $^{64/67}$Cu-SarTATE [103] and peptide receptor radionuclide therapy using ^{67}Cu-SarTATE in pediatric patients [104,105].

The decay properties of ^{67}Cu were also explored in order to improve the patient management in the hospital, going into two critical steps further: demonstrating the therapeutic efficacy of ^{67}Cu-PRIT (pre-targeted radioimmunotherapy) and illustrating the usefulness of pre-targeted ^{64}Cu-PET as a predictive indicator of response to ^{67}Cu-PRIT [106]. The use of γ-emissions of ^{67}Cu at 184.6 keV (48.7%) and rarely 93.3 keV (16.1%) for SPECT imaging, assuming that the branching ratios of ^{67}Cu's γ-emissions are much higher than those of ^{177}Lu, is expected to provide higher imaging sensitivity [107,108]. It is relevant to underline the recent work on SPECT imaging with Derenzo phantom, to study SPECT image quality using ^{67}Cu [10]. The use of a medium energy (ME) collimator is typically recommended for ^{177}Lu, since its γ-ray has an energy of 208 keV, while for the 140 keV γ-ray emitted by ^{99m}Tc, the gold-standard radionuclide for SPECT imaging, the low-energy high-resolution (LEHR) collimator is recommended. As the γ-emission energy of ^{67}Cu (185 keV, Table 1) falls between these two aforementioned energies, the appropriate collimator for SPECT imaging with ^{67}Cu was found to be the ME collimator. Although there is a reduced image quality for ^{67}Cu and ^{177}Lu, these radionuclides are still considered adequate for tumor identification, suggesting that the post-treatment dosimetry is possible for ^{67}Cu-labeled radiopharmaceuticals as it is for ^{177}Lu [10].

4. Discussion

Comparing with the accelerator-produced mode, the routine production of ^{67}Cu via the (n,p) reaction in a nuclear reactor is less profitable due to the relatively low yields and expensive ^{67}Zn target material. Only a limited number of nuclear reactors offer the fast neutron flux higher than 10^{14} n cm^{-2}s^{-1} and only a few higher than 10^{15} n cm^{-2}s^{-1}, which would be preferred for the more efficient production of ^{67}Cu [40]. For this reason, the reactor route is currently not expected to contribute significantly to the availability of ^{67}Cu, though reactor produced ^{67}Cu might be useful as a tracer in research phase on the development of new radiopharmaceuticals or to expand research globally [43]. In contrast, the number of LINACS and cyclotrons with high energy protons is increasing [109,110], thus rapidly filling the ^{67}Cu availability gap. It is worth to mention that the scarce availability of intense deuteron-beams with energy higher than 9 MeV is curtailing the use of the ^{70}Zn(d,x)^{67}Cu reaction. A more thorough detailed insight into the economy of charged-particle induced reactions is given in Table 5, where the target material costs were estimated considering the target thicknesses, a hypothetical beam spot area of 1 cm^2 and an average price of enriched ^{68}Zn about USD 3/mg, ^{70}Zn about USD 13/mg and ^{64}Ni about USD 30/mg. Table 5 also reports the estimated cost of ^{67}Cu activity for each case. However, these estimates did not include the influence of recovery and reuse of the irradiated enriched target material. Obviously, the ^{67}Cu cost will decrease if the same enriched material will be re-used for several production cycles. The number of these cycles has to be carefully studied, to assure a final ^{67}Cu product accomplishing the regulatory requirements.

The calculations revealed that when the target material is fully enriched in the desired isotope (i.e., 100% enrichment, unless for the ^{64}Ni case that is reported from [24] with a 98% enrichment), the convenient route to obtain ^{67}Cu (without ^{64}Cu coproduction) is by using deuteron beams and ^{70}Zn targets. On the other hand, if some ^{64}Cu coproduction is acceptable for clinical applications, the ^{68}Zn(p,2p)^{67}Cu reaction seems to be a better option, since it provides a larger ^{67}Cu yield at a lower price per GBq (mCi) in comparison with the ^{70}Zn(d,x)^{67}Cu route. The proton-induced reaction on a multi-layer target, composed of ^{68}Zn and ^{70}Zn, maximizes the ^{67}Cu yield at a reasonable price per GBq (or per mCi);

however, there is a coproduction of ^{64}Cu. This route has been previously inhibited by the lack of commercial availability of electron accelerators and the need for an enriched thick target that must be recycled. Electron accelerators are now becoming commercially available and methods for target recycling have been worked out which make this route of production more cost effective. It is important to note that the dose of ^{67}Cu radioactivity required for one treatment is about 3.7 GBq (100 mCi) [43], but it can vary depending on the specific case and radiopharmaceutical. The ^{64}Cu coproduction and its impact on the dose delivery has to be carefully considered for each radiopharmaceutical, taking into account the specific biodistribution and timing for wash-out [111]. In general, the limit for the RNP is 99% and the dose increase due to contaminant radionuclides is 10%; however, considering the ^{64}Cu and ^{67}Cu decay characteristics (i.e., the emission of β^- particles and Auger electrons), preclinical studies with $^{64/67}$Cu-radiopharmaceuticals are encouraged to determine the potential impact of this combined therapy.

Table 5. ^{67}Cu activity cost ($ (USD)/GBq) by using proton-, deuteron-, and alpha-beams on ^{70}Zn, ^{68}Zn, and ^{64}Ni enriched materials, considering I = 30 µA, T_{IRR} = 24 h and enriched targets.

Beam	Target	Energy Range (MeV)	Thickness (mm)	Target Cost ($)	^{67}Cu @ EOB (GBq) (mCi)	^{67}Cu Cost ($/GBq) ($/mCi)	^{64}Cu
Protons	^{70}Zn	25–10	1.22	11,284	3 (81)	3761 (139)	-
	^{68}Zn	70–35	6.43	13,758	17.5 (473)	786 (29)	Yes
	^{70}Zn + ^{68}Zn	70–55 + 55–35	3.26 + 3.27	30,186 + 7000	26.2 (709)	1420 (52)	Yes
Deuterons	^{70}Zn	26–16	0.58	6500	4.1 (110)	1323 (49)	-
Alpha	^{64}Ni	30–0	0.16	3300	1 (27)	4272 (158)	-

It is worth noting that pure ^{67}Cu is available using the photo-induced production route [10]. Its major drawback is the need of a large, massive enriched ^{68}Zn target, having not only an economic impact on the initial investment but also a technological one on the radiochemical processing and target recovery. In addition to this route, it is important to mention the possibility of using an online mass separator to select ^{67}Cu, to later apply the chemical separation of Cu-isotopes from the collected ^{67}X radionuclides.

Mass separation, in contrast to the commonly used ion exchange and extraction chromatography, is expected to increase the availability of certain "exotic" radionuclides, among them ^{67}Cu. This novel approach will be studied within the recently granted EU project (PRISMAP) [112]; however, the efficacy of this process is yet to be shown.

5. Conclusions

This review reveals the international effort to supply ^{67}Cu, a promising theranostic radionuclide. The increasing availability of intense particle accelerators and the optimization of the associated technologies (targetry and radiochemical processing) are making ^{67}Cu closer to the clinics. The positron emitter counter-parts of ^{67}Cu, i.e., ^{60}Cu, ^{61}Cu, and ^{64}Cu, can be produced in cyclotrons. In particular, ^{64}Cu is now widely available for clinical use, promoting the development of innovative Cu-labeled radiopharmaceuticals. The improved availability of ^{67}Cu would speed up further radiopharmaceutical applications for therapy. Should ^{67}Cu be produced in sufficient quantities and quality, its clinical use would spread worldwide. Therefore, in addition to a detailed analysis of the possible nuclear reactions to produce ^{67}Cu, the radiochemical procedures to extract and purify Cu from the bulk material were also described in this work. Recent developments in the photoproduction of ^{67}Cu, and in the possibility of having accelerators providing intense 70 MeV proton beams and/or intense 30 MeV deuteron beams, are grounds for a future reliable supply of ^{67}Cu.

In most of the nuclear reactions leading to ^{67}Cu, the ^{64}Cu is produced in various radioactivity ratios. What is the impact of this impurity on a patient's dosimetry? Further studies on the possible use of $^{67/64}$Cu labeled-radiopharmaceuticals are encouraged, to find out the possible therapeutic benefits for the patient exploiting, at the same time, the β^-

radiation emitted by ^{67}Cu and ^{64}Cu decay and the shorter range Auger-electrons emitted by ^{64}Cu. Encouraging technological achievements, namely working out methods to isolate the small amount of ^{67}Cu from the large mass of zinc in the target and to recover the target material in the accelerator-based production, show that, soon, ^{67}Cu will be available daily in the United States and in Europe for research purposes. The relatively long half-life of ^{67}Cu makes production in large facilities and the shipping of the purified product to clinical centers possible.

Author Contributions: Conceptualization, L.M., G.P., P.M., I.C. and R.M.; formal analysis, L.M. and G.P.; writing—original draft preparation, L.M., G.P., P.M., I.C. and R.M.; writing—review and editing, L.M., G.P., P.M., I.C., C.S.C. and R.M. All authors have read and agreed to the published version of the manuscript.

Funding: This work was partially supported by the IAEA/CRP code F22053; the Polish Ministry of Science and Higher Education, grant no. 3639/FAO/IAEA/16/2017/0; the CERAD project, financed under the Smart Growth Operational Programme 2014–2020, Priority IV, Measure 4.2. POIR.04.02.00–14-A001/16; COME project INFN-CSN3; LARAMED and TERABIO premium projects funded by the Italian Ministry for University and Research (MIUR).

Institutional Review Board Statement: Not applicable.

Informed Consent Statement: Not applicable.

Acknowledgments: We would like to acknowledge PRISMAP, the European medical radionuclide programme, for the production of high purity radionuclides (radioactive isotopes) by mass separation. This project has received funding from the European Union's Horizon 2020 research and innovation programme under grant agreement no. 101008571.

Conflicts of Interest: The authors declare no conflict of interest.

References

1. Srivastava, S.C. A Bridge Not Too Far: Personalized Medicine with the Use of Theragnostic Radiopharmaceuticals. *J. Postgrad. Med. Educ. Res.* **2013**, *47*, 31–46. [CrossRef]
2. Mirzadeh, S.; Mausner, L.F.; Srivastava, S.C. Production of No-Carrier Added ^{67}Cu. *Int. J. Radiat. Appl. Instrum. Part A Appl. Radiat. Isot.* **1986**, *37*, 29–36. [CrossRef]
3. Kolsky, K.L.; Joshi, V.; Meinken, G.E. Improved Production, and Evaluation of Cu-67 for Tumor Radioimmunotherapy. *J. Nucl. Med.* **1992**, *35*, 259.
4. Smith, A.; Alberto, R.; Blaeuenstein, P.; Novak-Hofer, I.; Maecke, H.R.; Schubiger, P.A. Preclinical Evaluation of ^{67}Cu-Labeled Intact and Fragmented Anti-Colon Carcinoma Monoclonal Antibody MAb35. *Cancer Res.* **1993**, *53*, 5727–5733. [PubMed]
5. International Atomic Energy Agency. *Therapeutic Radiopharmaceuticals Labelled with Copper-67, Rhenium-186 and Scandium-47*; IAEA-TECDOC-1945; IAEA: Vienna, Austria, 2021.
6. Jalilian, A.R.; Gizawy, M.A.; Alliot, C.; Takacs, S.; Chakarborty, S.; Rovais, M.R.A.; Pupillo, G.; Nagatsu, K.; Park, J.H.; Khandaker, M.U.; et al. IAEA Activities on ^{67}Cu, ^{186}Re, ^{47}Sc Theranostic Radionuclides and Radiopharmaceuticals. *Curr. Radiopharm.* **2021**, *14*, 306–314. [CrossRef]
7. NNDC.BNL NuDat 3.0. Available online: https://www.nndc.bnl.gov/nudat3/ (accessed on 25 January 2022).
8. Hao, G.; Singh, A.N.; Liu, W.; Sun, X. PET with Non-Standard Nuclides. *Curr. Top. Med. Chem.* **2010**, *10*, 1096–1112. [CrossRef]
9. Takacs, S. Therapeutic Radionuclides. Available online: https://www-nds.iaea.org/medical/therapeutic_2019.html (accessed on 25 January 2022).
10. Merrick, M.J.; Rotsch, D.A.; Tiwari, A.; Nolen, J.; Brossard, T.; Song, J.; Wadas, T.J.; Sunderland, J.J.; Graves, S.A. Imaging and Dosimetric Characteristics of ^{67}Cu. *Phys. Med. Biol.* **2021**, *66*, 035002. [CrossRef]
11. IAEA. *Alternative Radionuclide Production with a Cyclotron*; IAEA Radioisotopes and Radiopharmaceuticals reports No. 4; IAEA: Vienna, Austria, 2021.
12. NDS-IAEA Recommended Cross Sections for 68Zn(p,2p)^{67}Cu Reaction. Available online: https://www-nds.iaea.org/medical/zn867cu0.html (accessed on 9 September 2021).
13. Medvedev, D.G.; Mausner, L.F.; Meinken, G.E.; Kurczak, S.O.; Schnakenberg, H.; Dodge, C.J.; Korach, E.M.; Srivastava, S.C. Development of a Large Scale Production of ^{67}Cu from ^{68}Zn at the High Energy Proton Accelerator: Closing the ^{68}Zn Cycle. *Appl. Radiat. Isot.* **2012**, *70*, 423–429. [CrossRef]
14. Qaim, S.M. *Medical Radionuclide Production: Science and Technology*; De Gruyter: Berlin, Germany; Boston, MA, USA, 2019.
15. NDS-IAEA Recommended Cross Sections for 70Zn(p,a)^{67}Cu Reaction. Available online: https://www-nds.iaea.org/medical/zn067cu0.html (accessed on 9 September 2021).

16. Pupillo, G.; Mou, L.; Martini, P.; Pasquali, M.; Boschi, A.; Cicoria, G.; Duatti, A.; Haddad, F.; Esposito, J. Production of ^{67}Cu by Enriched ^{70}Zn Targets: First Measurements of Formation Cross Sections of ^{67}Cu, ^{64}Cu, ^{67}Ga, ^{66}Ga, ^{69m}Zn and ^{65}Zn in Interactions of ^{70}Zn with Protons above 45 MeV. *Radiochim. Acta* **2020**, *108*, 593–602. [CrossRef]

17. Qaim, S.M.; Hussain, M.; Spahn, I.; Neumaier, B. Continuing Nuclear Data Research for Production of Accelerator-Based Novel Radionuclides for Medical Use: A Mini-Review. *Front. Phys.* **2021**, *9*, 639290. [CrossRef]

18. Porile, N.T.; Tanaka, S.; Amano, H.; Furukawa, M.; Iwata, S.; Yagi, M. Nuclear Reactions of Ga-69 and Ga-71 with 13–56 MeV Protons. *Nucl. Phys.* **1963**, *43*, 500–522. [CrossRef]

19. IAEA ISOTOPIA. Available online: https://www-nds.iaea.org/relnsd/isotopia/isotopia.html (accessed on 2 April 2021).

20. NDS-IAEA Experimental Nuclear Reaction Data (EXFOR). Available online: https://www-nds.iaea.org/exfor/exfor.htm (accessed on 25 January 2022).

21. Mou, L.; Pupillo, G.; Martini, P.; Pasquali, M. A Method and a Target for the Production of ^{67}Cu 2019. Available online: https://patentscope.wipo.int/search/en/detail.jsf?docId=WO2019220224 (accessed on 18 November 2021).

22. Kozempel, J.; Abbas, K.; Simonelli, F.; Bulgheroni, A.; Holzwarth, U.; Gibson, N. Preparation of ^{67}Cu via Deuteron Irradiation of ^{70}Zn. *Radiochim. Acta* **2012**, *100*, 419–423. [CrossRef]

23. Nigron, E.; Guertin, A.; Haddad, F.; Sounalet, T. Is ^{70}Zn(d,x)^{67}Cu the Best Way to Produce ^{67}Cu for Medical Applications? *Front. Med.* **2021**, *8*, 1059. [CrossRef]

24. Takács, S.; Aikawa, M.; Haba, H.; Komori, Y.; Ditrói, F.; Szűcs, Z.; Saito, M.; Murata, T.; Sakaguchi, M.; Ukon, N. Cross Sections of Alpha-Particle Induced Reactions on NatNi: Production of ^{67}Cu. *Nucl. Instrum. Methods Phys. Res. Sect. B Beam Interact. Mater. At.* **2020**, *479*, 125–136. [CrossRef]

25. Shigeo, T. Reactions of Nickel with Alpha-Particles. *J. Phys. Soc. Jpn.* **1960**, *15*, 2159–2167. [CrossRef]

26. Antropov, A.E.; Zarubin, P.P.; Aleksandrov, Y.A.; Gorshkov, I.Y. Cross Sections Measurements of (p,n),(Alpha,Pn), (Alpha,Xn) Reactions on Nuclei of Middle Atomic Weight. In Proceedings of the 35th Conference on Nuclear Spectroscopy and Nuclear Structure, Leningrad, Russia, 16–18 April 1985; p. 369.

27. Levkovskij, V.N. *Cross Sections of Medium Mass Nuclide Activation (A = 40–100) by Medium Energy Protons and Alpha-Particles (E = 10–50 MeV)*; Inter-Vesi: Moscow, Russia, 1991; ISBN 5-265-02732-7.

28. Skakun, Y.; Qaim, S.M. Excitation Function of the ^{64}Ni(α,p)^{67}Cu Reaction for Production of ^{67}Cu. *Appl. Radiat. Isot.* **2004**, *60*, 33–39. [CrossRef]

29. Mausner, L.F.; Mirzadeh, S.; Schnakenberg, H.; Srivastava, S.C. The Design and Operation of the Upgraded BLIP Facility for Radionuclide Research and Production. *Int. J. Radiat. Appl. Instrum. Part A Appl. Radiat. Isot.* **1990**, *41*, 367–374. [CrossRef]

30. Stoner, J.; Gardner, T.; Gardner, T. A Comparison of DOTA and DiamSar Chelates of High Specific Activity ELINAC Produced ^{67}Cu. *J. Nucl. Med.* **2016**, *57*, 1107.

31. Ehst, D.A.; Smith, N.A.; Bowers, D.L.; Makarashvili, V. Copper-67 Production on Electron Linacs—Photonuclear Technology Development. *AIP Conf. Proc.* **2012**, *1509*, 157–161.

32. Hovhannisyan, G.H.; Bakhshiyan, T.M.; Dallakyan, R.K. Photonuclear Production of the Medical Isotope ^{67}Cu. *Nucl. Instrum. Methods Phys. Res. Sect. B Beam Interact. Mater. At.* **2021**, *498*, 48–51. [CrossRef]

33. Aliev, R.A.; Belyshev, S.S.; Kuznetsov, A.A.; Dzhilavyan, L.Z.; Khankin, V.V.; Aleshin, G.Y.; Kazakov, A.G.; Priselkova, A.B.; Kalmykov, S.N.; Ishkhanov, B.S. Photonuclear Production and Radiochemical Separation of Medically Relevant Radionuclides: ^{67}Cu. *J. Radioanal. Nucl. Chem.* **2019**, *321*, 125–132. [CrossRef]

34. Polak, P.; Geradts, J.; Vlist, R.V.A.N.D.E.R.; Lindner, L. Photonuclear Production of ^{67}Cu from ZnO Targets. *Radiochim. Acta* **1986**, *40*, 169–174. [CrossRef]

35. Starovoitova, V.; Foote, D.; Harris, J.; Makarashvili, V.; Segebade, C.R.; Sinha, V.; Wells, D.P. Cu-67 Photonuclear Production. *AIP Conf. Proc.* **2011**, *1336*, 502–504. [CrossRef]

36. Gopalakrishna, A.; Suryanarayana, S.V.; Naik, H.; Dixit, T.S.; Nayak, B.K.; Kumar, A.; Maletha, P.; Thakur, K.; Deshpande, A.; Krishnan, R.; et al. Production, Separation and Supply Prospects of ^{67}Cu with the Development of Fast Neutron Sources and Photonuclear Technology. *Radiochim. Acta* **2018**, *106*, 549–557. [CrossRef]

37. NNDC.BNL NuDat 2.8. Available online: https://www.nndc.bnl.gov/nudat2/ (accessed on 9 September 2021).

38. NIDC: National Isotope Development Center, Product Catalog Resources. Available online: https://isotopes.gov/sites/default/files/2021-02/Cu-67.pdf (accessed on 14 December 2021).

39. Mirzadeh, S.; Knapp, F.F. Spontaneous Electrochemical Separation of Carrier-Free Copper-64 and Copper-67 from Zinc Targets. *Radiochim. Acta* **1992**, *57*, 193–200. [CrossRef]

40. Mikolajczak, R.; Parus, J.L. Reactor Produced Beta-Emitting Nuclides for Nuclear Medicine. *World J. Nucl. Med.* **2005**, *36*, 184–190.

41. O'Brien, H.A. The Preparation of ^{67}Cu from ^{67}Zn in a Nuclear Reactor. *Int. J. Appl. Radiat. Isot.* **1969**, *20*, 121–124. [CrossRef]

42. Shikata, E. Research of Radioisotope Production with Fast Neutrons, (VI). *J. Nucl. Sci. Technol.* **1964**, *1*, 177–180. [CrossRef]

43. Smith, N.A.; Bowers, D.L.; Ehst, D.A. The Production, Separation, and Use of ^{67}Cu for Radioimmunotherapy: A Review. *Appl. Radiat. Isot.* **2012**, *70*, 2377–2383. [CrossRef]

44. IAEA. *Manual for Reactor Produced Radioisotopes*; IAEA-TECDOC-1340; International Atomic Energy Agency: Vienna, Austria, 2003; pp. 1–254, ISBN 92-0-101103-2.

45. Johnsen, A.M.; Heidrich, B.J.; Durrant, C.B.; Bascom, A.J.; Ünlü, K. Reactor Production of ^{64}Cu and ^{67}Cu Using Enriched Zinc Target Material. *J. Radioanal. Nucl. Chem.* **2015**, *305*, 61–71. [CrossRef]

46. Zinn, K.R.; Chaudhuri, T.R.; Cheng, T.P.; Morris, J.S.; Meyer, W.A.J. Production of No-Carrier-Added ^{64}Cu from Zinc Metal Irradiated under Boron Shielding. *Cancer* **1994**, *73*, 774–778. [CrossRef]

47. Vimalnath, K.V.; Rajeswari, A.; Jagadeesan, K.C.; Viju, C.; Joshi, P.V.; Venkatesh, M. Studies on the Production Feasibility of ^{64}Cu by (n,p) Reactions on Zn Targets in Dhruva Research Reactor. *J. Radioanal. Nucl. Chem.* **2012**, *294*, 43–47. [CrossRef]

48. Pupillo, G.; Sounalet, T.; Michel, N.; Mou, L.; Esposito, J.; Haddad, F. New Production Cross Sections for the Theranostic Radionuclide [67]Cu. *Nucl. Instrum. Methods Phys. Res. Sect. B Beam Interact. Mater. At.* **2018**, *415*, 41–47. [CrossRef]

49. Rowshanfarzad, P.; Sabet, M.; Reza Jalilian, A.; Kamalidehghan, M. An Overview of Copper Radionuclides and Production of [61]Cu by Proton Irradiation of [Nat]Zn at a Medical Cyclotron. *Appl. Radiat. Isot.* **2006**, *64*, 1563–1573. [CrossRef]

50. Ohya, T.; Nagatsu, K.; Suzuki, H.; Fukada, M.; Minegishi, K.; Hanyu, M.; Zhang, M.R. Small-Scale Production of [67]Cu for a Preclinical Study via the [64]Ni(α,p) [67]Cu Channel. *Nucl. Med. Biol.* **2018**, *59*, 56–60. [CrossRef]

51. Karimi, Z.; Sadeghi, M.; Hosseini, S.F. Experimental Production and Theoretical Assessment of [67]Cu via Neutron Induced Reaction. *Ann. Nucl. Energy* **2019**, *133*, 665–668. [CrossRef]

52. Mushtaq, A.; Karim, H.M.A.; Khan, M.A. Cu and [67]Cu in a Reactor. *J. Radioanal. Nucl. Chem.* **1990**, *141*, 261–269. [CrossRef]

53. Uddin, M.S.; Rumman-Uz-Zaman, M.; Hossain, S.M.; Qaim, S.M. Radiochemical Measurement of Neutron-Spectrum Averaged Cross Sections for the Formation of ^{64}Cu and [67]Cu via the (n,p) Reaction at a TRIGAMark-II Reactor: Feasibility of Simultaneous Production of the Theragnostic Pair ^{64}Cu/[67]Cu. *Radiochim. Acta* **2014**, *102*, 473–480. [CrossRef]

54. Yagi, M.; Kondo, K. Preparation of carrier-free [67]Cu by the [68]Zn(g,p) reaction. *Int. Appl. Radiat. Isot.* **1978**, *29*, 757–759. [CrossRef]

55. Chakravarty, R.; Rajeswari, A.; Shetty, P.; Jagadeesan, K.C.; Ram, R.; Jadhav, S.; Sarma, H.D.; Dash, A.; Chakraborty, S. A Simple and Robust Method for Radiochemical Separation of No-Carrier-Added [64]Cu Produced in a Research Reactor for Radiopharmaceutical Preparation. *Appl. Radiat. Isot.* **2020**, *165*, 109341. [CrossRef]

56. Schwarzbach, R.; Zimmermann, K.; Bluenstein, P.; Smith, A.; Schubiger, P.A. Development of a Simple and Selective Separation of [67]Cu from Irradiated Zinc for Use in Antibody Labelling: A Comparison of Methods. *Appl. Radiat. Isot.* **1995**, *46*, 329–336. [CrossRef]

57. Dasgupta, A.K.; Mausner, L.F.; Srivastava, S.C. A New Separation Procedure for [67]Cu from Proton Irradiated Zn. *Int. J. Radiat. Appl. Instrum. Part A Appl. Radiat. Isot.* **1991**, *42*, 371–376. [CrossRef]

58. Kim, J.H.; Park, H.; Chun, K.S. Effective Separation Method of ^{64}Cu from [67]Ga Waste Product with a Solvent Extraction and Chromatography. *Appl. Radiat. Isot.* **2010**, *68*, 1623–1626. [CrossRef]

59. Sen, N.; Chakravarty, R.; Singh, K.K.; Chakraborty, S.; Shenoy, K.T. Selective Separation of Cu from Large Excess of Zn Using a Microfluidic Platform. *Chem. Eng. Process.—Process Intensif.* **2021**, *159*, 108215. [CrossRef]

60. Dolley, S.G.; van der Walt, T.N.; Steyn, G.F.; Szelecsényi, F.; Kovács, Z. The Production and Isolation of Cu-64 and Cu-67 from Zinc Target Material and Other Radionuclides. *Czechoslov. J. Phys.* **2006**, *56*, 539–544. [CrossRef]

61. Dolley, S.G.; van der Walt, T.N. Isolation of Cu Radionuclides with Dithizone Impregnated Xad-8. *Radiochim. Acta* **2014**, *102*, 263–269. [CrossRef]

62. Roberts, J.C.; Newmyer, S.L.; Mercer-Smith, J.A.; Schreyer, S.A.; Lavallee, D.K. Labeling Antibodies with Copper Radionuclides Using N-4-Nitrobenzyl-5-(4-Carboxyphenyl)-10,15,20-Tris(4-Sulfophenyl) Porphine. *Int. J. Radiat. Appl. Instrum. Part A Appl. Radiat. Isot.* **1989**, *40*, 775–781. [CrossRef]

63. Ohya, T.; Nagatsu, K.; Hanyu, M.; Minegishi, K.; Zhang, M.R. Simple Separation of [67]Cu from Bulk Zinc by Coprecipitation Using Hydrogen Sulfide Gas and Silver Nitrate. *Radiochim. Acta* **2020**, *108*, 469–476. [CrossRef]

64. Ohya, T.; Nagatsu, K.; Suzuki, H.; Fukada, M.; Minegishi, K.; Hanyu, M.; Fukumura, T.; Zhang, M.-R. Efficient Preparation of High-Quality ^{64}Cu for Routine Use. *Nucl. Med. Biol.* **2016**, *43*, 685–691. [CrossRef]

65. McCarthy, D.W.; Shefer, R.E.; Klinkowstein, R.E.; Bass, L.A.; Margeneau, W.H.; Cutler, C.S.; Anderson, C.J.; Welch, M.J. Efficient Production of High Specific Activity [64]Cu Using a Biomedical Cyclotron. *Nucl. Med. Biol.* **1997**, *24*, 35–43. [CrossRef]

66. Le, V.S.; Howse, J.; Zaw, M.; Pellegrini, P.; Katsifis, A.; Greguric, I.; Weiner, R. Alternative Method for [64]Cu Radioisotope Production. *Appl. Radiat. Isot. Incl. Data Instrum. Methods Use Agric. Ind. Med.* **2009**, *67*, 1324–1331. [CrossRef]

67. EDQM. Radiopharmaceutical Preparations, General Monograph: 0125. In *European Pharmacopeia*; Council of Europe: Strasbourg, France, 2020.

68. Coenen, H.H.; Gee, A.D.; Adam, M.; Antoni, G.; Cutler, C.S.; Fujibayashi, Y.; Jeong, J.M.; Mach, R.H.; Mindt, T.L.; Pike, V.W.; et al. Consensus Nomenclature Rules for Radiopharmaceutical Chemistry—Setting the Record Straight. *Nucl. Med. Biol.* **2017**, *55*, v-xi. [CrossRef]

69. van So, L.; Pellegrini, P.; Katsifis, A.; Howse, J.; Greguric, I. Radiochemical Separation and Quality Assessment for the [68]Zn Target Based [64]Cu Radioisotope Production. *J. Radioanal. Nucl. Chem.* **2008**, *277*, 451–466. [CrossRef]

70. EDQM. Lutetium (177Lu) Solution for Radiolabelling, Monograph: 2798. In *European Pharmacopeia*; Council of Europe: Strasbourg, France, 2020.

71. Wadas, T.; Wong, E.; Weisman, G.; Anderson, C. Copper Chelation Chemistry and Its Role in Copper Radiopharmaceuticals. *Curr. Pharm. Des.* **2006**, *13*, 3–16. [CrossRef]

72. Wadas, T.J.; Wong, E.H.; Weisman, G.R.; Anderson, C.J. Coordinating Radiometals of Copper, Gallium, Indium, Yttrium, and Zirconium for PET and SPECT Imaging of Disease. *Chem. Rev.* **2010**, *110*, 2858–2902. [CrossRef] [PubMed]

73. Moi, M.K.; Meares, C.F.; McCall, M.J.; Cole, W.C.; DeNardo, S.J. Copper Chelates as Probes of Biological Systems: Stable Copper Complexes with a Macrocyclic Bifunctional Chelating Agent. *Anal. Biochem.* **1985**, *148*, 249–253. [CrossRef]

74. Boschi, A.; Martini, P.; Janevik-Ivanovska, E.; Duatti, A. The Emerging Role of Copper-64 Radiopharmaceuticals as Cancer Theranostics. *Drug Discov. Today* **2018**, *23*, 1489–1501. [CrossRef]

75. Anderson, C.J.; Ferdani, R. Copper-64 Radiopharmaceuticals for PET Imaging of Cancer: Advances in Preclinical and Clinical Research. *Cancer Biother. Radiopharm.* **2009**, *24*, 379–393. [CrossRef]

76. Bass, L.A.; Wang, M.; Welch, M.J.; Anderson, C.J. In Vivo Transchelation of Copper-64 from TETA-Octreotide to Superoxide Dismutase in Rat Liver. *Bioconjug. Chem.* **2000**, *11*, 527–532. [CrossRef]

77. Hao, G.; Singh, A.N.; Oz, O.K.; Sun, X. Recent Advances in Copper Radiopharmaceuticals. *Curr. Radiopharm.* **2011**, *4*, 109–121. [CrossRef]

78. Busch, D.H. The Complete Coordination Chemistry—One Practioner's Perspective. *Chem. Rev.* **1993**, *93*, 847–860. [CrossRef]

79. Sun, X.; Wuest, M.; Weisman, G.R.; Wong, E.H.; Reed, D.P.; Boswell, C.A.; Motekaitis, R.; Martell, A.E.; Welch, M.J.; Anderson, C.J. Radiolabeling and In Vivo Behavior of Copper-64-Labeled Cross-Bridged Cyclam Ligands. *J. Med. Chem.* **2002**, *45*, 469–477. [CrossRef] [PubMed]

80. Boswell, C.A.; Sun, X.; Niu, W.; Weisman, G.R.; Wong, E.H.; Rheingold, A.L.; Anderson, C.J. Comparative in Vivo Stability of Copper-64-Labeled Cross-Bridged and Conventional Tetraazamacrocyclic Complexes. *J. Med. Chem.* **2004**, *47*, 1465–1474. [CrossRef] [PubMed]

81. Marciniak, A.; Brasuń, J. Somatostatin Analogues Labeled with Copper Radioisotopes: Current Status. *J. Radioanal. Nucl. Chem.* **2017**, *313*, 279–289. [CrossRef]

82. Sun, X.; Wuest, M.; Kovacs, Z.; Sherry, A.D.; Motekaitis, R.; Wang, Z.; Martell, A.E.; Welch, M.J.; Anderson, C.J. In Vivo Behavior of Copper-64-Labeled Methanephosphonate Tetraaza Macrocyclic Ligands. *J. Biol. Inorg. Chem. JBIC Publ. Soc. Biol. Inorg. Chem.* **2003**, *8*, 217–225. [CrossRef]

83. di Bartolo, N.M.; Sargeson, A.M.; Donlevy, T.M.; Smith, S. v Synthesis of a New Cage Ligand, SarAr, and Its Complexation with Selected Transition Metal Ions for Potential Use in Radioimaging. *J. Chem. Soc., Dalton Trans.* **2001**, 2303–2309. [CrossRef]

84. Cooper, M.S.; Ma, M.T.; Sunassee, K.; Shaw, K.P.; Williams, J.D.; Paul, R.L.; Donnelly, P.S.; Blower, P.J. Comparison of ^{64}Cu-Complexing Bifunctional Chelators for Radioimmunoconjugation: Labeling Efficiency, Specific Activity, and in Vitro/in Vivo Stability. *Bioconjug. Chem.* **2012**, *23*, 1029–1039. [CrossRef]

85. Dearling, J.L.J.; Voss, S.D.; Dunning, P.; Snay, E.; Fahey, F.; Smith, S.V.; Huston, J.S.; Meares, C.F.; Treves, S.T.; Packard, A.B. Imaging Cancer Using PET—The Effect of the Bifunctional Chelator on the Biodistribution of a ^{64}Cu-Labeled Antibody. *Nucl. Med. Biol.* **2011**, *38*, 29–38. [CrossRef]

86. Maheshwari, V.; Dearling, J.; Treves, S.; Packard, A. Measurement of the Rate of Copper(II) Exchange for ^{64}Cu Complexes of Bifunctional Chelators. *Inorg. Chim. Acta* **2012**, *393*, 318–323. [CrossRef]

87. Boros, E.; Packard, A.B. Radioactive Transition Metals for Imaging and Therapy. *Chem. Rev.* **2019**, *119*, 870–901. [CrossRef]

88. Pasquali, M.; Martini, P.; Shahi, A.; Jalilian, A.R.; Osso, J.A.; Boschi, A. Copper-64 Based Radiopharmaceuticals for Brain Tumors and Hypoxia Imaging. *Q. J. Nucl. Med. Mol. Imaging* **2020**, *64*, 371–381. [CrossRef] [PubMed]

89. Paterson, B.M.; Donnelly, P.S. Copper Complexes of Bis(Thiosemicarbazones): From Chemotherapeutics to Diagnostic and Therapeutic Radiopharmaceuticals. *Chem. Soc. Rev.* **2011**, *40*, 3005–3018. [CrossRef] [PubMed]

90. Deshpande, S.V.; DeNardo, S.J.; Meares, C.F.; McCall, M.J.; Adams, G.P.; Moi, M.K.; DeNardo, G.L. Copper-67-Labeled Monoclonal Antibody Lym-1, a Potential Radiopharmaceutical for Cancer Therapy: Labeling and Biodistribution in RAJI Tumored Mice. *J. Nucl. Med.* **1988**, *29*, 217–225.

91. DeNardo, G.L.; Kukis, D.L.; Shen, S.; Mausner, L.F.; Meares, C.F.; Srivastava, S.C.; Miers, L.A.; DeNardo, S.J. Efficacy and Toxicity of ^{67}Cu-2IT-BAT-Lym-1 Radioimmunoconjugate in Mice Implanted with Human Burkitt's Lymphoma (Raji). *Clin. Cancer Res. Off. J. Am. Assoc. Cancer Res.* **1997**, *3*, 71–79.

92. Connett, J.M.; Anderson, C.J.; Guo, L.W.; Schwarz, S.W.; Zinn, K.R.; Rogers, B.E.; Siegel, B.A.; Philpott, G.W.; Welch, M.J. Radioimmunotherapy with a ^{64}Cu-Labeled Monoclonal Antibody: A Comparison with ^{67}Cu. *Proc. Natl. Acad. Sci. USA* **1996**, *93*, 6814–6818. [CrossRef]

93. Zimmermann, K.; Grünberg, J.; Honer, M.; Ametamey, S.; August Schubiger, P.; Novak-Hofer, I. Targeting of Renal Carcinoma with 67/64Cu-Labeled Anti-L1-CAM Antibody ChCE7: Selection of Copper Ligands and PET Imaging. *Nucl. Med. Biol.* **2003**, *30*, 417–427. [CrossRef]

94. Hughes, O.D.M.; Bishop, M.C.; Perkins, A.C.; Frier, M.; Price, M.R.; Denton, G.; Smith, A.; Rutherford, R.; Schubiger, P.A. Preclinical Evaluation of Copper-67 Labelled Anti-MUC1 Mucin Antibody C595 for Therapeutic Use in Bladder Cancer. *Eur. J. Nucl. Med.* **1997**, *24*, 439–443. [CrossRef]

95. Fani, M.; del Pozzo, L.; Abiraj, K.; Mansi, R.; Tamma, M.L.; Cescato, R.; Waser, B.; Weber, W.A.; Reubi, J.C.; Maecke, H.R. PET of Somatostatin Receptor Positive Tumors Using ^{64}Cu- and ^{68}Ga-Somatostatin Antagonists: The Chelate Makes the Difference. *J. Nucl. Med.* **2011**, *52*, 1110–1118. [CrossRef]

96. Cullinane, C.; Jeffery, C.M.; Roselt, P.D.; van Dam, E.M.; Jackson, S.; Kuan, K.; Jackson, P.; Binns, D.; van Zuylekom, J.; Harris, M.J.; et al. Peptide Receptor Radionuclide Therapy with ^{67}Cu-CuSarTATE Is Highly Efficacious Against a Somatostatin-Positive Neuroendocrine Tumor Model. *J. Nucl. Med.* **2020**, *61*, 1800. [CrossRef]

97. McInnes, L.; Zia, N.; Cullinane, C.; van Zuylekom, J.; Jackson, S.; Stoner, J.; Haskali, M.; Roselt, P.; van Dam, E.; Harris, M.; et al. A Cu-64/Cu-67 Bifunctional PSMA Ligand as a Theranostic for Prostate Cancer. *J. Nucl. Med.* **2020**, *61*, 1215.

98. Kelly, J.M.; Ponnala, S.; Amor-Coarasa, A.; Zia, N.A.; Nikolopoulou, A.; Nikolopoulou, A.; Williams, C.; Schlyer, D.J.; Schlyer, D.J.; Dimagno, S.G.; et al. Preclinical Evaluation of a High-Affinity Sarcophagine-Containing PSMA Ligand for ^{64}Cu/[67]Cu-Based Theranostics in Prostate Cancer. *Mol. Pharm.* **2020**, *17*, 1954–1962. [CrossRef] [PubMed]

99. Keinänen, O.; Fung, K.; Brennan, J.M.; Zia, N.; Harris, M.; van Dam, E.; Biggin, C.; Hedt, A.; Stoner, J.; Donnelly, P.S.; et al. Harnessing ^{64}Cu/ [67]Cu for a Theranostic Approach to Pretargeted Radioimmunotherapy. *Proc. Natl. Acad. Sci. USA* **2020**, *117*, 28316–28327. [CrossRef] [PubMed]

100. Strickland, G.T.; Beckner, W.M.; Leu, M.-L. Absorption of Copper in Homozygotes and Heterozygotes for Wilson's Disease and Controls: Isotope Tracer Studies with [67]Cu and [64]Cu. *Clin. Sci.* **1972**, *43*, 617–625. [CrossRef] [PubMed]

101. Denardo, G.L.; Denardo, S.J.; Kukis, D.L.; O'Donnell, R.T.; Shen, S.; Goldstein, D.S.; Kroger, L.A.; Salako, Q.; Denardo, D.A.; Mirick, G.R.; et al. Maximum Tolerated Dose of [67]Cu-2IT-BAT-LYM-1 for Fractionated Radioimmunotherapy of Non-Hodgkin's Lymphoma: A Pilot Study. *Anticancer Res.* **1998**, *18*, 2779–2788. [PubMed]

102. O'Donnell, R.T.; DeNardo, G.L.; Kukis, D.L.; Lamborn, K.R.; Shen, S.; Yuan, A.; Goldstein, D.S.; Carr, C.E.; Mirick, G.R.; DeNardo, S.J. A Clinical Trial of Radioimmunotherapy with [67]Cu-21T-BAT-Lym-1 for Non-Hodgkins Lymphoma. *J. Nucl. Med.* **1999**, *40*, 2014. [PubMed]

103. Bailey, D.; Schembri, G.; Willowson, K.; Hedt, A.; Lengyelova, E.; Harris, M. A Novel Theranostic Trial Design Using ^{64}Cu/[67]Cu with Fully 3D Pre-Treatment Dosimetry. *J. Nucl. Med.* **2019**, *60*, 204.

104. A Phase I/IIA Study of ^{64}Cu-SARTATE and [67]Cu-SARTATE for Imaging and Treating Children and Young Adults with High-Risk Neuroblastoma. Available online: https://www.mskcc.org/cancer-care/clinical-trials/20-218 (accessed on 11 January 2022).

105. [67]Cu-SARTATETM Peptide Receptor Radionuclide Therapy Administered to Pediatric Patients with High-Risk, Relapsed, Refractory Neuroblastoma. Available online: https://clinicaltrials.gov/ct2/show/NCT04023331 (accessed on 11 January 2022).

106. Biggin, C.; van Dam, E.; Harris, M.; Parker, M.; Schembri, G. Estimating External Exposure from Patients after Treatment with Cu-67 SARTATE. *J. Nucl. Med.* **2019**, *60*, 1624a.

107. Willowson, K.; Harris, M.; Jeffery, C.; Biggin, C.; Hedt, A.; Stoner, J.; Bailey, D. Development of [67]Cu Quantitative SPECT for Clinical Dosimetry. *J. Nucl. Med.* **2018**, *59*, 1748.

108. Hao, G.; Mastren, T.; Silvers, W.; Hassan, G.; Öz, O.K.; Sun, X. Copper-67 Radioimmunotheranostics for Simultaneous Immunotherapy and Immuno-SPECT. *Sci. Rep.* **2021**, *11*, 3622. [CrossRef]

109. Haddad, F.; Ferrer, L.; Guertin, A.; Carlier, T.; Michel, N.; Barbet, J.; Chatal, J.-F. ARRONAX, a High-Energy and High-Intensity Cyclotron for Nuclear Medicine. *Eur. J. Nucl. Med. Mol. Imaging* **2008**, *35*, 1377–1387. [CrossRef]

110. Esposito, J.; Bettoni, D.; Boschi, A.; Calderolla, M.; Cisternino, S.; Fiorentini, G.; Keppel, G.; Martini, P.; Maggiore, M.; Mou, L.; et al. Laramed: A Laboratory for Radioisotopes of Medical Interest. *Molecules* **2019**, *24*, 20. [CrossRef]

111. de Nardo, L.; Pupillo, G.; Mou, L.; Esposito, J.; Rosato, A.; Meléndez-Alafort, L. A Feasibility Study of the Therapeutic Application of a Mixture of [64/67]Cu Radioisotopes Produced by Cyclotrons with Proton Irradiation. *Phys. Med. Biol.* **2022**; *in press*.

112. PRISMAP—Building a European Network for Medical Radionuclides. Available online: https://www.arronax-nantes.fr/en/radionuclide-production/news/prismap-building-a-european-network-for-medical-radionuclides/ (accessed on 10 January 2022).

MDPI

St. Alban-Anlage 66

4052 Basel

Switzerland

Tel. +41 61 683 77 34

Fax +41 61 302 89 18

www.mdpi.com

Molecules Editorial Office

E-mail: molecules@mdpi.com

www.mdpi.com/journal/molecules